Gene Theme Scene

By

Clarisse Clemons Ferrara, MD

ISBN: 9781502794864

DEDICATION

I, Clarisse Clemons Ferrara, MD daughter of Clarence Clemons Chief of Research of the Cinti, Ohio EPA yes, the Environmental Protection Agency married my Mother Henrietta Hyatt Clemons teacher with a Masters in Administration and Supervision, dedicate this book to my grandfather, Rev Lucian Hyatt, as he asked my young parents "when are you getting married" the Power of being the Pastor helped as, they were married in their mutual birthday month of March. Two years later I became their first born and two years later my Brother Clarence now the owner of the Muddy Waters Equestrian Park, Cinti. Ohio, joined our family. Mother, Father, Clarence and I were born in Kentucky. Eventually, my sister Amy, now a Pharmacist with Kroger, completed my list of siblings.

Father's sister the lovely and wise Geraldine Clemons Downey continues to bless me in unmeasureable increments, since being able to hear her voice continues to be a weekly ray of sunshine for me on this earth. My father's younger brother Uncle Henry is survived by my Aunt Diane. My Mother's twin sisters Aunt Golda Mae/Aunt Goldie Hammond as well as my Uncle Lucian Hyatt, Jr completed my mother's list of siblings.

My Husband Joseph Ferrara, MD Cornell Medical Graduate and Board Certified Surgeon and I welcomed our twins Angelo Joseph and Angela Jolyn 7/23/93, to join his children with his first wife (Kate Von Braunsberg) Ann, Joey(Lauren), who were blessed with Joseph Salvalore Jr on 6/24/16, Maja(Frank) who also welcomed Antonio Joseph to this world, 3/30/15. I thank my in-laws, Salvatore Ferrara and Rose Giambalvo Ferrara, not only for the gift that is my husband, but also her brother Jim Giambalvo, MD(Liz), and the ability to brag that my mother in law gave birth to a Doctor (my husband), a Dentist my brother in law Peter and a lawyer, my sister in law Vickie.

My patient James Riley, Jr died before his twenty-third birthday from a synovial cell sarcoma that had metastasized to his lungs at the time of diagnosis. He is survived by his mother, Lucia, sister Cheryl Pepsii Riley and his children Jamal, Sheretta, Andrea and Tabitha, who became Mrs. King 7/19/14.

Our Previous President Jimmy Carter was diagnosed with a malignant melanoma with metastasis to his brain, his treatment with Keytruda is listed in the skin cancer chapter. Hopefully when the NIH publishes a new version of "What is a Mutation" carcinogens such a hexavalent chromium will be included since the field of Epigenetics is based on such environmental toxins and the resultant carcinogenic mutations, frankly, all disorders from mental health to webbed feet are identifiable via genomic sequencing.

Further dedication is to future molecular geneticist, as the road to increased numbers of gene based targeted therapies and immunomodulators need a path guided by Brian Druker, MD-like wisdom to realize this Helix to Health in our lifetime.

Included in this dedication section is information on the need for further research on Genetic Mutations as well as the need for hybrid training so that all Neurosurgeons, Neurologists and Interventional Radiologists can be trained in procedures such as intraarterial stent retraction as treatment in embolic stroke cases, this article is now in the appendix 2 are near the end of this book.

Joseph Vacanti, MD of MGH will be offered the opportunity to train Surgeons to reproduce the Replacing Body Parts Clinical Trial so that humans can use cadaver, plastic or pig parts all washed and seeded with the recipients' cells to create organs that will not be rejected even without the need for immunosuppression.

What is a gene mutation and how do mutations occur?

A gene mutation is a permanent alteration in the DNA sequence that makes up a gene, such that the sequence differs from what is found in most people. Mutations range in size; they can affect anywhere from a single DNA building block (base pair) to a large segment of a chromosome that includes multiple genes.

Gene mutations can be classified in two major ways:

> Hereditary mutations are inherited from a parent and are present throughout a person's life in virtually every cell in the body. These mutations are also called germline mutations because they are present in the parent's egg or sperm cells, which are also called germ cells. When an egg and a sperm cell unite, the resulting fertilized egg cell receives DNA from both parents. If this DNA has a mutation, the child that grows from the fertilized egg will have the mutation in each of his or her cells.
>
> Acquired (or somatic) mutations occur at some time during a person's life and are present only in certain cells, not in every cell in the body. These changes can be caused by environmental factors such as ultraviolet radiation from the sun, or can occur if a mistake is made as DNA copies itself during cell division. Acquired mutations in somatic cells (cells other than sperm and egg cells) cannot be passed on to the next generation.

Genetic changes that are described as de novo (new) mutations can be either hereditary or somatic.

In some cases, the mutation occurs in a person's egg or sperm cell but is not present in any of the person's other cells. In other cases, the mutation occurs in the fertilized egg shortly after the egg and sperm cells unite. (It is often impossible to tell exactly when a de novo mutation happened.) As the fertilized egg divides, each resulting cell in the growing embryo will have the mutation. De novo mutations may explain genetic disorders in which an affected child has a mutation in every cell in the body but the parents do not, and there is no family history of the disorder.

Somatic mutations that happen in a single cell early in embryonic development can lead to a situation called mosaicism. These genetic changes are not present in a parent's egg or sperm cells, or in the fertilized egg, but happen a bit later when the embryo includes several cells. As all the cells divide during growth and development, cells that arise from the cell with the altered gene will have the mutation, while other cells will not. Depending on the mutation and how many cells are affected, mosaicism may or may not cause health problems.

Most disease-causing gene mutations are uncommon in the general population. However, other genetic changes occur more frequently. Genetic alterations that occur in more than 1 percent of the population are called polymorphisms. They are common enough to be considered a normal variation in the DNA. Polymorphisms are responsible for many of the normal differences between people such as eye color, hair color, and blood type. Although many polymorphisms have no negative effects on a person's health, some of these variations may influence the risk of developing certain disorders.

For more information about mutations:

The Centre for Genetics Education provides a fact sheet discussing variations in the genetic code.

More basic information about genetic mutations is available from GeneEd.

Additional information about genetic alterations is available from the University of Utah fact sheet "What is Mutation?"

Next: How can gene mutations affect health and development?

Published: July 6, 2015

NCI to Support Epigenetics Studies of Cancer

Nov 30, 2009

NEW YORK (GenomeWeb News) – The National Cancer Institute has issued a new grant program that will support population-based studies aimed at creating understanding of how epigenetics may be used to study the development of cancer.

The program, "Epigenetic Approaches in Cancer Epidemiology," is funded through R01 grants that may vary in size and duration depending on the project, and R21 grants of up to $275,000 over two years or up to $200,000 per year.

NCI is interested in funding studies that determine the roles of physical, chemical, and infectious agents as well as behavioral factors on the types and levels of DNA methylation, histone modifications, and miRNA changes in human populations of different races and ethnicities.

The institute also will support research into the roles epigenetic changes may play in the risk of cancer in human populations, and studies that seek to identify genetic, environmental, and host susceptibility factors that modify the risk of cancer associated with different races and ethnic groups.

Gene Theme Scene
Presents

The Progression of Cancer Screening to Genetic Detection and Genomic Targeted Therapies. A Multi-Institutional Assignment to Promote Additional Gene Based Research and to Identify Environmental Toxins, Dietary Carcinogens, Infectious Disease Irritants, and Germ Line Mutations. NIH Chromosomal Facts and a Digest of the Biography of Cancer (Emperor of All Maladies by Siddhartha Mukherjee, MD) are included and condensed by Clarisse Clemons Ferrara, MD.

Table of Contents

Forward Facts

Congratulations to Cellceutix for the recent FDA approval of Kevetrin by the Harvard Cancer Center's Dana-Farber Cancer Center Institute. Now the challenge is to educate insurance companies, physicians, and patients that up to 50% of malignancies have the P53 mutation in combination with other genomic sequence variant signatures. Presently Foundation One sequences for 300 malignant genomic signatures, now Leiomyosarcoma patients can receive Kevetrin when the P53 mutation is present.

I am renewing the declaration of war against neoplasm that all humans can fight. Genetically we share approximately 98.5% of basic genetic sequences. Although germ line mutations we inherit or the somatic mutations we develop due to Environmental Toxins, Dietary Carcinogens, Infectious Disease Irritants as well as the 10-15% of known genetic variants are detectable, even prenatally. All pathology is represented by a genomic variant of either known or unknown significance. We plan regional Centers that offer complete genomic sequencing services as well as provide training programs for gene based researchers, especially since certain mutations have not been matched with a gene based targeted therapy or an immunomodulator therapy such as the SPOP gene mutation for prostate cancer.

Oregon State Oncologist BRIAN DRUCKER, MD fought and won the battle to have CIBA/GEIGY finance the clinical trial process to bring Imatinib to his Chronic Myelogenous Leukemia patients, now, Novartis markets Gleevec.

Since Imatinib is the prototype of Gene Based Kinase Inhibitor Targeted Therapy, I will concentrate on inspiring our youth to become Gene– Based Researchers by explaining the mechanism of action of Gleevec.

Chronic Myelogenous Leukemia is the result of a 9-22 translocation (Philadelphia Chromosome). This means that the breakpoint Cluster Region (BCR) gene from Chromosome 22 is fused with the Abelson (ABL) gene on Chromosone 9.

This Abelson gene has a domain that can add a phosphate group to the BCR-ABL fusion gene (mutated tyrosine kinase), Gleevec blocks this repeated phosphorylation process. Other examples of Tyrosine Kinases are Platelet Derived Growth Factor (PDGF), Epidermal Growth Factor (EGF), Fibroblast growth Factor (FGF), Insulin Receptors and peptide growth factor (type of ligand), (just a few of the tyrosine kinases).

Gleevec (Imatinib) is a tyrosine kinase inhibitor for Ph+ Chronic Myelogenous Leukemia (CML) in order to explain kinase inhibition normal tyrosine kinase function needs to be understood. Kinase enzymes modify other proteins by adding a phosphate group (phosphorylation) PROTEIN +ATP+ tyrosine kinase = PHOSPHORYLATED PROTEIN + ADP

These tyrosine kinases take part in many functional changes. Note: the human genome contains ~500 protein kinase genes which is only 2% of all genes.

The fact is that approximately 30% of all human proteins may be modified by these protein kinases. One such activity is regulation of cellular pathways involved in signal transduction. In Ph+ CML leukemic cells the BCR-ABL tyrosine kinase enzyme is stuck in the "on" position and keeps on adding phosphate groups. Imatinib competitively binds the ATP site of the BCR-ABL kinase domain so the phosphate group cannot be added.

Since Imatinib competitively binds to the ATP site of the BCR-ABL kinase domain the addition of the phosphate group is blocked, preventing proliferation of CML (mutated white blood cells).

We need additional gene based researchers to develop multiple generation therapies that are needed because sometimes secondary Mutations occur creating "shifts" that reopen active conformational sites.

A combination of ongoing Immunomodulator and gene based therapies are needed. Therfore we plan to hire multiple teams available to recheck the genomic sequence variant signatures as indicated. The More You Know the More You Know there is More to Know and Early Detection is the Key to Your Longevity!

List of Some of the Additional Gene Based Therapies on the Market Matched to the Specific Genetic Mutations Targeted

Note: all are in the Small Molecule Class except for five that are in the Monoclonal Antibody Class those are: Bevacizumab, Cetuximab, Panitumumab, Ranibizumab, and Trastuzumab

Gene Based Protein Kinase Inhibitor Pharmaceutical Companies	Genetic Mutation Targeted
Imatinib Novartis	Abl-Bcr, c-kit
Dasatinib Bristol/Myers/Squibb	Abl-Bcr, c-kit, Src
Nilotinib Novartis	Abl-Bcr
Bosutinib Pfizer	Abl-Bcr, Src
Axitinib Pfizer	VEGFR, PDGFR, c-kit
Pegaptanib Pfizer, OSI	VEGF
Pazopanib Glaxo/Smith/Kline	VEGFR, PDGFR, c-kit
Ranibizumab Genentech	VEGF
Sorafenib Bayer, Onyx	VEGFR, PDGFR, BRAF, c-kit
Sunitinib Pfizer, Sugen	VEGFR, PDGFR
Aflibercept Bayer, Regeneron	VEGF
Afatinib Boehringer/Ingelheim	EGFR, ErbB2
Gefitinib AstraZeneca	EGFR
Panitumumab Amgen	EGFR
Vandetanib AstraZeneca	EGFR, VEGFR
Cetuximab Imclone, Bristol/Myers/Squib	ErbB1
Erlotinib Genentech/Roche	ErbB1, EGFR
Lapatinib Glaxo/Smith/Kline	ErbB1and2, HER-2
Trastuzumab Genentech	ErbB2

Tofacitinib Pfizer	JAK
Ruxolitinib Incyte	JAK
Vemurafenib Roche	BRAF
Crizotinib Pfizer	ALK, HGFR, MET
Regorafenib Pfizer	RET, VEGFR, PDGFR

Ruxlitimib	JAK
Incyte	
Bevacizumab	VEGF
Genetech	
Crizotinib	ALK, MET
Pfizer	
Vemurafenib	BRAF
Roche	

Abbreviations

Bcr-Abl	Breakpoint Cluster Region-Abelson
C-Kit	Binds to Stem Cell Factor/Receptor Tyrosine Kinase
Src	Sarcoma—Proto-Oncogene Encoding a Tyrosine Kinase
VEGFR	Vascular Endothelial Growth Factor Receptor
VEGF	Vascular Endothelial Growth Factor
PDGFR	Platelet Derived Growth Factor Receptor
BRAF	Murine Sarcoma Viral Oncogene Homolog B (B-Raf is a Growth Signal Transduction Kinase)
EGFR	Epidermal Growth Factor Receptor member of the Erb Receptor Tyrosine Kinases
Erb1-4	Erb family of Four Receptor Kinase Inhibitors Her1 (ErbB1, EGFR), Her2 (ErbB2, Neu), Her3 (ErbB3), Her4 (ErbB4)
HER-2	Human Epidermal Growth Factor Receptor 2
JAK	Janus Kinase (JAK) are Pseudo-Kinase Domain
Alk	Anaplastic Lymphoma Kinase Receptor Cluster Differentiation 246 (CD246) encoded by the ALK gene
MET	Mesenchymal Tyrosine Kinase (Hepatocyte Growth Factor) due to the MET gene receptor
RET	Receptor Tyrosine Kinases proto-oncogene encodes Glial Cell Line-Derived Neurotrophic Factor (GDNF)

W

Chapter One - 2655 B. C. to 168 A. D. Era Inferences

To digest the history of cancer we will start with Queen Atossa's 'time travel' to 2655-2600 B.C. - the era of Imhotep of Egypt. This 27th Century B.C. period would provide anatomical observations including aggressive types of cases that Imhotep states in hieroglyphics.

"There is no treatment!"

Queen Atossa's chiseled likeness

Hippocrates of Cos, Greece (460-377 B.C.) identifies Queen Atossa's 'tumor' as Karkinos, the Greek word for Crab; thus giving her illness a name. According to the thermology world, Hippocrates placed clay on the body to see what area dried first indicating increased heat generation due to illness.

Other authors imply that the clay was ingested for medicinal reasons. Siranus of Ephesus, a second century B.C. Greek gynecologist, was Hippocrates first biographer.

Aristotle of the fourth century B.C. also wrote about Hippocrates. Cleopatra in 50 B.C. reportedly used clay on her skin for cosmetic purposes.

In 500 B.C., the actual time period of Queen Atossa Achaemenes of Persia, Iran, my review contrary to the Emperor of All Maladies account that she ordered a slave to perform a mastectomy revealed that she lived to bear four sons to Darius; and that her diagnosis was probably inflammatory mastitis, probably incised by her physician Demos cedes who reportedly wanted Atossa to return to Greece.

Hippocrates

In 47 AD the Greco-Roman Philosopher, Celcius wrote an encyclopedia of medicine in which officially used the word cancer which is the Latin equivalent of crab. Fast forward to 168 A.D. when Claudius Galen described a breast tumor' vascular tributaries as the appearence of crab legs extending from its body. Galen hypothesized a universal cause; systemic overdose of black bile, trapped melancholy boiling out as a tumor. The Egyptians believed that the body has four humors - blood, phlegm, yellow bile and black bile. An alignment with Hippocrates thinking that health was a state of balanced humors. This theory was passed on by the Romans and was embraced by Galen's medical teaching which remained unchallenged throughout the middle ages for over 1300 years. During this period, autopsies were prohibited thus limiting knowledge.

Varied attempts were made to purge Atossa's black bile including frog's blood, lead plates, holy water, crab paste and caustic chemicals according to Siddartha Muhkagee, MD a senior Oncologist at Columbia University Medical Center and the Pulitzer Prize winner for non-fiction.

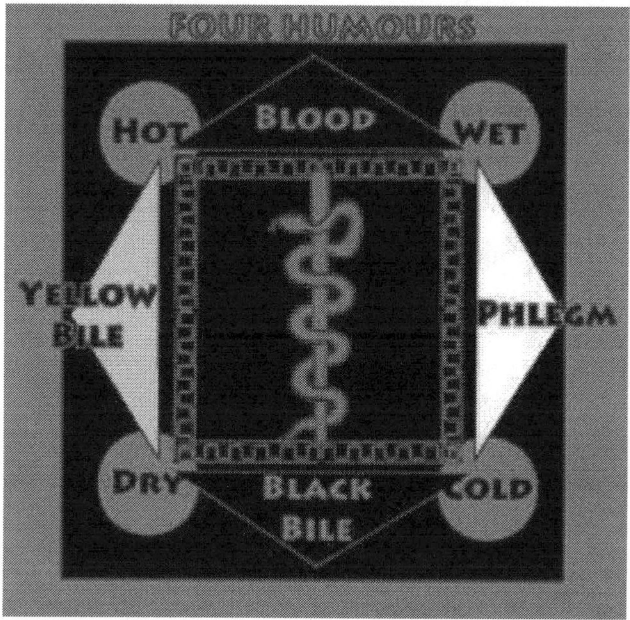

A diagram of the main principle behind medical treatment of the age

http://www.bbc.co.uk/schools/gcsebitesize/history/shp/ancient/greekknowledgerev1.shtml

Chapter Two -
Era of Universities from 1533 to 1793

In 1793 Matthew Baillie a Scottish physician and pathologist was a conformational anatomical Authority after publishing the Morbid Anatomy of Some of the Most Important Parts of the Human Body. Baillie In 1538 Andreas Vesalius at the University of Paris decided to create his own anatomical map. Vesalius published these anatomic drawings but did not find Galen's black bile.

First use of the word Autopsy, the Greek word for, "to see for yourself", depicted Vesalius work. Described cancers of the lungs, stomach and testicles. Like Vesalius, Baillie drew what he actually saw and Galen's channels of black bile, the humors of tumors that had gripped the minds of doctors for centuries vanished from the picture.

Treatment of Queen Atossa in the fifteenth century during which Galileo and Newton continued the development of a greater understanding of the human body, now would include the use of post-mortum anatomic evidence to study disease.

By 1628 autopsies performed by Harvey accelerated the understanding of blood circulation. 1751 Giovanni Morgagni of Padua was the first to use autopsies to relate the patient's illness to the pathologic findings.

These advances laid the foundation of scientific oncology, the study of cancer. In 1760, John Hunter a Scottish surgeon and Baillie's maternal uncle (who was one of the black bile theorist), began removing cancer surgically, initially on animals and cadavers in his own house. He found that when the tumor was moveable with the part, he should exercise great caution to determine the presence of additional sites. Dr. Hunter only had opium and alcohol to achieve near oblivion in his patients. Therefore the leap back to life was fraught with complications.

The pain of surgery and the high incidence of infection, led to a miserable death for those who survived the operation in many cases. The Scottish Surgeon John Hunter (1728-1793) suggested that some cancer could be cured by surgery and described how the surgeon would decide which cancers to operate on based on tumor invasion of nearby tissue or not. John Hunter offered sympathy in the case of invasive cancer since those cancers were considered non-resectable.

Chapter Three - Nineteenth Century Leaps

Rudolph Virchow 1821-1902, founder of cellular pathology, with the use of the microscope, documented his study of diseased tissues, therefore providing the scientific basis for modern pathology. As Morgagni had linked autopsy findings seen with the unaided eye to clinical illness, so Virchow correlated the microscopic pathological findings.

This method allowed a better understanding of the cellular picture of cancer and the evaluation of surrounding tissues for post-operative diagnostic decisions. In 1845 Scottish physician John Bennett used leaches and purging to treat leukemia. At autopsy, blood laden with white cells were thought at the time to have caused death due to an infection instead of leukemia. Four months later in 1845 Rudolph Virchow had a patient with symptoms similar to John Bennett's case.

Both patients had an absence of a source of infection which made Dr. Virchow think that the blood itself was abnormal and he called this blood 'Weissesblut' which means white blood, a literal description of the millions of white cells he had seen under the microscope.

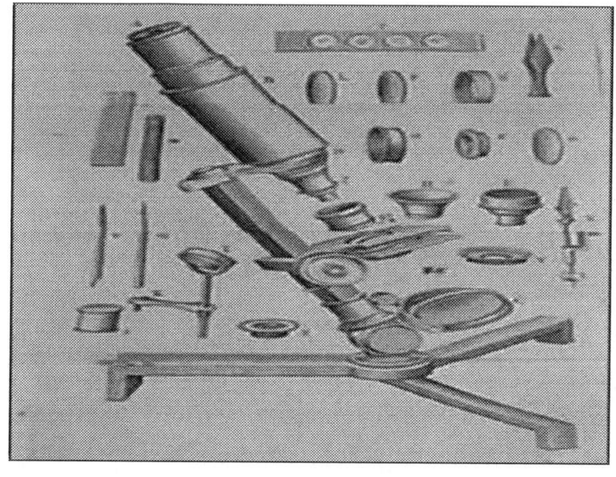

Early microscopes aided medical scientists by giving them new insight into tumors during autopsy.

http://www.antique-microscopes.com/photos/improved.htm

In 1847 Professor Virchow changed the name to Leukemia using 'leukos' the Greek word for 'white'. Professor Virchow at the University of Wurzburg launched a project of describing human disease in cellular terms. Between 1846 and 1867 anesthesia was publicly demonstrated at the Massachusetts General Hospital less than ten miles from where Sidney Farber's basement lab would be located a century later. William Morton unveiled a glass vaporizer with a quart of ether fitted with an inhaler.

One patient stated, that although he knew the operation was proceeding, pain was nullified by ether. Until the mid-nineteenth century, post-surgical infections were common and universally lethal.

In 1865, Joseph Lister, a Scottish surgeon who had learned the subtle unseen particle principle, based on an experiment performed by Louis Pasteur, confirmed what Pasteur wisely noticed that meat broth sealed in a sterilized vacuum jar would not permit turbidity. Pasteur stated that microorganisms that fell from the air caused turbidity.

Lister concluded that applying a dressing composed of a material capable of destroying the life of the floating particle should be used.

This chemical was carbolic acid originally known to be used by the sewage disposers. August, 1867 Lister used a combination of anesthesia and carbolic acid as an antiseptic and was able to routinely operate on cancer patients. By the mid 1870's.

Viennese surgeon, Theodur Bilroth, while a professor in Berlin, launched a systematic study that defined clean and safe surgical approaches to abdominal surgery. By the mid 1890's Bilroth had operated on 41 patients with gastric carcinoma. 19 of the 41 patients survived the surgery. Uterine, ovarian, breast, prostate, colon and lung tumors were surgically resected (in early cases) before they invaded other organs. The operations were considered cures in a significant fraction of patients. William Stewart Halsted entered surgery at a transitional moment in history.

Bloodletting, leaching and purging were still common procedures. Surgical sutures were made of catgut and surgeons walked around with scalpels dangling from their pockets.

If Queen Atossa's time travel had taken her to 1877, she might have encountered William Halstead, MD after his flight from this gruesome medical world of purges, bleeders, pus pails and Quacks.

After Halstead's European surgical training he refined the techniques of Bilroth, Volkmann, Hans Chiari, and Anton Woller.

When Halsted returned to NY, in 1880 he operated on patients at Roosevelt Hospital, Columbia, Bellevue and Chambers. (see page 62 of E of M) Queen Atossa, if seen in Halsted's Johns Hopkins clinic in the 1890's would have undergone a radical mastectomy including excision of the tumor with removal of the deep chest muscles, lymph nodes of the axillae, and clavicle. In the early nineteenth century, she would have consulted a radiation oncologist that would obliterate her tumor locally using X-RAY's.

Note: the history of Emil Grubes use of X-rays as well as Roetgen, the founder of X-RAYs is included in future pages. Therefore, in keeping with approximate chronologic oncological facts, Thomas Hodgkin, 1798-1866 published a paper in 1832 that described "Morbid Appearances of Absorbent Glands and Spleen. Before the continuation of Dr. Halsted's history, is the inclusion of the relevant Thomas Hodgkin a Quaker Physician that described the pathologic appearances of lymphoma. His descriptions were acknowledged, in 1898, (thirty years after Hodgkin's death).

Doctor Karl Sternberg, the Australian pathologist, also found bi-lobed nuclei (owl's eyes) in a field of lymphocytes.

The owl's eyes were lymph cells that had become malignant, now representative of Hodgkin's Lymphoma.

Now we will continue with facts relevant to William Halstead, MD. Surgical anesthetic Cocaine was used in 1884 at Volkmann's clinic. Doctor Halsted sampled Cocaine for five years after which he entered the Butler Sanitarium in Providence where he was treated for his cocaine habit.

Later, Doctor Halsted was recruited to Johns Hopkins by physician William Welch. By 1898, Dr. Halsted's radical mastectomy was taught. After 76 radical mastectomies had been performed, 40 survived more than three years. The other 36 died from disease that had not been "uprooted" from the body.

Dr. William Halsted
http://esgweb1.nts.jhu.edu/hmn/W02/annals.html

Chapter Four - Dr Halsted, Roentgen's X-Rays, Fleming's Penicillin, End of Nineteenth Century and Twentieth Century Advances

4-a Dr. William Halsted's Impact on the Field of Surgery

In 6I907 Halsted presented his data to the American Surgical Association in Washington D.C. By 1930 Surgeons mirrored Dr. Halsted's radical mastectomies. Chicago based gynecological surgeon Alexander Brunschwig, carried the radical surgical theme further, by devising an extensive operation for cervical cancer called a complete pelvic exenteration. A surgeon in New York in the thirties, George Pack, was nicknamed Pack the Knife. In 1933 Surgeons had a chilling discussion of stomach cancer during which the Arabian proverb was stated that 'he is no physician who has not "slain many patients".

To arrive at that sort of logic, (the Hippocratic Oath turned upside down) ignores the fact that untreated metastatic cancer is lethal even without a surgeons touch or introduced "air".

Radical surgery thus drew the blinds of circular logic around the value of the most extreme process for nearly a century.

In 1904 Young with Halsted's assistance excised the entire prostate in an operation called 'radical prostatectomy'.

In the early 1900's another of Halsted's students H. Cushing became a Neurosurgeon. Everts Graham, at the Barnes Hospital in St. Louis pioneered operations to remove cancerous lung tissue in 1933. He strived to excise the entire known lesion. The surgical community frowned on procedures that did not attempt to remove involved tissue. Salvaging procedures were considered a" make shift operation".

To indulge in such makeshift operations was to succumb to the old flaw of" makeshift kindness" that a generation of surgeons had tried so diligently to banish.

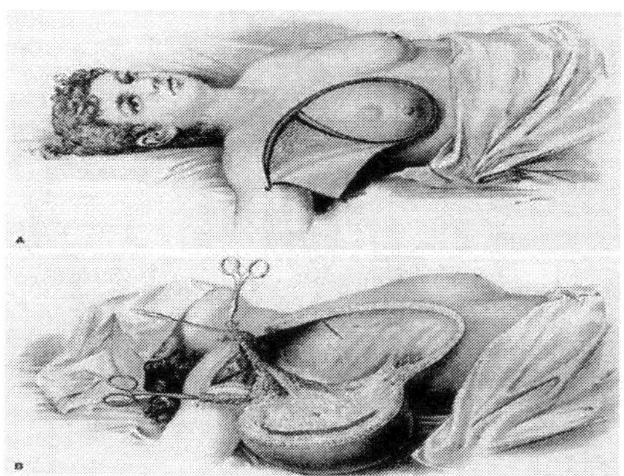

An example of Halsted's radical surgeries
http://edgeofthewest.wordpress.com/2008/07/30/mastectomy-through-the-ages/

4-b Roentgen, Becquerel, Pierre and Marie Curie (Develop the First Diagnostic Imaging)

In 1885 a few months after Halsted unveiled the radical mastectomy in Baltimore, William Roentgen, a lecturer at the WÜrzberg Institute in Germany was working with an electronic vacuum tube that shot electrons from one electrode to another when he noticed a leakage of radiant energy so powerful that it penetrated cardboard producing a white phosphorescent glow on a barium screen.

Roentgen's wife Anna's hand was placed between the source of his rays and a photographic plate, producing an image of her bones. In 1896, after Roentgen's discovery, Henri Becquerel, the French chemist, discovered that uranium emits invisible rays with properties similar to those of X-RAYS.

In Paris, friends of Becquerel, a young physicist named Pierre Curie who was married to Marie Curie, a chemist, began to scour the world for more chemical radiant energy.

Both were originally interested in magnetism. Pierre Curie used a miniscule quartz crystal to craft an instrument called an electrometer capable of measuring small dosages of radiation.

In 1902, a derivative of tons of pitchblende waste yielded purified-radium, that glows with a hypnotic blue light in the dark.

This elements instability was found to be a chimera of matter and energy-matter decomposing into energy.

Radium by virtue of its potency revealed an unexpected property of transmission of radiation through human tissue with resultant deposit of energy deep into the tissues. Radiation burned Marie Curie's bone marrow leaving her permanently anemic, eventually causing her death due to Leukemia in 1934. Radium and X-RAYS in excess can shatter strands of DNA and generate mutations that cause damage. Cells respond to these mutations by dying or more often by ceasing to divide.

X-RAYS preferentially kill the most rapidly proliferating cells in the body such as cells in the skin, nails, gums, and blood.

4-c Emil Grubbe Founder of Radiation Oncology

Grubbe, was a 21 year old Medical student in Chicago, when he determined the value of X-RAYS to inhibit malignant breast tissue cells on 3/26/86.

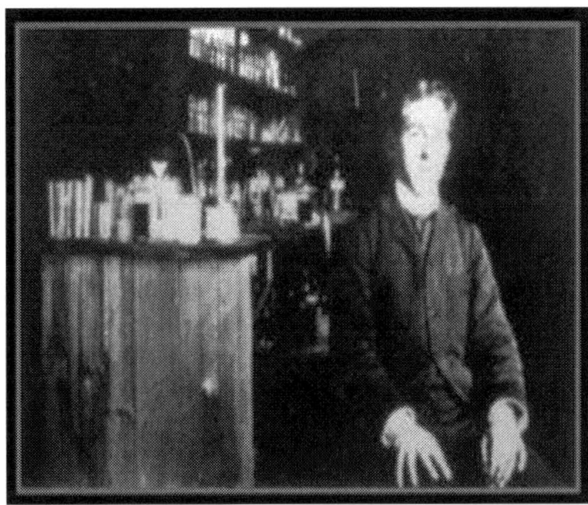

Grubbe suffered from radiation dermatitis and eventually endured an amputation of his left hand. In 1960 metastatic squamous cell skin carcinoma, caused the death of Dr. Grubbe, the father of radiation oncology, who is also credited with Continuing Medical Education for Physicians.

Dr. Emil Grubbe

http://jorhodeshomeopathy.co.uk/jo-blogs/2008/10/like-cures-like.html

Since X-RAYS could only be directed locally, radiation was of limited use for a cancer that had metastasized. Radiation itself produced cancers since DNA damage creates various known genetic mutations. Radium girls who painted watch dials with fluorescent radium sued U.S. Radium Company in 1927, since many developed leukemia and other types of cancers.

4-d Pathologist, Chemists and Pharmacologist

In 1856 William Perkin boiled nitric acid in benzene to produce a violet color creating a new chemical called aniline mauve. A German textile chemist synthesized phenols, alcohols, bromides, alkaloids, alazarine, and amides. Frederick Wohler found that boiling ammonium cyanate in inorganic salt would create urea.

In 1878 Paul Ehrlich found that analine derivatives stained organelles of the cell. Compound 606 (Arsphenamin) is considered the first modern chemotherapeutic agent used to treat Treponema Pallidum that causes Syphilis. Ehrlich met with successes with trypan red which he named Salvarsan (for salvation) which treated Trypanosomiasis, a parasite, and other microbial diseases.

From 1904 to 1908 Ehrlich tried amides, analines, sulfa derivatives, arsenics, bromides, and alcohols to destroy cancer cells-inevitably killing normal cells as well. 1908 Ehrlich won the Nobel Prize for the 'principle of specific affinity'. Then in 1915 Ehrlich developed an association with Bayer / Hoechst who became the producers of war gases such as thioglycol, a dye intermediate formed by boiling hydrochloride, creating mustard gases.

In 1917, Ehrlich's mustard gases lethal properties were evident, including (Nitrogen Mustard's respiratory damage, skin blisters, and blindness).

In 1919 American Pathologists Ed Krumblieur analyzed those that survived the Ypres mustard gas bombing. His findings included bone marrow depletion, leaving victims anemic and in need of multiple blood transfusions. They were prone to infections since their white blood cell counts were below normal. The sixteenth century physician, Paracelsus, once opined that every drug is a poison in disguise, thus cancer therapy found its roots in this obverse logic that every poison is a drug in disguise.

On 12/2/1943 the Johns Harvey Ship that was stockpiled with seventy tons of mustard gas, was blown up during WWII.

Burnt garlic and horseradish smell was in the breeze. 83 died in the first week due to mustard gas toxins which spread quickly over the next week.

Nearly a thousand died over the next few months due to complications of the mustard gas. The chemical warfare unit was housed within the wartime office of scientific research and development.

Goodman and Gilman of Yale secured research funding to study the white blood cell destroying properties of war gases. In 1942 Goodman and Gilman convinced Dr. Lindskowski to treat 48 lymphoma patients with intravenous mustard.

4-e Alexander Fleming, the Penicillin Mold, and other Microbial Related Advances

Note: since infectious disease irritants such as helicobacter pylori, hepatitis c, and human papilloma virus are known carcinogens, included here (not in the original text of EoM), is a history of the research that lead to the use of penicillin that has positively impacted mankind to the present day.

In 1928 Alexander Fleming, while in England, noticed a halo of inhibition of bacterial growth around a contaminated blue green mold on a Staphylococcus plate culture.

Although, Sir Alex Fleming discovered the penicillin mold in 1928, the Nobel Prize was not awarded until 1945. In the span between the two world wars Fleming's focus was the mold, *Penicillium notatum*. During the first years after his discovery, he grew and distributed the original mold unsuccessfully. He was unable to get help from a chemist with enough skill to make a form stable enough for mass production. 1941 to 1944 in Brooklyn, New York, Jasper Kane and other Pfizer scientists developed the practical deep tank fermentation method for the production of large stable quantities of pharmaceutical grade penicillin.

In 1952, Australian Biochemie, now called Sandoz, developed the first acid-stable penicillin for oral administration which is called Penicillin V. In the 1940's Selmon Waksman purified a diverse series of antibiotics by systemically scouring soil bacteria. One antibiotic came from a rod shaped microbe called Actinomyces. Waksman called it Actinomycin D, which was found to work by binding and damaging DNA.

Sidney Farber found that Actinomycin D resolved leukemia, lymphomas and breast cancers in mice. In 1955, Dr. Farber launched a series of trials to evaluate the efficacy of Actinomycin D.

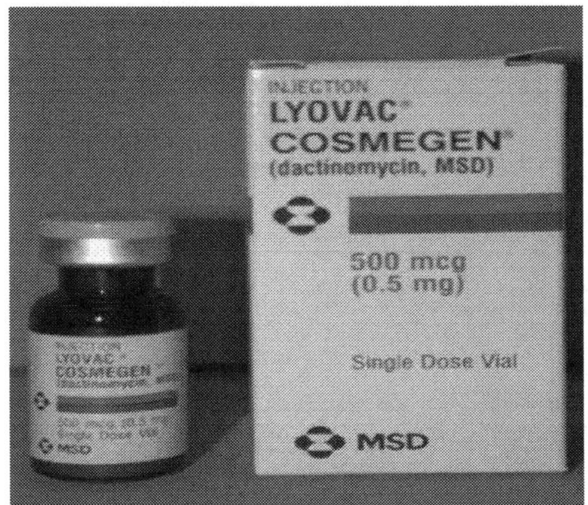

Dactinomycin

http://pharm.kku.ac.th/thaiv/depart/clinic/DispensingPharmacy/cancer/antitumor.htm

This clinical trial became a pharmacist's nightmare. It was so toxic that even if a small amount leaked out of the veins the skin would necrose. Actinomycin D had no effect on leukemia, but, one form of cancer that did show a response was Wilm's Tumor, a rare variety of kidney cancer, seen in children. Wilm's was the first metastatic solid tumor to respond to Actinomycin D. Goodman and Gilman published their findings in 1946 months before Sidney Farber's paper on the anti -folate, (Aminopterin), appeared in the press. In 1951, Trudy Elion found a variant anti-folate (6-mercaptopurine).

In 1948, Cornelius Rhodes pursued a collaboration between Elion's labs at Burrows Welcome that used 6-mercaptopurine in clinical trials with children with lymphocytic leukemia.

4-f Polio Vaccinations, Sidney Farber, Mary Lasker, and the History of the American Cancer Society.

In the 1920's Sidney Farber witnessed the polio epidemic. Initially, during the acute phase of polio, the virus paralyzed the diaphragm making breathing nearly impossible. In the 1930's the iron lung was the "artificial respirator" of this era.

In 1927 Franklin Roosevelt (paralyzed from the waist down due to the polio virus), launched a polio hospital and research center called the

Warmspring Foundation in Georgia. The National Foundation for Infantile Paralysis was an advocacy group formed to advance research and publicize polio. Eddie Cantor, an actor, created the March of Dimes campaign to coordinate the national fundraising efforts. Within weeks 2,680,000 dimes poured into the White House.

In the late 1940's John Enders with funds from these campaigns succeeded in culturing the polio virus in the lab. Sabin and Salk, building on Ender's work, prepared the first polio vaccine.

Farber wanted a similar campaign for leukemia. While Farber was in the middle of his Aminopterin trial Bill Koster toured Sidney Farber's laboratory. Bill Koster's group was the Variety Club of New England that had been founded in 1927, originally in Philadelphia by producers, directors, actors, entertainers and film theater owners.

1948 Farber and Koster founded the Children's Cancer Research Fund. Later called the Jimmy Fund. The child's real name is Einar Gustafason, a Swedish-American child diagnosed with Intestinal lymphoma who was hospitalized in the Boston Children's Hospital in May of 1948.

$231,000 was contributed during some of the original Jimmy Fund efforts.

If Queen Atossas time travel lead her to the US in 1918, she and Mary Woodward Lasker (a teenager back in 1918) would have witnessed the lethal Spanish Flu epidemic that took 50 million lives world wide and 600,000 American lives that year making it the deadliest pandemic in history at the time. By 1939 sulfa drugs were in existence, niacin (B3), was known to prevent pellagra. Tryptophan, leucine, and lysine and deficiencies were known. Vitamin C, (ascorbic acid), was found to prevent scurvy.

Mary Lasker was infuriated at the lack of knowledge of the American Medical Association regarding cancer. According to Business Week Magazine, Mary Lasker became the "fairy godmother" of medical research

Mary Lasker receiving the Presidential Medal of Freedom

http://wn.com/List_of_notable_Presidential_Medal_of_Freedom_recipients

In 1943, Mary Lasker visited the office of Doctor Little who was the director of the American Society for the Control of Cancer in New York. She was quoted as saying to a reporter, "I am opposed to heart attacks and cancer in the way one is opposed to sin". In 1943 the Reader's Digest, based on Mary Lasker's request, ran a series of articles on the screening and detection of cancer. In months $300,000 in donations were raised.

In 1945 the American Society for the Control of Cancer's Board had members who were businessmen, movie producers, pharmaceutical executives, lawyers and Mary Woodward Lasker's special contacts. This "lay group" as it was called, renamed the organization the American Cancer Society (ACS). The previous fund-raisers, The Women's Field Army was edged out and replaced by a fund-raising machine.

In 1944 donations were $832,000; 1945 the donations were $4,292,000; and in 1947 $12,045,000.

The previous director, Doctor Little, resigned by force in 1949 and Jim Adams, the president of the Standard Corporation and a member of the 'lay group' stated that the ACS should consist of laymen, with only four professional and scientific members.

Mary Lasker and Sidney Farber met in Washington, D.C. in the 1940's soon after his anti-folates shot Farber into international fame. In 1948 after Farber's anti-folate paper was published John Heller the director of the NCI wrote to Lasker about Sidney Farber.

1951 Albert Lasker, husband of Mary Lasker was diagnosed with colon cancer that had taken his life by 5/30/52. At the time, Mary Lasker and the Laskerettes as they were fondly called pointed out that one in four Americans would succumb to cancer.

Now, the American Cancer Society states that 50% of men and 33% of all women will develop cancer in their lifetime.

Sidney Farber

http://www.childrenshospital.org/gallery/index.cfm?G=49

Sidney Farber underwent a colectomy for a lesion of his colon for which he used a colostomy bag for the remainder of his life.

4-g Chemical Warfare, Hodgkins Lymphoma, National Cancer Institute Research

The MIT engineer Vaniveur Bush was the director of the OSRD (Office of Scientific Research and Development), in 1941. This organization developed the atomic bomb that was used on August 7, 1945. The morning after the Hiroshima bombing, the New York Times reported that Los Alamos, New Mexico's nuclear product was a result of pre-war physics/chemistry knowledge from Europe. National Cancer Institute Searches for Cancer Chemotherapies

In 1954, the U.S. Senate authorized the National Cancer Institute (NCI), to build a program to find chemotherapeutic drugs in a targeted manner.

In 1955 this effort was called the Cancer Chemotherapy National Service Center (CCNSC). During the interval of time from 1954-64, (CCNSC) tested 822,700 synthetic chemicals, 115,000 fermentation products and 17,200 plant derivatives that were used to treat 1,000,000 mice yearly with various chemicals in an attempt to find an ideal chemotherapeutic agent.

In the 1950's Henry Kaplan trained at the NCI (National Cancer Institute) and used X-RAYS to treat lymphoma in animals. In 1953 he persuaded physicists and engineers at Stamford to create a linear X-RAY accelerator to create a beam of radiation able to reach the inside of human tumors. In 1956, Henry Kaplan personally wielded a large block of lead with a minuscule pinhole that could direct millions of radiant energy (beams of X-RAYS) that were used to blast the malignant cells of Hodgkin's Lymphoma.

In the 1930's, the Swiss radiologist named Rene Gilbert was able to reduce the size of a Hodgkin's tumor with radiation. Independently, Vera Peters broadened the radiation field to regional lymph nodes. Independent of Peters, Henry Kaplan realized that the extended field of radiation improved the survival of Hodgkin's patients. Henry Kaplan's lymphography is a procedure considered to be a precusor of future radiologic evaluations of lymph nodes.

Hodgkin's Lymphoma is a regional illness and when early diagnosis is accomplished radiation is able to improve the five year survival and in some cases surpass the five year survival interval.

Kaplan performed an exploratory laparotomy with biopsies to ensure that the case was regional without abdominal lymph node involvement.

In 1963 George Canellos, a senior fellow at NCI, with Tom Frei listed the cytotoxic agents on one side of the board and a few of the cancers on the other actually deriving mathematical equations.

Cytoxin	Breast
Vincristine	Ovarian
Procarbazine	Lung
	Lymphomas
Nitrogen mustard (mustard gas)	
Actinomycin D (from soil bacteria)	

6- Mercaptopurine resulted from testing thousands of molecules in a drug screening effort to find the handful that possessed cytotoxic properties. Indiscriminant inhibitors of cellular growth was the common bond of agents such as nitrogen mustard that damages DNA. Vincent DeVita, M.D., was a fellow at the NCI in 1963. Vincent DeVita led the first test of the intensive combination MDMP. Methotrexate, later substituted by Procarbazine (Oncovin, Vincristine), Nitrogen Mustard, Oncovin, Procarbazine and Prednisone. The MOPP and VAMP Chemo/Combination caused nausea and infection.

Chemotherapy caused permanent sterility in men and women. The first known case of PCP (Pneumocystis carinii pneumonia) was a patient receiving MOPP therapy. The same PCP pneumonia in immune compromised homosexual men in 1981 was seen during the arrival of the HIV epidemic in America.

Several people after being treated with Hodgkin's lymphoma developed a drug-resistant leukemia that was caused by the MOPP chemotherapy.

In 1961 Donald Pinkle, a protégé of Sidney Farber, was recruited to Danny Thomas' St. Jude Hospital in Memphis, Tennessee. Dr. Pinkle decided to infuse methotrexate by a spinal tap into the cerebrospinal fluid, since the blood brain barrier had prevented chemotherapy from reaching brain metastases via the blood stream. The brain was also irradiated. Antibiotics and transfusions were used, and Dr. Pinkle's leukemia treatments were spread out over the course of two 2.5 years.

From 1968 to 1979 Pinkle's team ran eight trials, 278 patients. One fifth relapsed, 80% remained disease free. The longest was a six year remission at the time of the publication of the trial. “The iron is hot and it is time to pound without cessation”(Sidney Farber to Mary Lasker, September, 1965). September 1968 the Jimmy Fund turned twenty one.

The real Jimmy, Einar Gustafson, then thirty years old, was living in Maine with his wife and three children.

Chapter Five - Cancer Etiology Theories

The viral cause of cancer was fathered by a chicken virologist named Peyton Rous whose lab was located at the Rockefeller Institute in New York. In the 1920's the only known causes of human cancer were environmental carcinogens such as radium which caused leukemia in Marie Curie. Other carcinogens in the 1920's were organic chemicals such as paraffin and dye products that caused solid tumors. Percival Pott believed scrotal cancer, endemic to chimney sweeps, was caused by soot and smoke. These observations led to the somatic mutation hypothesis of cancer. Soot, paraffin and radium possessed the capacity to trigger the malignant potential of the cell. In 1910, Rous, the chicken virologist injected the tumor of one chicken into another chicken. He later wrote "this neoplasm is true to the original neoplasm".

The Rous Sarcoma Virus is called RSV. It was isolated after passing cells from the tumor through a series of filters with fine sieves. RSV was the first viral particle known to cause cancer. In 1935 Rous's colleague Richard Schope reported a papilloma virus that caused tumors in cotton tailed rabbits.

In 1958, Dennis Burkitt discovered an aggressive lymphoma that occurred endemically in a malaria ridden belt of sub-Saharan Africa. The distribution of the disease suggested an infectious cause.

The new virus was called Epstein Barr Virus or EBV which is also known to cause Infectious Mononucleosis. In 1952 the polio vaccine was a success. In 1962 Life Magazine stated that cancer "may be infectious". TB and Smallpox victims were confined and hysteria ensued. Reporters questioned if cancer was infectious why not quarantine patients to prevent its spread?

In 1960, Sidney Farber insisted that the NCI start a special cancer virus program. In 1966 Peyton Rous was awarded the Nobel Prize in Physiology and Medicine. December 10, 1966 in Stockholm Sweden Rous stated that a few viruses have a connection with the production of neoplasm. Rous realized that inherent in the cells a genetic mutation could exist (although he later went on to become lambastic saying that numerous facts exclude the somatic mutation theory). He believed that the viral theory was the unifying hypothesis. The viral theorist purported that radium and soot caused insults that activated endogenous viruses.

In 1972, Woodward, from the Chicago Tribune said that if Richard Nixon could end the War in Vietnam and defeat the ravages of cancer then he would have had a Lincolnesque niche greater than putting a man on the moon.

By 12/23/71 Nixon signed a bill giving the NCI 400 million dollars, 1972, 500 million, in 1973, 600 million and in 1974, a total of 1.5 Billion over the next five years. Sidney Farber had a cardiac arrest in his office and Mary Lasker wrote that the world would never be the same, 3/30/73.

Geoffery Keynes

http://www.modern-humanities.info/people/Keynes_Geoffrey.htm

Chapter Six - Surgical Changes, Medical Authority, and Women's Decisions

6-a Surgical Changes

In 1927 if Queen Atossa consulted with Geoffrey Keynes, she would have had a simple mastectomy followed by radiation. Dr Keynes wrote a technical report to his department at St. Bartholomew's Hospital in London about the removal of malignant lumps followed by radiation that he had performed since 1924. In 1953 a colleague of Keynes affiliated with the Cleveland Clinic, gave a lecture on the history of breast cancer, focusing on Keynes observations on minimal surgery for the breast.

George Barney Crile, had pioneered the use of blood transfusions. Keynes had used an apparatus devised by the elder Dr. Crile to transfuse blood in the First World War. Crile realized that breast cancer was inherently localized or inherently systemic.

Therefore Crile began to operate in a manner similar to Keynes's simple mastectomy. In 1928, four years after Keynes began his lumpectomies in London, two statisticians Jerzy Neyman and Egan Peareson provided a statistical method to evaluate a negative statistical claim.

'Power' is a measure of the ability of a test or trial to reject a hypothesis. A scientist's capacity to reject a hypothesis depends on how intensively he has tested the hypothesis and thus on the number of samples that have been independently tested.

Thus, even Crile, forty years after Keynes' discovery couldn't run a trial to dispute Halsted's radical mastectomy since the hierarchical practice of medicine, the gospel of the Surgical Profession, as Crile mockingly called it, was ideally arranged to resist change and to perpetuate orthodoxy.

Finally, Bernard Fisher ran a controlled clinical trial to test the radical mastectomy against the simple mastectomy and lumpectomy and radiation. By the late 1960's the cancer patient realized the fallibility of medical "judgements".

6-b Medical Authority and Womens Decisions

In 1973, Jane Roe (a pseudonym) sued the state for blocking her ability to end her pregnancy in a medical clinic, launching the Roe vs Wade case between the state, medical authority and women's bodies. Dr. Crile added a new generation of women in 1973, to refuse to submit to a radical mastectomy. Dr. Fisher wrote 'In God we Trust, all others must have data'.

In ten years the trial recruited 1,765 patients in 34 centers in the U.S. and Canada.

In 1981 the group treated with the radical mastectomy paid heavily in morbidity but accrued no benefits in survival, recurrence or mortality. From 1891 to 1981 was approximately 100 years after radical mastectomies. Dr. Bernard Fisher's Clinical Trial has now made the radical mastectomy rarely if ever performed today. Fisher was the chairman of the National Surgical Breast and Bowel Project that led large scale trials in Breast Cancer.

Chapter Seven - Chemotherapeutic and Hormonal Research

7-a Chemotherapeutic Research

Cisplatin is a drug that has the molecular structure of a central planar platinum atom with four extensions making it a symmetric chemical structure. In 1965 Barnett Rosenberg, a Biochemist, found that a platinum electrode had reacted to the salt in his solution to generate a new growth-arresting molecule that had diffused throughout the liquid. That chemical, cisplatin was used to treat testicular cancer with a new regimen BVP, Bleomycin, Vinblastine and Cisplatin (the P is for platinum).

In 1975, Dr. Einhorn had treated 20 patients with the BVP regimen and found dramatic and sustained responses; although the nausea caused patients to vomit as many as 12 times a day! By 1978 cisplatin was used in the treatment of thousands of cancer patients.

Funding designated by the National Cancer Act had stimulated the drug discovery program to the extent that thousands of chemicals were being tested for their cytotoxic capability, although the biology of cancer was still poorly understood. Taxol was purified from the bark of a hundred Pacific yew trees (endophytic fungi) to create synthesized paclitaxel.

In 1969 the blood- red colored Adriamycin was used therapeutically and caused the orange-red tinge hue of patient's skin. Eventually it was found that cytotoxic doses could irreversibly damage the heart. Etoposide was derived from the fruit of a poisonous may apple.

Bleomycin is a mold-derived antibiotic that was found to cause damage to the lungs (i.e. searing of the pulmonary tissue). Dr. Zubrod, the chief of the NCI, proposed additional solid tumor research. In the 1970's Burkitt's lymphoma, the aggressive tumor discovered in Africa was treated with seven drugs including a molecular cousin of nitrogen mustard.

In the 1970's the National Cancer Institute had many doctors who opposed the Vietnam War. Many Doctors applied to the NCI, since enrollment in a Federal Program gave an exempt status from the draft. By 1979 the NCI had recognized at least twenty comprehensive cancer centers. Human trials needed the approval of each hospital's institutional review board to test the many regimens and combinations that evolved from increased physician-researcher input.

7-b Hormonal Research

In 1927 the Urology department chairman Charles Huggins, who was trained as a general surgeon/physiologist, was interested in glandular secretions. Dr. Huggins found that surgical castration arrested the growth of prostate cancer. Edward Doisey was a biochemist that studied the hormonal factors in the estrus cycle of females. In 1929, Edward Doisey found the hormone estrogen which was extracted from hundreds of gallons of pregnant mare's urine. By the mid 1940's laboratories and pharmaceutical companies raced to synthesize analogs of estrogen.

The two most widely used analogs were primarin (pregnant mare urine) and DES (diethylstilbestrol). Both were marketed to treat the symptoms of menopause.

Dr. Charles Huggins

http://www.brimr.org/Reprints/HugginsQuotes.htm

Doctor Huggins realized that he could Use this estrogen in men to inhibit the production of testosterone and arrest the growth of prostate cancer-as he called it 'chemical castration'. In 1968 Dr. Huggins and Chemist, E. Jensen, found the 'estrogen receptor', the molecule responsible for binding estrogen and relaying its signal to the cell. Jensen found that some breast cancer cells were with or without this receptor (estrogen receptor positive (ER positive) and estrogen receptor negative (ER negative) breast Cancers). In 1962 a British chemist at ICI, Imperial Chemical industries, filed a patent for Tamoxifen (patent ICI46474).

A hormone biologist, Arthur Wadpole while attemping its use as an estrogen agonist noted that instead of turning the estrogen on as designed, Tamoxifen turned the estrogen receptor off, becoming an 'estrogen antagonist'.

Doctor Mary Cole, an oncologist became the clinical collaborator since she had interest in breast cancer. In 1969, ten of 46 patients of Dr. Cole's responded to Tamoxifen. In 1973 Craig Jordan a biochemist in Massachusetts at the Worcester Foundation, while searching contraceptives, used a molecular staining technique to stain breast cancer cells. Cancer cells that expressed the estrogen receptor, were responsive to tamoxifen, and cells that did not express the receptor did not respond to tamoxifen. Min Chiu Li had treated choriocarcinoma with methotrexate even after the tumor was not visible, and because of this he had been expelled from the NCI.

He felt that he had to treat beyond the disappearance of the tumor to the point of disappearance of all traces of human chorionic gonadotropin (the hormone produced by tumor cells -found in the serum of those who had choriocarcinoma). Later the NCI, found that patients treated by Dr. Li had the best survival.

In the1980's Queen Atossa's royal status may have landed her in the care of Dr. Paul Carbone, who was inspired by Dr Min Chiu Li to add chemotherapy after surgery, to decrease the rate of relapse from breast cancer. The word 'adjunctive' is Latin for 'to help'. Chemotherapy was found to eradicate residual malignant cells present after surgical resection. In 1972 Italian oncologist, Giovanni Boradonna, used the research information provided by DeVita, Canelos and Cordone; his concoction of Cytoxan, a cousin of nitrogen mustard, methotrexate (a variant of Farber's aminopterin) and flurouracil (an inhibitor of DNA synthesis).

This combination regime (CMF) was used as "adjuvant chemo", thus, the NCI conducted clinical trials to evaluate the efficacy of this combination. In 1973 Dr Boradonna had almost four hundred women to the "trial." Half had the CMF treatment and half did not. By 1977 Bernie Fisher recruited one thousand eight hundred ninety one women with estrogen receptor positive (ER-postive) breast cancer that had limited metastasis to the axillary lymph nodes. In 1985, while re-analyzing the relapse and survival data the effect of Tamoxifen was found to be dramatic. Tamoxifen had prevented 55 relapses in the over 50 year old group.

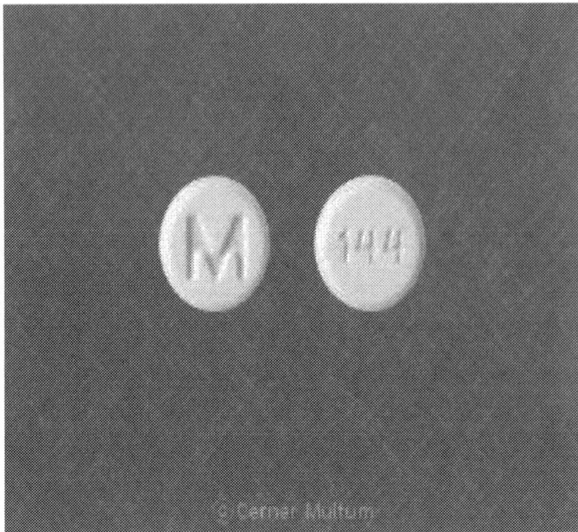

Tamifloxen, used in therapy for breast cancer to prevent relapse

http://www.drugs.com/tamoxifen.html

Palliate is the Latin word meaning "to cloak". In 1974 Yale New Haven had the first Hospice Facility in the US. In 1985, John Cairns, a Harvard Biologist used a State by State Statistical Record of Cancer Related Deaths sub-classified by the type of Cancer involved. These Registries had existed since WW11, and are called the Cancer Registry. Cairns estimated that between thirty five and forty thousand lives were being saved. This number included lives saved due to strategies that included pap smears, and mammograms, that detect cancer in its early stages.

Cairns believed that to spend the effort on treatment rather than vaccination or screening is to deny all precedent since the death rates due to malaria, typhus, scurvy, pellagra, scurvy and other scourges of the past have dwindled in the U.S. Since humankind has developed preventions for these diseases.

Percival Pott, a surgeon, noticed that chimney sweep adolescents developed scrotal wounds that were initially thought to be syphilitic. The wounds were treated using toxic mercury-based chemicals. Syphilis was mocked as 'one night with Venus, a thousand nights with Mercury."

Dr. Pott later thought that chimney soot lodged in the skin caused scrotal cancer. Pott wrote that this illness was man-made so his solution was to remove known carcinogens to avoid repetitive malignant transformations in the chimneysweeps of London. In 1788, The Chimney Sweep Act was passed in Parliament after the plight of London's chimney sweeps was publically exposed by English reformers.

Chapter Eight - 100% of Doctor Lung Cancer Death due to Smoking

1761 John Hill in a pamphlet entitled 'Caution against the Immoderate Use of Snuff' argued that oral tobacco, (snuff), could cause lip, mouth, and throat cancer. In 1948 Ernst Wynder saw the autopsy of a smoker with tar stained bronchi and soot stained lungs. Wynder and Doctor Graham, a thoracic surgeon and smoker himself set up a case controlled study. Dr. London Hill's Department was called the "Statistical Unit" known for plaques and inscriptions about the mosquito as the carrier of malaria, the sandfly as the carrier for black fever, and the tsetse fly as the vector of sleeping sickness.

In the 1940's, the Oxford geneticist, Edmond Ford, studied moths in the marshes not far from Oxford. In 1951 British efforts to nationalize health care resulted in a registry of all doctors, Approximately 60,000 names.

On 10/31/51 Hill and Doll sent a survey to these doctors about their smoking habits and 41,024 wrote back. Hill and Doll then divided the master list of doctors into smokers and non-smokers noting that every time a doctor died they would be notified with a description of the cause of death. From October 1951 to March of 1954, 789 deaths were reported including the cause of death. 36 of these physicians died of lung cancer, all 36 of these physicians were smokers!

Richard Doll and Bradford Hill published the above prospective study in 1956, the same year that it was reported that 45% of Americans were smokers. Clarence Cook Little was discovered by the tobacco lobbyists in 1954. Little believed that all diseases including cancer were hereditary. Smoking, little argued, unveiled the inherent genetic aberration.

Everts Graham the above mentioned thoracic surgeon died of metastatic lung cancer in 1937. President Kennedy's Surgeon General, Luther Perry, released the smoking-lung cancer bomb shell in 1964.

advertisements. Clarence Little argued that the efficacy of cigarette filters was immaterial because "there was nothing harmful to be filtered anyway". When B. Banzhoff filed a case against a TV station, he approached the American Cancer Society and the American Lung Association for support and in all cases he was rebuffed. In 1968, William Tallman an actor and former smoker, announced that he was

dying of lung cancer. He clearly stated, "If you do smoke, quit, don't be a loser!"

1/1/71 the last tobacco TV commercial was aired. Tallman did not live to see the termination of Advertisements since his lung cancer metastasized to his liver, bones, and brain.

Dr. Luther Terry, Surgeon General who announced that cigarettes caused lung cancer.

The FTC (Federal Trade Commission) was conceived to regulate

From 1954 to 1984, 300 product liability cases had been launched against tobacco companies.

Sixteen cases went to trial. None were settled out of court, the tobacco industry all but declared absolute victory, until Marc Edell, a New Jersey attorney insisted that he needed to know what cigarette makers knew about smoking risks. The jury later awarded Anthony Cippolone $400,000 in damages after internal knowledge was identified.

In 1994, the State of Mississippi filed suit against several tobacco companies seeking to recover over a billion dollars in health care costs. Florida, Texas and Minnesota also filed suit against the tobacco companies. In 1998, 46 states signed the MSA (Master Settlement Agreement) with four of the largest cigarette companies.

Tobacco smoking is a major preventable cause of death in India and China.

Richard Peto, an epidemiologist at Oxford and close collaborator of Richard Dolls (until Dolls death in 2005) estimated that one million smoking related deaths will occur in India. China noted that in 2010, lung cancer is a leading cause of death due to smoking in Chinese men.

Mesothelioma cancer clusters appeared in certain professions such as insulation installers, fire fighters, shipyard workers, and Chrysolite (white asbestos) miners. The rare profession and the rare tumor made the causal agent more identifiable.

Chapter Nine - Carcinogenic Mutations

In 1971 Diethylstilbesterol (DES) taken by mothers, caused vaginal and uterine cancers in daughters exposed in utero. Late 1960's a bacteriologist, Bruce Ames was working on mutations in Salmonella that possessed genes that allowed it to digest galactose. Ames could now test thousands of chemicals that increased the mutation rate (mutagens). Mutagens tended to be carcinogens as well, i.e. dye derivatives, XRAYS, Benzene compounds, nitroguanidine derivatives, even tobacco smoke constituents caused mutations.

Ames concluded carcinogens caused alterations in genes. In the late 1960's, Baruch Blumberg a biologist in Philadelphia, PA, was interested in genetic anthropology. He wanted to link genetic variations in humans to susceptibility for diseases.

Blumberg originally thought the AU was a gene but later found that AU was a piece of viral protein now known as the hepatitis B virus. Chronic infection with HBV (hepatitis B virus) causes liver cancer that is endemic in parts of Asia and Africa.

Decades of hepatitis C lead to liver cancer also. In 1969, Japanese researchers and Blumberg's group learned that viruses were transmitted through blood transfusions and that by screening blood before transfusion, looking for the AU antigen as one of the early bio-markers one could block the transmission of Hepatitis B.

Chronic infection with HBV is a carcinogen capable of being transmitted from one host to another. By 1979 Blumberg had devised a vaccination for HBV .The hepatitis B vaccination was commercially available by 1981. In 1982 Warner and Marshall identified Helicobacter Pylori.

The word 'pyloris' is Latin for 'gate-keeping', based on its location at the outlet of the stomach. Bismuth was one of the multi -drug treatments for H. Pylori. Barry Marshall's experiment on himself, to create a pre-cancerous state in his own stomach, was the way that H. Pylori was found to be the cause of peptic ulcerations that are pre-cancerous. DES, Asbestos, Radiation, Hepatitis Virus and H. Pylori, a stomach bacterium all cause cancer, an abnormal dys-regulatation of cell division that leads to a malignant transformation.

In 1962 Sidney Farber wrote to Etta Rosensohn, "the greatest need we have today in the human cancer problem, except for a universal cure, is a method of detecting the presence of cancer before there are any clinical signs or symptoms." George Papanicolaou was a Greek cytologist at Cornell University in NY. In 1928 he published "New Cancer Diagnosis" in a race betterment eugenics conference.

From 1928 to 1950 Dr. George Papanicolaou stated that the purpose of his " pap" smear was to detect the antecedents of cancer, by sampling asymptomatic women. He could capture the disease at its earliest stages, pushing the diagnostic clock back.

In 1952 Doctor Papilloma convinced the NCI to launch the largest clinical trial in Shelby County, Tennessee. 150,000 women were tested with the Pap smear. 555 women had invasive cervical cancer. 557 women had pre-invasive (pre-cancerous) changes, now referred to as cervical atypia.

In 1913 a Berlin surgeon named Albert Solomon took amputated breasts and had them x-rayed to see the shadowy outlines of cancer. He saw microscopic sprinkles of calcium lodged in cancerous tissue. In 1954, lawmakers in New York unveiled a program to provide a subscriber based health insurance to a group of employees in New York. This program called HIP was the ancestor of HMO organizations.

In 1981 eight years after the Mammography Study had been launched there were 31 deaths in the mammography group and 52 deaths in the control group. The American Cancer Society's Breast Cancer Detection Demonstration Project found that between 1976 and 1992 several parallel trials were conducted. Researchers named the randomized trials the National Breast Screening Study (NBSS).

Chapter Ten - HIV Victims in the Eighties not as Fortunate as Tim Brown Whose Donor 61 has a CCR5 Gene Mutation that Prevents the Production of Membrane Receptor that Allows HIV to enter Cells

In the 1980's, the Autologous Bone Marrow transplant and the STAMP (Solid Tumor Autologous Marrow Program were born. In 1981 the journal Lancet reported eight cases of Kaposi's sarcoma in eight homosexual men. One man also had PCP (Pneuocystis carinii pneumonia). The CDC (Center for Disease Control) in Atlanta from June to August of 1981 collected additional cases of PCP, Kaposi's sarcoma, Cryptococcal Meningitis, and rare lymphomas.

These patients were homosexual men whose immune systems were severely compromised. In 1982 the term Acquired Immunodeficiency Syndrome (AIDS) became the official terminology for this syndrome.

In 1983 Luc Montagnier's group found an RNA virus in a lymph node that could incorporate its genes into DNA and lodge them in the human genome, a retro-virus, first called Immune Deficiency Associated Virus (IDAV). In the late 1980's, marrow transplantation cost from $50,000 to $400,000 per patient.

In 1991, a patient with metastatic cancer that needed an autologous marrow transplant, was turned down for insurance payment, by Healthnet because this investigational procedure was not listed as one of the covered procedures based on the HMO's clinically proven protocols.

From 1988 to 2002 eighty six cases were filed against HMO's to pay for bone marrow transplants. In 1993, Charlotte's Law mandated coverage of bone marrow transplants by HMO's. Presently the American Society Clinical Oncology (ASCO) has yearly conferences that discuss the latest clinical trials in Oncology.

In 2000 Werner Bezwoda admitted to falsifying parts of his bone marrow studies and admitting to a serious breach of scientific honesty an integrity. 1986 Bailor's Article "Cancer Undefeated" discussed the overall mortality for lung cancer that had increased from 1970 to 1994. Lung cancer deaths for women over the age of 50 had increased 400%.

Chapter Eleven - The Genetic Info of the Time

In 1879 Walther Flemming, a biologist was studying abnormal cell division of salamander eggs. He stained the cells with aniline, a dye stain that highlighted a blue threadlike substance located deep within the cell's nucleus that became a cerulean shade before cell division. Flemming called his blue-staining bodies 'chromosomes' (colored bodies). The fourteen chromosomes of the salamanders were his focus. In 1914, Bovary published his chromosomal theory in a paper entitled 'Concerning the Origin of Malignant Tumors'.

In 1909, Carl Landsteiner implicated a viral cause for polio. In the 1920's viruses that cause cowpox and human herpes had been grown in laboratories. The mechanistic understanding of the cancer cell had been suspended in limbo between viruses and chromosomes.

In 1860, Gregor Mendel, (1822-1884), known as the founder of the science of genetics, via his study of pea plant traits, described "inherited particles of inheritance-a gene with traits" but no explanation about a gene's structure or functions were understood in his lifetime.

1915 Morgan proposed a crucial advance to Mendel's theory of inheritance, that genes are 'borne' on the chromosomes, therefore transmission of chromosomes during cell division allowed genes to move from a cell to its progeny.

In 1926 Avery at Rockefeller University in New York found that genes could move from one generation to the next by being carried on the chromosome. In 1944 Avery said genes are carried by one chemical, DNA, Deoxyribonucleic Acid-later to be described by Watson and Crick as the 'double helix'. George Beadle, Thomas Morgan's Student found that genes carried instructions to build proteins.

Beadle and Tatum found that a gene works by providing the blueprint to build a protein. Francis Crick in Cambridge discovered that the genesis of proteins from genes requires an intermediary step, a molecule called RNA, ribonucleic acid. DNA-RNA-Protein became the central dogma of molecular biology. In 1950, Monod's identification of RNAs ability to "copy genes" illuminated the internal mechanism of the normal cell.

In 1872 Hilario De Gouvea, a Brazilian Ophthalmologist, had a patient with retinoblastoma. After he removed the eye, the patient married a woman with no history of cancer in her family.

Two of their children had retinoblastomas in both eyes and died. The inherited factor of the retinoblastoma, as he called it, lived in the genes of his patient, and caused retinoblastomas that caused the death of his patient's daughters.

In 1910 Thomas Morgan, the fruit fly geneticist, noticed that occasionally the fruit flies would develop mutations that would be "carried" from one generation to the next. In 1928 Herman Joseph Mueller, a student of Morgan's, noticed X-RAYS could increase the rate of mutations in fruit flies. He produced hundreds of mutant flies over a few months.

Radiation was known to cause Marie Curie's leukemia and the tongue cancer of the radium watch makers. Could genetic alterations be the "unitary cause" of cancer since cancer is a disease of mutations? If Mueller and Morgan, student and mentor, put their scientific skills together, they would have uncovered this essential link between mutations and malignancy. Instead Mueller and Morgan went from close colleagues to pitted and embittered rivals. Morgan refused to give Mueller full recognition for his theory of mutagenesis.

Mueller in turn in 1932 was so sensitive and paranoid, that in 1932 after moving to Texas he took sleeping pills in a non-fatal suicide attempt.

In 1933, Morgan received the Nobel Prize in Physiology and Medicine for his work in fruit fly genetics. 1946 Mueller received the Nobel Prize also. Morgan wrote that one day in the future the convergence of medicine and genetics will mean that a doctor will call his geneticist friend for a consultation!

Unfortunately at the time this was laughable and it wasn't until the nineteen seventies that researchers and oncologists returned to the language of genes and mutations as the 'unitary cause' of cancer. Cancer researches knew that X-RAYs, soot, cigarette smoke, asbestos, Rous Sarcoma Virus, and the Brazilian case of the family with hereditary retinoblastoma all caused cancer.

In the early 1950's the Rous Sarcoma Virus occupied the limelight. Peyton Rous's personality was persuasive and inflexible and acquired a paternal attachment to his virus. He acknowledged the 1950 study of Doll and Hill that clearly showed that smoking was associated with increased incidence of cancer but Rous still thought 'Viruses were the answer'.

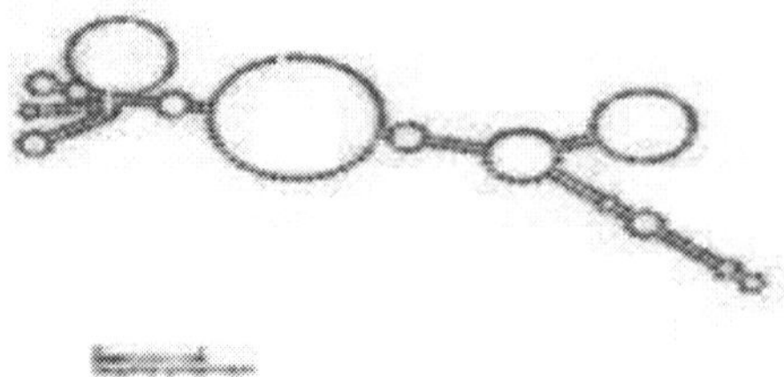

RSV is known to cause cancer in humans
http://www.freebase.com/view/en/rous_sarcoma_virus

In the nineteen fifties there were three camps of thought, the virologist's, the epidemiologist's thoughts of exogenous chemicals, and Theodore Bovari's successors, of the genes internal to the cell causing cancer.

Chapter Twelve - The Contemporaries of Renato Dulbecco Correlate the Viral Invasion of Genes

In 1951, Howard Temin, a post-doctoral researcher at the California Institute of Technology in Pasadena decided to study the Rous Sarcoma Virus in Renato Dulbecco's laboratory at Cal Tech. In 1958 Temin's eighth year in Dulbecco's lab he added the Rous Sarcoma virus to a layer of normal cells in a Petri dish which caused the cells to grow uncontrollably with pathological mitosis. The Rous Sarcoma Virus had physically attached itself to the cell's DNA and thereby altered the cell's genetic makeup, its genome.

Temin's statement was speculative since formal proof of the structural attachment of the Rous Sarcoma Viral genes into the cellular genome would not come until years later. In normal cells the conversion of DNA into RNA is called transcription.

The virus, had to possess the reverse capability called reverse transcription to infect the cell. Temin hired a Japanese post-doctoral student named Satoshi Mizutani, to purify this reverse transcription enzyme from the virally infected cells.

Mizutani, a chemist made chemical extracts from the virally-infected cells, he found enzymatic activity in the cellular extracts of the Rous Sarcoma Virus that was capable of converting RNA into DNA.

When he added RNA to this cellular extract, he could identify the creation of a DNA copy, REVERSING TRANSCRIPTION. Temin had his proof. The Rous Sarcoma Virus could write genetic information backwards, it is a retro-virus.

In May 27, 1970 Temin presented at the 10th International Cancer Congress at MIT in Boston. David Baltimore also picked up on the hint of the RNA to DNA conversion activity. He had befriended Temin in the 1940's while at a science summer camp in Maine. The 1970 Nature Magazine published the reports of Temin and Baltimore who had both identified the RNA→DNA enzymatic activity from the virus particle now called the retrovirus! Postulations by Temin and Baltimore were that the genes of the retrovirus existed as RNA outside the cells. When these viruses infect cells they make DNA copies of their RNA and attach the copies to the cells genes.

This DNA copy, called a pro-virus makes RNA copies. The RNA virus is regenerated using the cellular genome. Therefore RNA→DNA→ RNA, RNA→DNA→RNA, ad infinitum.

Sol Spiegelman, a Columbia University Virologist, heard Temin suggest that an RNA virus could enter a cell, make a DNA copy of its genes, and attach itself to a cells genome. This he believed could activate a viral gene. Therefore both an internal aberration and exogenous infection could be responsible for cancer. Spiegelman's conjecture about a human retrovirus was half right and half wrong. He was looking for the right kind of virus in the wrong kind of cell.

Retroviral activity explains HIV. This was the strange illness Spiegelman heard about before his death in 1983 (due to pancreatic cancer). This illness, erupting in homosexual men and blood transfusion recipients, this retrovirus, was identified one year after Sol Spiegelman's death. Clarified as a human immunodeficiency virus, HIV. Temin argued that since the Rous Sarcama virus could cause cancer by inserting a viral gene into cells, this proved that genetic alterations could cause cancer.

Chapter Thirteen - Foundations of Viral Mediated Genetic Carcinogenesis

In the 1970's Martin, Vogt, and Duesberg, made mutants of the Rous Sarcoma Virus and pinpointed the RSV cancer causing ability, to a single gene in the virus called SRC (sarc), a diminutive of sarcoma. This SRC gene was termed an oncogene (a gene capable of causing cancer). Erickson decided to decipher the function of the SRC gene and found that this oncogene encodes a protein whose function is to modify other proteins, by attaching a phosphate group to these proteins, a molecular tag.

Even in normal cells enzymes that attach phosphate groups to other proteins, are called kinases, behaving as molecular switches within the cell. The attachment of the phosphate group to a protein acted like an 'on switch' activating the protein's function. After a kinase turned on another kinase which turned on another and so forth. .

The signal was amplified at each step of the chain reaction and many molecular switches were thrown into their on positions. The confluence of many activating switches produce a powerful internal signal to a cell to change, moving from a non-dividing to a dividing state. SRC is a proto-typical kinase on hyperdrive.

The protein made by the viral SRC gene phosphorylated anything and everything around it including many crucial proteins in a cell. The SRC genome unleashes an indiscriminate volley of phosphorylation activating dozens of molecular switches.

In SRC's case the activated series of proteins eventually impinged on proteins that controlled cell division. The SRC forcibly induced a cell to change its state from non-dividing to dividing ultimately inducing accelerated mitosis, the hallmark of cancer.

In 1989, Mike Bishop, University of California, San Francisco (UCSF) virologist, became pre-occupied with the evolutional origin of this viral process, and Harold Varmis, a National Institute of Health Researcher joined forces with him, to be awarded the Nobel Prize for their discovery of the cellular origin of retro-viral oncogenes. The Varmis/ Bishop Proto-oncogene theory explained how radiation, soot, cigarette smoke and RSV (Rous Sarcoma Virus) could all initiate mutations that activate precursor oncogenes in the cell.

Chapter Fourteen - Environmental Toxins Affect the Structure and Function of Genes

Chemicals that cause mutations in DNA produce cancers because they alter cellular proto-oncogenes. This was Bruce Ames' correlation between carcinogens and mutagens. Smokers and non-smokers have the same proto-oncogenes in their cells. Smokers develop cancer at a higher rate because tobacco increases the mutation rate of these genes.

Thus the conclusion that other endogenous proto-oncogenes are present in the human cellular genome was rightfully affirmed. Geneticists have two ways to "see genes". Number one- structural pieces of DNA lined up along chromosomes. Number two-functional pieces imagined by Mendel as inheritance of traits from one generation to the next. One is structural and one is functional.

From 1970 to 1980 a Chicago hematologist, Janet Rowley, saw a human cancer gene in a physical form after she stained patterns of chromosomes in order to locate chromosomal abnormalities in the cancer cell. The cells she was studying were chronic myelogenous leukemia cells.

Janet Rowley

http://www.mdanderson.org/newsroom/news-releases/2010/cancer-genetics-pioneer-wins-margaret-kripke-legend-award.html

The Philadelphia Chromosome was named by Noell and Hungerford while identifying the fact that humans have 23 matched pairs of chromosomes, (a total of 46 chromosomes), in the case of the Philadelphia Mutation, one copy of the 22nd chromosome was missing a segment. Later Rowley found the segment of the 22nd chromosome on the tip of chromosome 9 and the tip of chromosome 9 was attached to the tip of chromosome 22. This genetic event was termed 'translocation'. In case after case, Janet Rowley found the same '9/22 translocation'.

Chromosomal abnormalities had been known since Von Hanseman and

Bovari. Finally Dr. Rowley had demonstrated an 'organized chromosomal chaos', with specific and identical mutations in a particular form of cancer! These chromosomal translocations create new genes called chimeras.

In the early 1970's, in Houston, Alfred Knudson studied Retinoblastoma, the eye cancer that De Gouveda had identified in Brazil. Knudson grouped children with the sporadic form separate from the familial form of

Retinoblastoma. The sporadic form of RB, appears between ages 2 to 4 years, the inherited or familial form of RB appears 2 to 6 months after birth.

The sporadic form of RB requires two genetic changes, the familial form requires one genetic change in the Retinoblastoma gene.

RB suppresses cell proliferation, since RB is a cancer suppressor gene which is the functional opposite of SRC, (an anti -oncogene). Some children carry one tumor suppressor mutation in the germ line and are highly susceptible to this RB tumor because only one somatic event is necessary.

Other children even though they do not carry the mutation in the germ line can acquire the RB tumor gene as a result of two somatic events. Sporadic Retinoblastoma develops at a later age because two independent mutations have to accumulate in the same cell. Knudson understood genes in a statistical sense.

By the late 1970's Varmis, Bishop and Knudson could describe the core molecular aberration of the cancer cell by understanding the SRC pro-oncogene and the anti-oncogene, the suppressor gene that like RB (Retinoblastoma) suppresses cell division by supplying the brakes to cellular proliferation. Doctor Robert Weinberg worked in the Dulbecco lab at the Salk Institute, isolating DNA from Monkey viruses to study their genes. In 1979 Temin and Baltimore discovered reverse transcriptase. Varmis and Bishop officially announced the presence of celluar SRC.

In 1972, Weinberg moved to MIT to study cancer causing viruses. In 1982 Weinberg, Barbacid, and Wigler isolated the RAS gene from human cancer cells. RAS is a gene present in all cells.

In normal cells RAS gene encodes modulatation of the on and off switch. In cancer cells the mutated RAS protein signals a cell to divide permanently in the locked on position. In 1983 Erickson received the

General Motors Prize for his research on SRC. Tom Frei was honored for advancement in the cure of Leukemia. By the mid 1980's, geneticists had demonstrated that the RB gene loci is chromosome 13. Webster Canence's Cinti. Lab, Brenda Gallie's Lab in Toronto, Weinberg's Lab in Boston, as well as Thad Dryja joined the hunt for RB.

A RB Ophthalomologist, turned Geneticist found that every normal cell has two copies of chromosone thirthteen, the RB gene (one in each copy of chromosome 13).

In 1985, Dryja and Friand, later in 1986 Friand, Weinberg and Dryja wrote a Nature Article on the isolation of the tumor suppressor gene. RB is also isolated from lung, bone, esophageal, breast and bladder cancer. RB like Ras is expressed in every dividing cell! Calling it RB underestimates the prowess of the gene.

The RB suppressor gene encodes a protein RB whose function is to bind to several other proteins and keep them tightly sealed, preventing them from activating cell division.

When a cell decides to divide it tags RB with a phosphate group, a molecular signal that inactivates the gene and forces the protein to release it's partner RB, thus acting as a gatekeeper for cell division, (key molecular openings), a series of key molecular floodgates.

The RB Suppressor gene encodes a protein whose function is to bind to several other proteins keeping them tightly sealed in their pocket, preventing them from activating cell division.

When a cell decides to divide, it tags RB with a phosphate group-molecular signal that inactivates the gene and forces the protein to release its partners. RB thus acts as a gatekeeper for cell division opening a series of key molecular floodgates each time cell division is activated and closing them sharply when cell division is completed. Mutations in RB inactivate this function. The cancer cell perceives its gates are perpetually open and is unable to stop dividing. The cloning of RAS oncogene and Retinoblastoma and its tumor suppressor anti-oncogene was transformative to cancer genetics.

From 1983-1993 additional human cancer genes were identified, myc, neu, fas, ret, akt (all oncogenes), and, p53, pVHL, APC (all tumor suppressors). Mutations of the tumor suppressors increase the risk of developing various types of cancers.

The Lynch Syndrome was described by Henry Lynch, an oncologist in Nebraska whose family of patients had a family history of colon, ovarian, stomach and biliary cancer from generation to generation.

In Li-Fraumeni Syndrome there were recurrent bone, visceral sarcomas, leukemias, and brain tumors. 1980-1990's with the use of molecular genetic techniques, cancer geneticists began to clone and identify cancer linked genes.

Mary Claire King and later Marc Skolnick who worked together at Myriad Genetics Center in Utah, found BRCA -1, (the cancer predisposing gene alteration), frequently represented in the population. BRCA-1 is a gene that strongly predisposes humans to breast and ovarian cancer.

Note: after speaking to Falicitas Lacbawan, M.D., she stated that the cost is $3,500 to have this test. Dr. Lacbawan is an NIH trained clinical geneticist, and molecular pathologist. The cost has decreased now that Myriad no longer monopolizes BRACA testing. Robert Koch postulated that for an agent to be the cause of a disease it must, 1) be present in the diseased organism, 2) be capable of being isolated from the diseased organism, 3) be able to recreate the disease in a secondary host when transferred from the diseased organism.

C- myc is an oncogene discovered in leukemia cells. Phillip Leder's team at Harvard patented the Onco Mouse. Even though an aggressive oncogene had been 'stitched' into the mouse breast loci, the mice did not develop cancer until late in life after pregnancy. This fact indicated that the hormones of pregnancy are required to achieve full transformation of the breast cancer oncogene

Chapter Fifteen - Hallmarks of Cancer

In 1998 Physician-Scientist Bert Vogelstein at Johns Hopkins determined the signaling pathways for the RAS, MEK, and ERK proteins that cause uncontrolled division (pathological mitosis). 1990's Surgeon Scientist Judah Folkman, at Children's Hospital in Boston, demonstrated that activated signal pathways within cancer cells (Ras), could induce neighboring blood vessels to grow, Dr. Folkman called this 'tumor angiogenesis'.

Stan Korsmeyer found mutated genes that blocked cell death, thus influencing cancer cells to resist death signals. Other pathways provide neoplastic cells with motility to migrate from one tissue to another initiating metastasis.

Judah Folkman
http://news.harvard.edu/gazette/story/2008/02/cancer-research-pioneer-judah-folkman-dies-suddenly-at-74/

Other gene cascades increase cell survival in hostile environments, so that cancer cells traveling though the bloodstream can invade other organs and not be rejected or destroyed in environments not designed for their survival. Cascades of aberrant signals originating in mutant genes fanned out within the cancer cell, promoting survival, accelerating growth, enabling motility, recruiting blood vessels, enhancing nourishment, drawing oxygen, and sustaining cancer's life. These gene cascades are perversions of signaling pathways used by the body in normal circumstances.

The motility genes are normally used when immunological cells move to a site of infection. Tumor oncogenesis exploits the same pathways that are used to create blood vessels to heal wounds.

In the early 1990's the genesis of cancer in terms of molecular changes in genes, was the model cancer biologist sought for in each case. In 2000, Weinberg and Hanahan, published Hallmarks of Cancer, to summarize the rules of cancer's pathways, behaviors and genes, rules that govern the transformation of normal human cells into malignant cancers.

The molecular, biochemical and cellular traits are acquired capabilities shared by perhaps all types of human cancer.

Hallmarks of Cancer

1. Self-sufficiency in growth signals
2. Insensitivity to growth-inhibiting antigrowth signals
3. Evasion of programmed cell death (apoptosis)
4. Limitless replicative potential
5. Sustained Angiogenesis
6. Tissue invasion and metastasis.

Stem Cell Cryobanks

Umbilical Blood rich in stem cells can be used for a bone marrow transplant in the future when stored in cryobanks.

In 2003, the mortality from lung, breast, colon and prostate cancer had dropped 1%. Between 1990 and 2005 the cancer death rate had dropped 15%. Mary Lasker died of heart failure at age 93 in 1994. In 1994 a year after Mary Lasker's death, Ed Harlow, a cancer geneticist stated that our knowledge of molecular defects in cancer, has come from the dedication of twenty years of the best molecular biology research. Yet, at the time information did not translate to effective treatment nor to any understanding of why many of the current treatments succeed, or why others fail. In 2004 Bronchoalveolar Cell Lung cancer was treated with Carboplatin and Taxol, augmented with radiation.

Could carboplatin "fix" a mutated gene? How would Taxol know which cells carried the mutations in order to kill them? How would the mechanistic explanation of illness connect with the medical interventions? Antifolates, such as Farber's Aminopterin, interrupt the metabolism of folic acid, blocking a crucial nutrient needed for cell division. Nitrogen mustard and cisplatin chemically react with DNA. DNA-damaged cells cannot replicate their DNA, therefore, unable to divide. Vincristine thwarts the ability of a cell to construct the molecular scaffold required for cells to divide. Cisplatin targets cellular growth of hair, blood, skin and GI mucosa.

Malignacies occur because of an accumlation of mutations in the cellular DNA.

1. Cancer Mutations activate internal proto-oncogenes and inactivate tumor suppressors, thus unleashing the accelerators and brake

suppressors that operate during cell division targeting the hyperactive genes, while sparing their modulated normal precursors as a way to attach cancer more discriminately.

2. Proto-oncogenes and Tumor Suppressor Genes lie at the hub of cellular signaling pathways. These pathways exist in normal cells but are tightly regulated. The dependence of a cancer cell on permanently activated pathways is a second potential vulnerability.

3. The relentless cycle of mutation, selection and survival creates a cancer cell that has acquired properties besides uncontrolled cell growth, i.e. the capacity to resist cell death signals, the capacity to metastasize throughout the body, and the capacity to incite the growth of blood vessels. The acquired dependence of a cancer cell on these processes is a third potential vulnerability.

Thus new cancer treatments, modalities needed to target genes, pathways and acquired capabilities, were represented by Tamoxifen, which attacks the breast cell dependence on estrogen. This was the most targeted treatment until the late 1980's.

In 1986, acute promyelocytic leukemia (APL) that was first identified in the 1950's as a distinct form of adult leukemia was researched. Retinoic acid, an oxidized form of Vitamin A was identified in 1985 by a team of researchers from Shanghai, China led by Zhen Yi Wang who traveled to France to meet the hematologist, Laurent Degos.

Since both Wang and Degos identified retinoic acid, called cis-retinoic acid, and trans-retinoic acid, they found that the molecules were similar, although the atoms were arranged differently in the two chemicals. Trans-retinoic acid with chemotherapy caused Wang and Degos to conclude that 75% of patients treated with trans-retinoic acid and chemotherapy would have prolonged remission to APL.

In 1984 Rowley identified chromosome 15 fused with a fragment of chromosome 17 in acute promyelocytic leukemia patients. This created an activated chimeric oncogene with the proliferation of promyelocytes.

In 1990 scientists from France, Italy and America found the APL oncogene that encodes the protein that is tightly bound by trans-retinoic acid. This binding to trans-retinoic acid extinguishes the APL oncogene signal. In 1982 Lakshmi Padhly a scientist from Banbary working in Weinberg's lab reported the isolation of an oncogene from a rat tumor called neuroblastoma. This new oncogene was named neu. Neu, was tethered to the cell membrane with a large fragment that hung outside the cell membrane making drug access easier.

Chapter Sixteen - Her-2 Testing and Timely Treatment Battles Won and Lost

In 1986, Jeff Orebin, and Mark Greene demonstrated the fact that anti-neu antibodies arrested growth of neu cancer cells. In the early 1980's Weinberg's team discovered the human homolog of the neu gene noting its resemblance to another growth-modulating gene, human epidermal growth factor receptor. The researchers called the gene Her-2. Genetech, short for Gene Engineering Technology, hired researchers in the 1970's that invented a technology termed Recombinant DNA. This technology allowed genes to be manipulated by engineers.

Axel Ullrich rediscovered the Her-2 neu oncogene tethered to the cell membrane.

Dennis Slomon, the UCLA oncologist, performed a series of studies on human leukemia virus, HLV-1, the retrovirus shown to cause human cancer. Slomon had been collecting and storing cancer tissue from surgical specimens. Slomon proposed collaboration with Ullrich, who sent Slomon, the DNA probes for Her-2 from Genetech so he could test his cancer samples with Her-2. This new Her-2 was able to save Barbara Bradfield's who was still alive in 1993.

She was a gynecologist who had to fight her HMO to pay for testing of the Her-2 marker, especially since Herceptin which was named for the Her-2 marker (the word 'Intercept' and the word 'Inhibitor'), was a treatment she sought to receive. Herceptin was still investigational, therefore testing Marta Nelson's tumor was not paid for until political activists picketed outside Genetech in October, 1994.

Doctor Nelson's tumor was finally tested and she was found to be an ideal candidate for Herceptin; but before she could receive Herceptin, (awaiting approval), she had drifted into a coma and died nine days after the tumor was tested.

She was forty one years of age at the time of her death (she was diagnosed at the age of 33).

In 1998, Herceptin was FDA approved.

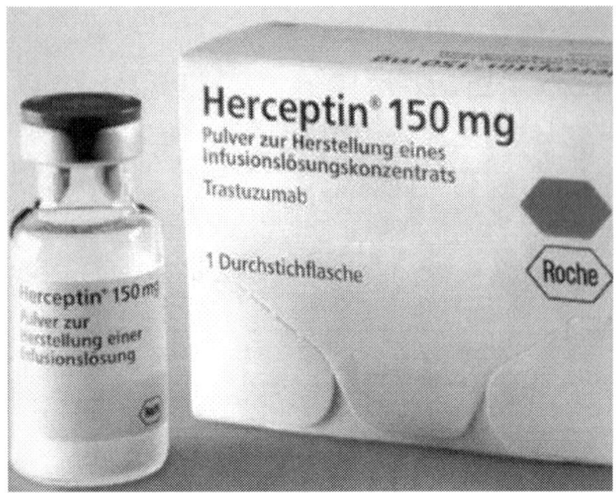

Herceptin

http://www.targetedcancerdrugs.com/herceptin.htm

Chapter Seventeen - Doctor Druker's Gene Based Imatinib (Gleevec) is a Targeted Therapy for Chronic Myelogenous Leukemia (CML)

While examining chronic myelogenous leukemia cells in 1973, Janet Rowley identified the chromosomal fusion 9:22 chimeric gene. In 1982, a Dutch Researcher identified a gene on chromosome 9 and called it ABL (Abelson). In 1984, while working in Maryland, ABL's partner on chromosome 22 was found, he then called the new gene BCR (Breakpoint Cluster Region). The fusion of these genes created the chronic myelogenous leukemia oncogene.

In the mid 1980's, Ciba-Geigy chemists located in Basel, Switzerland were developing drugs that would inhibit kinases. The human genome has about 500 kinases about 90 belong to the subclass that contain SRC and BCR/ABL.

Every kinase attaches a phosphate tag to a unique set of proteins in the cell. Kinases are molecular master switches, turning on or off certain pathways, thus providing the cell a co-ordinate set of internal signals to grow, shrink, move, stop or die. The CIBA /GEIGY team hoped to discover drugs that could activate or inhibit kinases selectively in cells, thus manipulating the cells master switches.

This team was led by Alex Matter, a Swiss physician/biochemist. In 1986, Matter was joined by Nick London in this hunt for selective kinase inhibitors.

Jurg Zimmerman, a chemist on Matter's team created thousands of variants of the parent molecules handing them off to a cell biologist, Elizabeth Brukgundr who 'weeded out' those that were insoluble or toxic and sent them back to Zimmerman. If it did not fit he changed it again like a key in the hands of a locksmith.

In the late 1980's Nick London travelled to Dana Farber Cancer Institute in Boston and met Brian Druker, M.D., a faculty member at Dana Farber who was about to launch an independent lab in Boston.

Dr. Druker whom I met in the 1990's and later spoke to in regard to a relative with cancer, was particularly interested in chronic myelogenous leukemia. This leukemia could be identified by the BCR/ABL kinase mutation. Dr. Druker had researched the fact that present in the CIBA

/GEIGY freezer in Switzerland, was the chemical kinase inhibitor with special affinity for the BCR/ABL gene that he needed for his chronic myelogenous leukemia patients. Dr. Druker proposed a collaboration between the Dana Farber Cancer Institute and the CIBA/ GEIGY pharmaceutical company in Basel, Switzerland.

Although that agreement fell apart, Brian Druker, MD was persistent, in 1993 he left Boston to start his own laboratory at the Oregon Health and Science University in Portland, Oregon. Doctor Druker, recontacted Dr. London who had found a molecule that might bind BCR-ABL with high specificity and selectivity.

The molecule was called CGP57148. Dr. Druker went to the legal team at OHSU and had the lawyers to humor him by signing the agreement. Two weeks later he had a package from Basel with a small collection of kinase inhibitors to test in his lab.

Back in 1993 one of the suggested treatments for CML was an allogenic bone marrow transplant, which was the protocol pioneered in Seattle, back in the 1960's, by Donnall Thomas.

Dr. Brian Druker

http://www.ohsu.edu/xd/about/news_ev e nts/news/2008/cancergift102908.cfm

When Dr. Lydon's kinases arrived in Dr. Druker's hands he added CGP57148 to Petri dishes with chronic myelogenous leukemia cells and overnight, the CML cells involuted in response. Dr. Druker implanted CML cells in mice and after treatment with the CGP57148 the tumors regressed in days. Thankfully, normal mouse cells were undamaged. Dr. Druker then took bone marrow from humans with CML and added CGP57148 (also called imatinib or Gleevec), to the human marrow cells in the petri dish. The mutated Chronic Myelogenous Leukemia Cells died leaving the normal blood cells!

These facts were published in Nature Medicine, with Dr Druker, as first author, London, the senior author, Buchdunder, and Zimmerman as key contributors.

Dr. Druker expected CIBA/GEIGY to be thrilled at these results, but since only a few thousand patients develop CML each year, the cost of one to two hundred million to launch this clinical trial seemed prohibitive, even to the pharmaceutical industry. Plus, the fact that CIBA /GEIGY had partnered with Sandoz to form a new company called Novartis, made the prospect of spending millions on a new molecule to benefit "only thousands" gave Novartis cold feet.

Dr. Druker found himself in the position of pushing a pharmaceutical company to propel it's own discovery into a clinical trial! So from 1995 to 1997 Dr. Druker flew back and forth between Basel and Portland trying to convince Novartis to continue the clinical development of the CML drug.

Dr. Druker was known to have said, "Either get this drug into clinical trials, or license it to me." Dr. Druker assembled a team of physicians to run a potential clinical trial of CGP57148 on CML patients.

Finally in 1998 Novartis relented and agreed to synthesized and release a few grams of CGP57148 (about enough for a trial on a hundred patients)

By the winter of 1998, Druker, Sawyers, and Talpaz had witnessed dozens of 'remissions'. Of the patients on Gleevec (the trade name for CGP57148) 53 showed complete response within days of starting this drug. Clearly with patients still in remission, Gleevec is still a success!

In 2005 a Bristol Myers Squib chemist working with the oncologists Sawyers and Talpaz had a patient that had responded for three years on Gleevec, after which this patient's CML had become resistant. These cases were rare, as the vast majority of CML patients remain in striking remissions to this day. But occasionally, Gleevec resistant leukemia cells recur.

Sawyer discovered that Gleevec resistance was explained by new mutations occurring that altered the structure of the BCR-ABL that creates a protein still able to drive the growth of the leukemia.

Now, no longer capable of binding to Gleevec, this mutated CML, escaped targeted therapy by changing the target. By 2005 chemists at Bristol Myers Squibb and Sawyer's team generated another kinase inhibitor to target Gleevec resistant BCR-ABL.

This drug DASATINIB is not a structural analog of Gleevec, it accessed BCR -ABL through a separate molecular crevice on the protein surface. This patient that had become resistant to the original Gleevec responded to DASATINIB and in 2009 remained in remission.

Chapter Eighteen - The Human Genome Project and the Cancer Genome Atlas

The Human Genome Project was completed in 2003. Fully sequencing the genome of several human cancer cells is more complex. This effort once completed is called the Cancer Genome Atlas. Presently the Cancer Genome Atlas includes the most common types of cancer. The equivalent of ten thousand Human Genome Projects has yet to be sequenced in terms of completing the Cancer Genome Atlas. The Human Genome Project presents the normal genome compared to cancer's abnormal genome to be juxtaposed and contrasted.

The Cancer Genome Atlas will present the entire genome signature of cancer, derived by sequencing the entire genome of several tumor types making sure that every single mutated gene pattern is identified.

The Cancer Genome Atlas Consortium has multiple interconnected teams and labs in several nations. Bert Vogelstein's group at Johns Hopkins, has assembled a Cancer Genome Sequencing facility. The human genome contains approximately 20,000 to 25,000 genes in total.

From gene to gene and now pathway to pathway, the knowledge must move from anatomy and physiology, to therapeutics, with an increase in the number of patients driven by screening, since, early detection is the key to longevity, and, the more you know, the more you know, there is more to know!

Open Letter to New York State Representatives in RE to the Gene Theme Scene Medical, PC

As the Founder of Cancer Screening Centers, a 501 c 3 organization, I have included the Continuing Medical Education Certificate granted by the Omnia-Provia Education Collaborative, Inc. accredited by the Accreditation Council for Continuing Education (ACCME) to provide continuing medical education for physicians, the full title is Evolving Strategies in Melanoma Management. Genomics and Genetics.

The second document, was issued from the University of Michigan Medical School Office of Continuing Medical Education on-line activity entitled Genetic Profiling in Early Stage Breast Cancer signed by Julie Wilson CME Credit Coordinator. From the end of 2009 to Oct 18, 2013, I have listened to Reach MD during my hour ride in and out of New York, from Long Island. Presently I listen to Doctor Radio produced by New York University, Langone Medical Center.

I am honored to state the fact that I had a return of my call by Deirdre Parker from the Office of the Scientific Director of the National Institutes of Health/ National Human Genome Research Institute. Subsequently, I received the spring 2014 Syllabus entiltled Current Topics in Genome Analysis.

The Fourteen Lectures that I will attend in the NIH Clinical Center Wednesday Mornings 9:30-11am. From Febuary the 26 to June 4 include the following:

2/26/14 The Genomic Landscape circa 2014 Eric Green, NHGRI
3/5/14 Biological Sequence Analysis 1 Andy Baxevanis, NHGRI
3/12/14 Biological Sequence Analysis 11 Andy Baxevanis, NHGRI
3/19/14 Genome Scale Sequence Analysis Tyra Wolfsberg, NHGRI
3/26/14 Regulatory and Epigenetic Landscapes of Mammalian Genomes Laura Elnitski, NHGRI
4/2/14 Applications of Genomics to Improve Public Health Colleen Mcbride, NHGRI
4/9/14 Introduction to Population Genetics Lynn Jorde, University of Utah
4/23/14 Genomic Approaches to Complex Genetic Diseases Karen Mohike, University of North Carolina
4/30/14 Identifying the Genomic Basis of Rare Diseases David Valle The Johns Hopkins University

5/7/14 Pharmacogenomics Howard McLeod, Moffitt Cancer Center 5/14/14 Large Scale Expression Analysis Paul Meltzer, NCI 5/21/14 Genomic Medicine Bruce Korf, University of Alabama at Birmingham
5/28/14 Next Generation Sequencing Technologies Elaine Mardis, Washington University
6/4/14 Genetics of Microbes and Microbiomes Julie Segre, NHGRI

As of October of 2013 via Dr. Klement and Foundation One I am able to offer genomic sequencing to patients with a pathology report that reveals a confirmed malignancy. Back in 2008 this complete genomic sequencing, cost 350,000 by Knome, in Boston MA.

The Foundation One version includes 300 known genomic sequence variant signatures known to be present in various cancers. My patient had an excision of an abdominal lesion found to be a Leiomyosarcarcoma, as the pathologic diagnosis. I requested that genomic sequencing be performed. I was told that the patient's doctor had to request this service in writing, therefore I did, afterwhich, I spoke to the Pathologist who then told me that no genomic sequencing services were available in house. This conversation occurred after 8/9/13, the date on my written request, fortunately for the patient Mark Wright had introduced me to Boris Epshteyn who advised me to contact the Newman Lakka Instititute for Personal Cancer Care, Giannoula Klement, MD, to whom I sent the pathology report. Subsequenty, I spoke to members of her staff, Jen DaGasee , RN that arranged to have Blocks of her Leimyosarcoma tissue mailed to Foundation One where 300 tumor marker variant signatures are sequenced.

The P53 mutation was identified and I called the Harvard Clinical Trial Coordinator. On 11/22/13 she was accepted into the Kevetrin Clinical Trial known to target P53 mutation driven, potential residual leiomyosarcomatous cells.

Therefore, my lifes work is to have Veterans and Civilians alike organized regionally, eventually, offering this service in all Veteran Hospitals and Cooperating Civilian Hospitals. Creating Project Manager Positions to coordinate each facilities Audio-Visual Department, to create a television worthy documentary about participants with their Physicians that undergo the Genomic Sequencing Process.

Educational lecture series with updated interventional radiological facts, including Procedures including the use of MR spectroscopy to identify active cells. So that the prostate biopsy includes the representative tissue, as presently performed by Herbert Lepore, MD at NY University

Langone Medical Center. I will also advocate for hybrid training of Surgeons to include the interventional radiological skills to place Aortic Stents as well as the ability to use available instumentalizations to remove cerebral arterial clot formations in time to save brain function.

To Conclude, Brian Druker Oncologist/Researcher at Oregon State University, who I had the pleasure of meeting at a lecture at Yale after Imatinib (Gleevec), the gene based targeted therapy, had met FDA approval as treatment for Chronic Myelogenous Leukemia, is greatly admired, because he fought the system, wisely to bring Gleevec to his CML patients.

The Clinical Trial that my patient has been entered at Harvard, is based on the presence of the P53 gene mutation, Kevetrin (Cellceutix Corporation) was FDA approved in 2014.

Additional Kinase Inhibitors and several Immunomodulators that boost our immunological response to malignant cell types, as a combination will greatly impact our ability to fight malignant cell types. This is needed to create second line therapy, when original cell types mutate.

My life work includes providing the education needed to increase the number of genomic based researchers to meet this demand.

Media attention in the form of a Television Docudrama Series, creating jobs and saving lives as an essential component. Dr. Klement, and I spoke in reference to this community outreach program that we are both dedicated to implementing in the near future. Meanwhile, the Newman Lakka Institute for Personal Cancer Care, continues to perform this service on a case by case basis. Physician education is mandatory to utilize the kinase inhibitors and immumunomodulators on the market now, as most patients with excised malignant lesions, are not sequenced.

I will create an educational stream of information via lives saved, that will identify all pathology, via genomic sequencing of veterans and civilians that have been exposed to Environmental Toxins, Dietary Carcinogens, Infectious Disease Irritants as well as 10-15% of Cancer that have a hereditary basis. Granting the Gene Theme Scene Medical PC will be greatly appreciated in this worthy effort.

Selected Fundamental Information on the Cell and DNA

Major parts of a cell

The nucleus contains most of the cell's genetic material

What is a cell?

Cells are the basic building blocks of all living things. The human body is composed of trillions of cells. They provide structure for the body, take in nutrients from food, convert those nutrients into energy, and carry out specialized functions. Cells also contain the body's hereditary material and can make copies of themselves.

Cells have many parts, each with a different function. Some of these parts, called organelles, are specialized structures that perform certain tasks within the cell. Human cells contain the following major parts, listed in alphabetical order:

Cytoplasm

Within cells, the cytoplasm is made up of a jelly-like fluid (called the cytosol) and other structures that surround the nucleus.

Cytoskeleton

The cytoskeleton is a network of long fibers that make up the cell's structural framework. The cytoskeleton has several critical functions, including determining cell shape, participating in cell division, and allowing cells to move. It also provides a track-like system that directs the movement of organelles and other substances within cells.

Endoplasmic reticulum (ER)

This organelle helps process molecules created by the cell. The endoplasmic reticulum also transports these molecules to their specific destinations either inside or outside the cell.

Golgi apparatus

The Golgi apparatus packages molecules processed by the endoplasmic reticulum to be transported out of the cell.

Lysosomes and peroxisomes

These organelles are the recycling center of the cell. They digest foreign bacteria that invade the cell, rid the cell of toxic substances, and recycle worn-out cell components.

Mitochondria

Mitochondria are complex organelles that convert energy from food into a form that the cell can use. They have their own genetic material, separate from the DNA in the nucleus, and can make copies of themselves.

Nucleus

The nucleus serves as the cell's command center, sending directions to the cell to grow, mature, divide, or die. It also houses DNA (deoxyribonucleic acid), the cell's hereditary material. The nucleus is surrounded by a membrane called the nuclear envelope, which protects the DNA and separates the nucleus from the rest of the cell.

Plasma membrane

The plasma membrane is the outer lining of the cell. It separates the cell from its environment and allows materials to enter and leave the cell.

Ribosomes

Ribosomes are organelles that process the cell's genetic instructions to create proteins. These organelles can float freely in the cytoplasm or be connected to the endoplasmic reticulum

What is DNA?

DNA, or deoxyribonucleic acid, is the hereditary material in humans and almost all other organisms. Nearly every cell in a person's body has the same DNA. Most DNA is located in the cell nucleus (where it is called nuclear DNA), but a small amount of DNA can also be found in the mitochondria (where it is called mitochondrial DNA or mtDNA).

The information in DNA is stored as a code made up of four chemical bases: adenine (A), guanine (G), cytosine (C), and thymine (T). Human DNA consists of about 3 billion bases, and more than 99 percent of those bases are the same in all people. The order, or sequence, of these bases determines the information available for building and maintaining an organism, similar to the way in which letters of the alphabet appear in a certain order to form words and sentences.

DNA bases pair up with each other, A with T and C with G, to form units called base pairs. Each base is also attached to a sugar molecule and a phosphate molecule. Together, a base, sugar, and phosphate are called a nucleotide. Nucleotides are arranged in two long strands that form a spiral called a double helix. The structure of the double helix is somewhat like a ladder, with the base pairs forming the ladder's rungs and the sugar and phosphate molecules forming the vertical sidepieces of the ladder.

An important property of DNA is that it can replicate, or make copies of itself. Each strand of DNA in the double helix can serve as a pattern for duplicating the sequence of bases. This is critical when cells divide because each new cell needs to have an exact copy of the DNA present in the old cell.

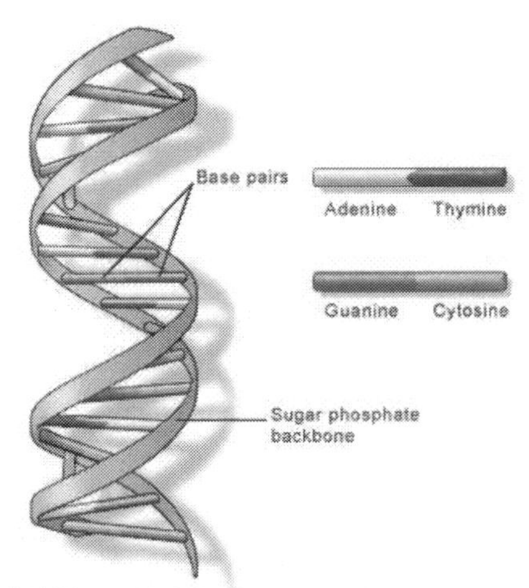

DNA is a double helix formed by base pairs attached to a sugar-phosphate backbone.

For more information about DNA:

The National Human Genome Research Institute fact sheet Deoxyribonucleic Acid (DNA) provides an introduction to this molecule

DNA Is a Structure That Encodes Biological Information

What do a human, a rose, and a bacterium have in common? Each of these things — along with every other organism on Earth — contains the molecular instructions for life, called **deoxyribonucleic acid** or **DNA**. Encoded within this DNA are the directions for traits as diverse as the color of a person's eyes, the scent of a rose, and the way in which bacteria infect a lung cell.

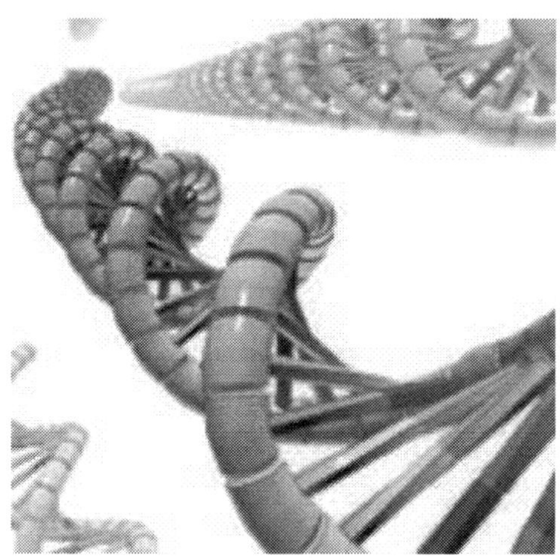

DNA is found in nearly all living cells. However, its exact location within a cell depends on whether that cell possesses a special membrane-bound organelle called a nucleus. Organisms composed of cells that contain nuclei are classified as **eukaryotes** , whereas organisms composed of cells that lack nuclei are classified as **prokaryotes**. In eukaryotes, DNA is housed within the nucleus, but in prokaryotes, DNA is located directly within the cellular cytoplasm, as there is no nucleus available.

But what, exactly, is DNA? In short, DNA is a complex molecule that consists of many components, a portion of which are passed from parent organisms to their offspring during the process of reproduction. Although each organism's DNA is unique, all DNA is composed of the same nitrogen-based molecules. So how does DNA differ from organism to organism? It is simply the order in which these smaller molecules are arranged that differs among individuals. In turn, this pattern of arrangement ultimately determines each organism's unique characteristics, thanks to another set of molecules that "read" the pattern and stimulate the chemical and physical processes it calls for.

What components make up DNA?

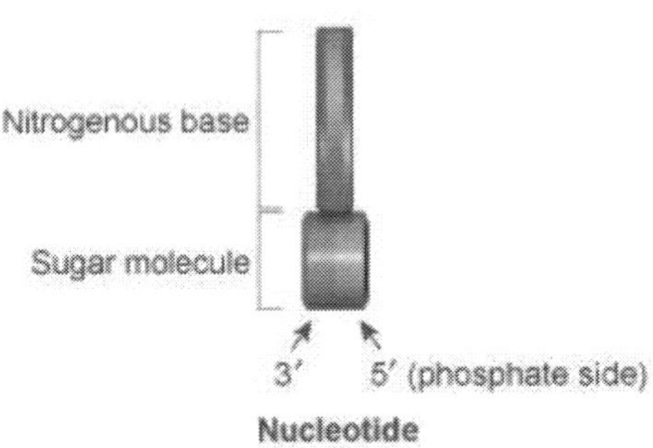

Figure 1: A single nucleotide contains a nitrogenous base (red), a deoxyribose sugar molecule (gray), and a phosphate group attached to the 5' side of the sugar (indicated by light gray). Opposite to the 5' side of the sugar molecule is the 3' side (dark gray), which has a free hydroxyl group attached (not shown).

At the most basic level, all DNA is composed of a series of smaller molecules called **nucleotides**. In turn, each nucleotide is itself made up of three primary components: a nitrogen-containing region known as a **nitrogenous base**, a carbon-based sugar molecule called **deoxyribose**, and a phosphorus-containing region known as a **phosphate group** attached to the sugar molecule (Figure 1). There are four different DNA nucleotides, each defined by a specific nitrogenous base: **adenine** (often abbreviated "A" in science writing), **thymine** (abbreviated "T"), **guanine** (abbreviated "G"), and **cytosine** (abbreviated "C") (Figure 2).

Figure 2: The four nitrogenous bases that compose DNA nucleotides are shown in bright colors: adenine (A, green), thymine (T, red), cytosine (C, orange), and guanine (G, blue).

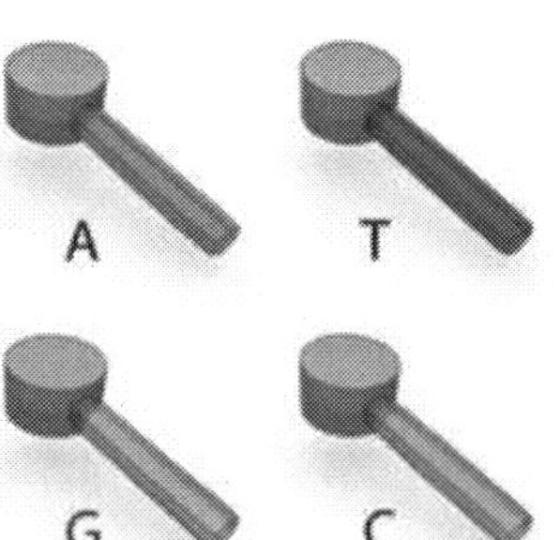

Although nucleotides derive their names from the nitrogenous bases they contain, they owe much of their structure and bonding capabilities to their deoxyribose molecule. The central portion of this molecule contains five carbon atoms arranged in the shape of a ring, and each carbon in the ring is referred to by a number followed by the prime symbol ('). Of these carbons, the 5' carbon atom is particularly notable, because it is the site at which the phosphate group is attached to the nucleotide. Appropriately, the area surrounding this carbon atom is known as the **5' end** of the nucleotide. Opposite the 5' carbon, on the other side of the deoxyribose ring, is the 3' carbon, which is not attached to a phosphate group. This portion of the nucleotide is typically referred to as the **3' end** (Figure 1). When nucleotides join together in a series, they form a structure known as a **polynucleotide**. At each point of juncture within a polynucleotide, the 5' end of one nucleotide attaches to the 3' end of the adjacent nucleotide through a connection called a **phosphodiester bond** (Figure 3). It is this alternating sugar-phosphate arrangement that forms the "backbone" of a DNA molecule.

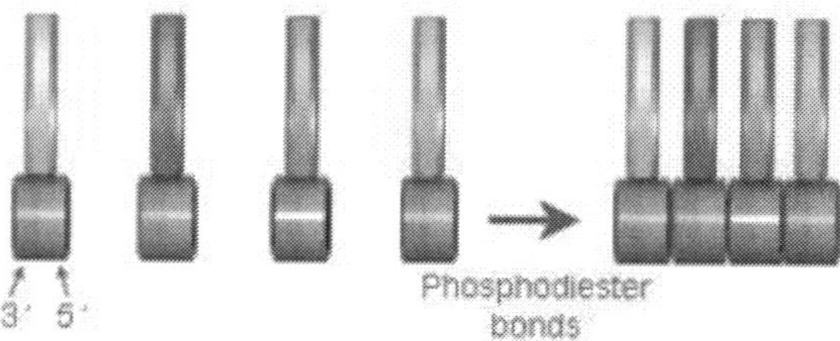

Figure 3: All polynucleotides contain an alternating sugar-phosphate backbone. This backbone is formed when the 3' end (dark gray) of one nucleotide attaches to the 5' phosphate end (light gray) of an adjacent nucleotide by way of a phosphodiester bond.

How is the DNA strand organized?

Although DNA is often found as a single-stranded polynucleotide, it assumes its most stable form when double stranded. Double-stranded DNA consists of two polynucleotides that are arranged such that the nitrogenous bases within one polynucleotide are attached to the nitrogenous bases within another polynucleotide by way of special chemical bonds called **hydrogen bonds**. This base-to-base bonding is not random; rather, each A in one strand always pairs with a T in the other strand, and each C always pairs with a G. The double-stranded DNA that results from this pattern of bonding looks much like a ladder with sugar-phosphate side supports and base-pair rungs.

Note that because the two polynucleotides that make up double-stranded DNA are "upside down" relative to each other, their sugar-phosphate ends are **anti-parallel**, or arranged in opposite orientations. This means that one strand's sugar-phosphate chain runs in the 5' to 3' direction, whereas the other's runs in the 3' to 5' direction (Figure 4). It's also critical to understand that the specific sequence of A, T, C, and G nucleotides within an organism's DNA is unique to that individual, and it is this sequence that controls not only the operations within a particular cell, but within the organism as a whole.

Figure 4: Double-stranded DNA consists of two polynucleotide chains whose nitrogenous bases are connected by hydrogen bonds. Within this arrangement, each strand mirrors the other as a result of the anti-parallel orientation of the sugar-phosphate backbones, as well as the complementary nature of the A-T and C-G base pairing.

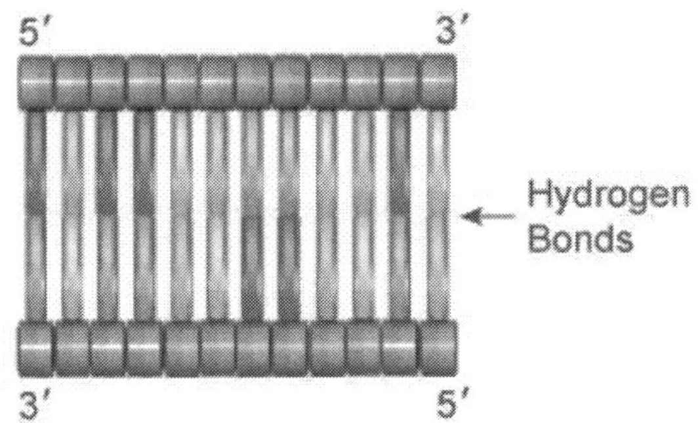

Figure 5: Rosalind Franklin's X-ray diffraction image of DNA. Images like this one enabled the precise calculation of molecular distances within the double helix.

Beyond the ladder-like structure described above, another key characteristic of double-stranded DNA is its unique three-dimensional shape. The first photographic evidence of this shape was obtained in 1952, when scientist Rosalind Franklin used a process called X-ray diffraction to capture images of DNA molecules (Figure 5). Although the black lines in these photos look relatively sparse, Dr. Franklin interpreted them as representing distances between the nucleotides that were arranged in a spiral shape called a **helix**.

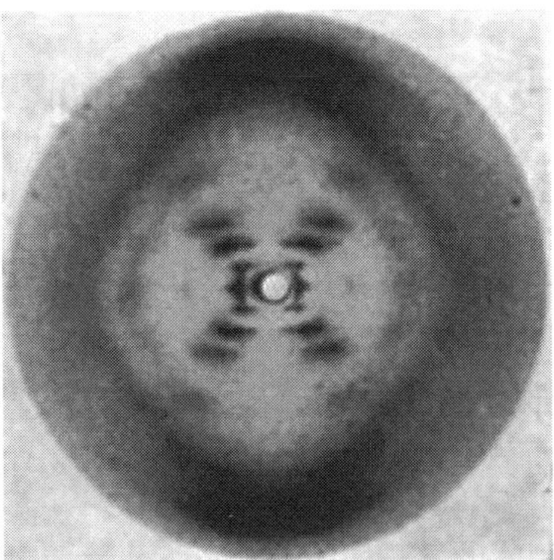

Around the same time, researchers James Watson and Francis Crick were pursuing a definitive model for the stable structure of DNA inside cell nuclei. Watson and Crick ultimately used Franklin's images, along with their own evidence for the double-stranded nature of DNA, to argue that DNA actually takes the form of a **double helix**, a ladder-like structure that is twisted along its entire length (Figure 6). Franklin, Watson, and Crick all published articles describing their related findings in the same issue of *Nature* in 1953.

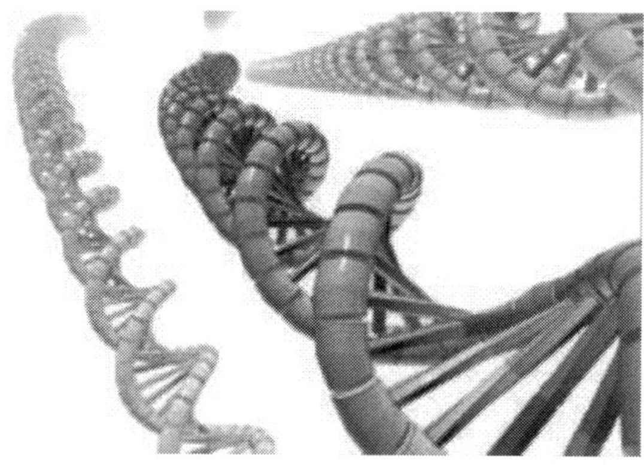

Figure 6: The double helix looks like a twisted ladder.

How is DNA packaged inside cells?

Figure 7: To better fit within the cell, long pieces of double-stranded DNA are tightly packed into structures called chromosomes.

Most cells are incredibly small. For instance, one human alone consists of approximately 100 trillion cells. Yet, if all of the DNA within just one of these cells were arranged into a single straight piece, that DNA would be nearly two meters long! So, how can this much DNA be made to fit within a cell? The answer to this question lies in the process known as **DNA packaging**, which is the phenomenon of fitting DNA into dense compact forms (Figure 7).

Chromatin

During DNA packaging, long pieces of double-stranded DNA are tightly looped, coiled, and folded so that they fit easily within the cell. Eukaryotes accomplish this feat by wrapping their DNA around special proteins called **histones**, thereby compacting it enough to fit inside the nucleus (Figure 8). Together, eukaryotic DNA and the histone proteins that hold it together in a coiled form is called **chromatin**.

Figure 8: In eukaryotic chromatin, double-stranded DNA (gray) is wrapped around histone proteins (red).

DNA can be further compressed through a twisting process called **supercoiling** (Figure 9). Most prokaryotes lack histones, but they do have supercoiled forms of their DNA held together by special proteins. In both eukaryotes and prokaryotes, this highly compacted DNA is then arranged into structures called **chromosomes**. Chromosomes take different shapes in different types of organisms. For instance, most prokaryotes have a single circular chromosome, whereas most eukaryotes have one or more linear chromosomes, which often appear as X-shaped structures . At different times during the life cycle of a cell, the DNA that makes up the cell's chromosomes can be tightly compacted into a structure that is visible under a microscope, or it can be more loosely distributed and resemble a pile of string.

Figure 9: Supercoiled eukaryotic DNA.

How do scientists visualize DNA?

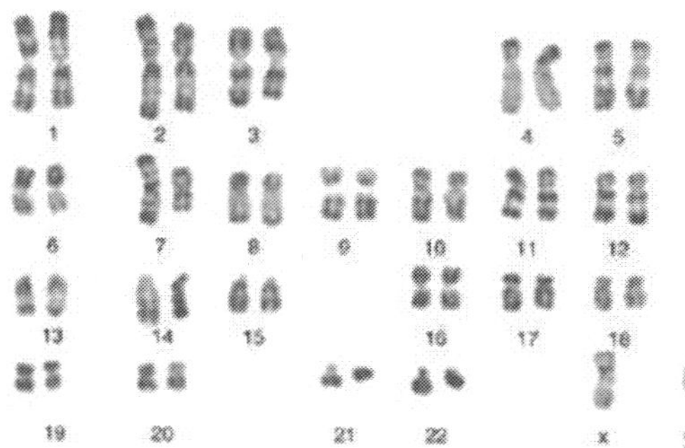

Figure 10: This karyotype depicts all 23 pairs of chromosomes in a human cell, including the sex-determining X and Y chromosomes that together make up the twenty-third set (lower right).

It is impossible for researchers to see double-stranded DNA with the naked eye — unless, that is, they have a large amount of it. Modern laboratory techniques allow scientists to extract DNA from tissue samples, thereby pooling together miniscule amounts of DNA from thousands of individual cells. When this DNA is collected and purified, the result is a whitish, sticky substance that is somewhat translucent.

To actually visualize the double-helical structure of DNA, researchers require special imaging technology, such as the X-ray diffraction used by Rosalind Franklin. However, it is possible to see chromosomes with a standard light microscope, as long as the chromosomes are in their most condensed form. To see chromosomes in this way, scientists must first use a chemical process that attaches the chromosomes to a glass slide and stains or "paints" them. Staining makes the chromosomes easier to see under the microscope. In addition, the banding patterns that appear on individual chromosomes as a result of the staining process are unique to each pair of chromosomes, so they allow researchers to distinguish different chromosomes from one another. Then, after a scientist has visualized all of the chromosomes within a cell and captured images of them, he or she can arrange these images to make a composite picture called a **karyotype** (Figure 10)

Selected Fundamental Information on the Synthesis of Proteins

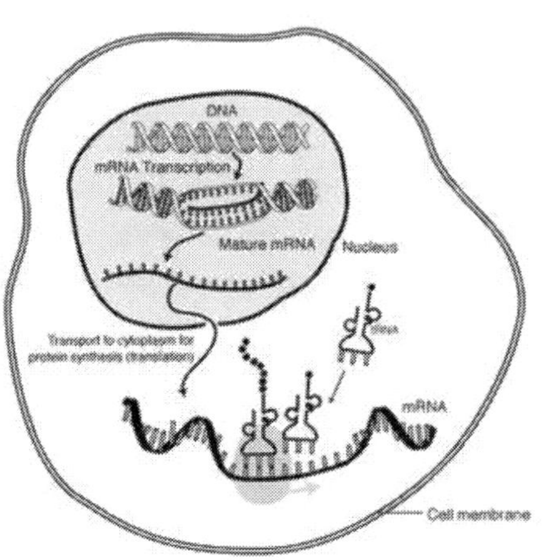

RNA is transcribed in the nucleus; once completely processed, it is transported to the cytoplasm and translated by the ribosome (shown in very pale grey behind the tRNA).

Main article: Transcription (genetics)

Diagram showing the process of transcription

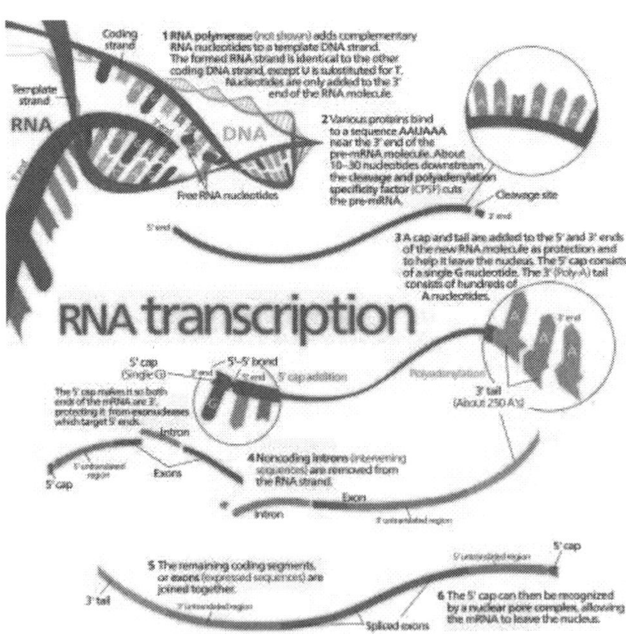

In transcription an mRNA chain is generated, with one strand of the DNA double helix in the genome as a template. This strand is called the template strand.

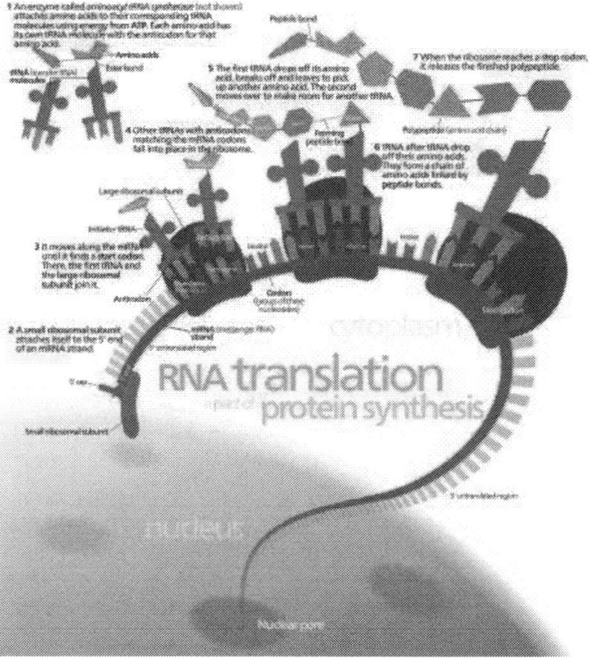

Diagram showing the process of translation

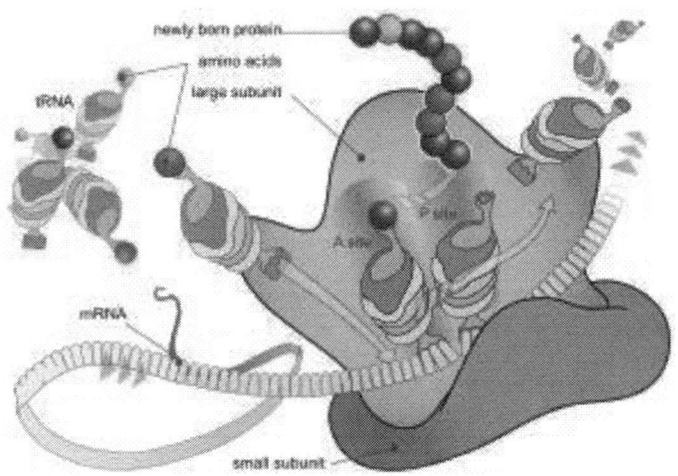

Diagram showing the translation of mRNA and the synthesis of proteins by a ribosome

The synthesis of proteins from RNA is known as translation. Translation occurs in the cytoplasm, where the ribosomes are located. Ribosomes are made of a small and large subunit that surround the mRNA. In translation, messenger RNA (mRNA) is decoded to produce a specific polypeptide according to the rules specified by the trinucleotide genetic code. This uses an mRNA sequence as a template to guide the synthesis of a chain of amino acids that form a protein. Translation proceeds in four phases: activation, initiation, elongation, and termination (all describing the growth of the amino acid chain, or polypeptide that is the product of translation).

In activation, the correct amino acid (AA) is joined to the correct transfer RNA (tRNA). While this is not, in the technical sense, a step in translation, it is required for translation to proceed. The AA is joined by its carboxyl group to the 3' OH of the tRNA by an ester bond. When the tRNA has an amino acid linked to it, it is termed "charged". Initiation involves the small subunit of the ribosome binding to 5' end of mRNA with the help of initiation factors (IF), other proteins that assist the process. Elongation occurs when the next aminoacyl-tRNA (charged tRNA) in line binds to the ribosome along with GTP and an elongation factor.

Khan Academy Genetics 101 Synopsis Plus

Our bodies are composed of approximately 100 trillion cells. Sperm cells and egg cells have 23 chromosomes via the process of meiosis during which the recombination makes us unique. During prophase 1, the homologous chromosomes that originally came from the mother and father of the individual whose cells are undergoing meiosis, have, homologs, that zip together. During which enzymes cut the chromosomes into pieces and seal newly combined strands back together in an action called crossing-over, causing the maternal and paternal chromosomes to be combined.

Our cells have 46 chromosomes via the process of mitosis, one set from each parent causing the genetic recipe that results in us!

Cells have numerous varieties and function, and the nucleus of these cells have 99.99% of our genes with the mitochondria containing a few additional genes totaling 20-25,000 genes.

Chromosomes are made up of short segments of DNA called genes. The DNA that is coiled into the nucleus of our cells is actually approximately six feet long! Each rung of the ladder shaped DNA molecule is made up of a pair of interlocking units called bases that are normally designated by the purine Adenine (A) the pyrimidine Thymine (T), the purine Guanine (G), and the pyrimidine Cytosine (C.) . Note the purine bases in DNA adenine and guanine have two rings, as does the purine caffeine! The pyrimidine bases in DNA cytosine and thymine are single ring structures.

Additional note: because all four bases are flat ring shaped molecules they stack up in DNA like coins.

This stacking arrangement makes the DNA molecule compact and strong. Deoxyribose differs from Ribose due to the absence of an oxygen atom at the 2'site. Nucleotides are composed of a nitrogenous base, a five-carbon ribose if RNA, or deoxyribose if DNA, and at least one phosphate group.

Nucleoside plus a phosphate group equals a Nucleotide.

DNA is organized into Chromosomal strands, the 46 Chromosomes that make up a human results from the 23 pairs from your Mother and 23 pairs from your Father.

Chromosomal short segments of DNA are called genes. Imagining DNA as a cookbook, genes become the recipes and within the DNA alphabet of A,T, C, G the recipes tell your cells how to function with gene regulators turning your cells on and off.

Cells use the recipes written in your genes to make proteins, some proteins give cells their shape and structure, while others carry out biologic functions. Using different combinations of the DNA alphabet creates different proteins just as different combinations of the same ingredients result in different meals. Every somatic (non egg or sperm germ cell) has the same set of instructions so that intricate systems of genetic switches turn genes on and off making sure that the right protein is made at the right time in the right cells. Even in the Fold It video game the P53 tumor suppressor gene is mentioned as being present in up to 50% of Cancers and in 2014 Kevetrin was approved as the gene based treatment that causes a reactivation of the P53 mutation, so that the tumor suppresor gene is reactivated. Note: Kevetrin has now passed the FDA approval process and as of 2014 is on the market to reactivate the P53 tumor suppressor gene.

There are three billion base pairs, variations of these base pairs are called ,SNPs (Single Nucleotide Polymorphisms) . When a genome is copied to make new cells a single base pair can be left out, added or substituted. Single Base Pair substitutions create SNPs. There are 10 Million SNPs in humans that create the differences between our fellow humans ability to metabolize drugs, become addicted, develop diseases as well as our talents and phenotypic expressions.

Homologous Genes are the 22 pairs of genes in our somatic cells. The 23rd pair of genes are the XX combination in the Germ (Sex) cell of your Mothers egg and your Fathers sperm, when you are a female. XY combination is the 23rd pair, when you are a male. Viable eggs have 23 pairs of chromosomes and functional sperm have 23 pairs of chromosomes also.

If genes were considered clothes, the the variations could be considered the brand of clothes, or the allele, aka, SNP. Therefore, SNPs are a measure of genetic similarity, since DNA is passed from parent to child, you inherit certain SNP versions from your parents. Therefore the number of SNPs you match with another person can be used to match how closely related you are. Note: most of the Y chromosome, is passed from father to son, and the genetic material in mitochondrial DNA is passed from mother to child, without recombination. These types of DNA can be used to trace your ancestry.

Phenotypes are observable traits, the field of epigenetics identifies the result of interactions between your genes and the environment. SNPs have been linked to health issues as well as your phenotype. Humans share 98.5% of our Genomic Sequence with other Humans.

The Gene Theme Scene Synopsis of the History of Genetics

500-428 BC Anaxagoras speculated that everything contains a portion of everything else.

460-378 BC Hippocrates and 389-322 BC Aristotle taught that the semen and menstrual blood interacted in the womb to form new life.

1744-1829 Jean Baptist Lamarck theory of evolution included platitudes shed by the organs that were carried in the bloodstream where they accumulated in the germ cells with a secondary environmental force that adapted to local environments.

1809-1882 Charles Darwin the Father of ten with his first cousin had a theory of pangenesis, mentioning, a subunit he called, gemmules (platitudes), as the pangens (particles of inheritance) proposed as his Pangenesis Theory of 1868.

1822-1884 Gregor Mendel aka Father of Genetics coined the term recessive and dominant traits of the pea plant experiments demonstrating the invisible factors of traits in predictable way.

1871 Miescher isolated the cell nucleus then treated them with pepsin recovering an acidic substance he called nuclein.

1889 Altmann attempted to isolate purified DNA.

1902 Garrod noticed that close relatives developed an inborm error of metabolism alkaptonuria.

1904 Sutton showed that grasshoppers had matched pairs of chromosomes.

1904 Boveri stated that sea urchins had to have chromosomes for embryonic development to occur.

1908 Hardy-Weinstein equilibriun model description of gene pool alleles.

1910 Morgan studied the Drosophila Fly affirming the Chromosome Therory of Heredity.

1911 Sturtevant created the procedure of linkage mapping.

1923 Griffin bacterial observations that DNA carried the genetic basis of the bacteria's pathogenicity.

1933 Morgan received Nobel Prize for Chromosomal Linkage Mapping.

1941 Beadle and Tatum studied Neurospora to demonstrate genetic codes for enzymes.

1943 Luria-Delbruck demonstated spontaneous mutations conferring resistance to a bacteriophage.

1944 Avery-MacLeod-McCarty provided additional evidence that DNA carries genetic information reponsible for pathogenicity

1947 Luria studied the repair process of the irradiated bacteriophage and the DNA repair process in humans.

1950 Chargaff determined the pairing process of nitrogenous bases.

1953 Watson-Crick-Franklin-Wilkins proposed the double helix model of DNA.

1955 Todd Nobel prize in Chemistry for his contributions in the synthesis of nucleotides and nucleotide coenzymes

1955 Tjio studied human embryonic lung tissue and determined the way of separating the 46 Chromosomes.

1957 Kornbergand Ochoa synthesized DNA after discovering that DNA polymerase 1 established requirements for invitro duplication of DNA (Nobel Prize 1959)

1961 Crick and Brenner frame shift mutations identifing the genetic code as a triplet code (codon) sepcifing a particular amino acid 1966 Nirenberg-Leder -Khorana found which triplets of RNA were translated into amino acids in yeast cells

1976 Sanger and Coulson determined the sequence of DNA using a four lane polyacrylamide gel electrophoresis (PAGE)

1980 Cohen and Boyer received the first patent for gene cloning of a plasmid This team also replicated HGH, Erythropoietin, and Insulin.

1982 FDA approved the first genetically engineered human insulin using recombination DNA methods by Genentech in 1978

1983 Mullis amplified DNA using a cloning procedure called Polymerase Chain Reaction

1990 King Laboratory in California discovered the first breast cancer gene BRCA1

1993 Sharp and Roberts were awarded the Nobel Prize for discovering that genes in DNA are made up of introns and exons (not all of the nucleotides on the RNA strand as a product of DNA transcription are used in the translation process since the intervening sequences in the RNA strand are first spliced out so that the RNA segment left behind after splicing would be translated to polypeptides

1994 Stratton and Wooster BRCA11 breast cancer gene

1996 Drucker (Novardis) Imatinib (Gleevec) tyrosine-kinase inhibitor gene based treatment for Philadelphia Chromosome Positive (Ph+) (CML) Chronic Myelogenous Leukemia

1997 Wilmut cloned Dolly the Sheep

2003 The Human Genome Project is officially completed

2004 Merck introduces a vaccine for Human Papillomavirus to protect against HPV strain 16 and 18 known to inactivate tumor suppressor genes (70% of cervical cancer is due to HPV strains 16and 18)

2007 Worobey analyzed HIV genetic mutations and traced them back to infections in the United States as early as the 1960's

2008 Advexin (FDA approval pending) utilizing an Adenovirus to carry a replacement gene coding for the P53 protein

2014 Kevetrin is FDA approved to reactivate the P53 tumor suppressor gene Stem Cell

Research for ALS (Amyotrophic Lateral Sclerosis) and MS (Multiple Sclerosis) NIH Fact Sheet on Human Pluripotent Stem Cell Research Guidelines

The Promise of Stem Cell Research

Human pluripotent stem cells are a unique scientific and medical resource. In 1998, scientists at the University of Wisconsin and at Johns Hopkins University isolated and successfully cultured human pluripotent stem cells. The pluripotent stem cells were derived using non-Federal funds from early-stage embryos donated voluntarily by couples undergoing fertility treatment in an in vitro fertilization (IVF) clinic or from non-living fetuses obtained from terminated first trimester pregnancies. Informed consent was obtained from the donors in both cases. Women voluntarily donating fetal tissue for research did so only after making the decision to terminate the pregnancy.

Because pluripotent stem cells give rise to almost all of the cells types of the body, such as muscle, nerve, heart, and blood, they hold great promise for both research and health care. This advance in human biology continues to generate enthusiasm among scientists, patients suffering from a broad range of diseases, including cancer, heart disease and diabetes, and their families. For example, further research using human pluripotent stem cells may help:

Generate cells and tissue for transplantation. Pluripotent stem cells have the potential to develop into specialized cells that could be used as replacement cells and tissues to treat many diseases and conditions, including Parkinson's disease, amyotrophic lateral sclerosis, spinal cord injury, burns, heart disease, diabetes, and arthritis.

Improve our understanding of the complex events that occur during normal human development and also help us understand what causes birth defects and cancer.

Change the way we develop drugs and test them for safety. Rather than evaluating the safety of candidate drugs in an animal model, drugs might be initially tested on cells developed from pluripotent stem cells and only the safest candidate drugs would advance to animal and then human testing.

The Potential of Adult Stem Cell Research

Questions have been raised about the usefulness of adult stem cells in research and treatment, especially as compared to pluripotent stem cells derived from embryos or fetal tissue. Indeed, there is enormous potential for research using such cells. Human adult stem cells have been isolated from tissues such as blood, brain, intestine, skin, and muscle. Furthermore, some adult stem cells have been shown to be more "plastic" than first thought—that is, some of these stem cells appear to be capable of developing into different kinds of cells than first predicted.

There is, however, considerable evidence that adult stem cells may have limited potential compared to pluripotent stem cells derived from embryos or fetal tissue. Human adult stem cells have not yet been isolated from all cell and tissue types, and they have not been shown to be capable of developing into all of the different cell and tissue types of the body. Furthermore, adult stem cells are difficult to obtain, since they are often present in only minute quantities. They are difficult to isolate and purify, and their numbers appear to decrease with age. Moreover, adult stem cells may have more DNA damage, and they appear to have a shorter life span than pluripotent stem cells. For all of these reasons, and because of the enormous potential of stem cell approaches to research and treatment, it is vitally important that scientists study and compare both pluripotent and adult stem cells.

The Need for Guidelines to Govern Research Using Pluripotent Stem Cells

The NIH is prohibited from using any appropriated funds for "... (1) the creation of a human embryo or embryos for research purposes; or (2) research in which a human embryo or embryos are destroyed, discarded, or knowingly subjected to risk of injury or death greater than that allowed for research on fetuses in utero under 45 CFR 46.208(a)(2) and section 498(b) of the Public Health Service Act (42 U.S.C. 289g(b))." Because of the enormous potential of human pluripotent stem cells to medical research, the NIH asked the General Counsel of the Department of Health and Human Services (DHHS) to determine whether research utilizing pluripotent stem cells is permissible under existing Federal law governing embryo and fetal tissue research. After careful consideration, the DHHS concluded that because human pluripotent cells are not embryos, current Federal law does not prohibit DHHS funds from being used for research utilizing these cells.

Recognizing the ethical and legal issues surrounding human pluripotent stem cell research and the need for stringent oversight of this class of research—oversight that goes beyond the traditional rigorous NIH scientific peer review process—the NIH issued a moratorium on the funding of this research until Guidelines could be developed and an oversight process could be implemented.

In April 1999, the NIH convened a working group of the Advisory Committee to the Director (ACD), NIH, to provide advice to the ACD relevant to guidelines and oversight for this research. The working group met in public session and included scientists, clinicians, ethicists, lawyers, patients, and patient advocates. During their deliberations, the group considered advice from the National Bioethics Commission, the public, and scientists. Draft guidelines for this research were published for public comment, and, after reviewing and considering all comments received, the *NIH Guidelines for Research Using Human Pluripotent Stem Cells* (*NIH Guidelines*) were published in the *Federal Register* and became effective on August 25, 2000. (Because the *NIH Guidelines* contained a few incorrect citations and other minor errors, a notice of correction (65 FR 69951) was published on November 21, 2000.) The revised *NIH Guidelines* and other information about stem cell research can be found at the URL: /news/pages/default.aspx.

Specifics of the Guidelines

The purpose of the *NIH Guidelines* is to set forth procedures to help ensure that NIH-funded research in this area is conducted in an ethical and legal manner. By issuing these *Guidelines*, the NIH aims to enhance both the scientific and ethical oversight of this important arena of research and the pace at which scientists can explore its many promises. These *Guidelines* will encourage openness, provide appropriate Federal oversight, help make certain that all researchers can make use of these critical research tools, and help assure full public access to the practical medical benefits of research using these cells.

The *Guidelines* prescribe the documentation and assurances that must accompany requests for NIH funding for research using human pluripotent stem cells derived from human embryos or fetal tissue. These include the following:

For studies using human pluripotent stem cells derived from human embryos, NIH funds may be used only if the cells were derived from frozen embryos that were created for the purposes of fertility treatment, were in excess of clinical need, and were obtained after the consent of the donating couple.

The *NIH Guidelines* prohibit the use of inducements, monetary or otherwise, for the donation of the embryo. There must also have been a clear separation between the fertility treatment and the decision to donate embryos for this research.

The *NIH Guidelines* require that the informed consent specify whether or not information that could identify the donor(s) will be retained.

Investigators who propose using NIH funds to conduct research using human pluripotent stem cells derived from fetal tissue are expected to follow both the *NIH Guidelines* and all Federal and state laws and regulations governing human fetal tissue and human fetal tissue transplantation research.

The *Guidelines* require that the donation of human embryos or fetal tissue be made without any restriction or direction regarding the individual(s) who may be the recipient of the cells derived from the human pluripotent stem cells.

They also require review and approval of the derivation protocol by an Institutional Review Board.

The informed consent should include statements that the embryos or fetal tissue will be used to derive human pluripotent stem cells for research that may include human transplantation research; that derived cells may be kept for many years; that the research is not intended to provide direct medical benefit to the donor; and that the donated embryos will not be transferred to a woman's uterus and will not survive the stem cell derivation process.

The informed consent must also state the possibility that the results of the research may have commercial potential, and that the donor will not receive any benefits from any such future commercial development.

The *Guidelines* also set forth the areas of research that are ineligible for NIH funding, including: 1) as required by law, research involving the derivation of pluripotent stem cells from human embryos; 2) research in which human pluripotent stem cells are utilized to create or contribute to a human embryo; 3) research utilizing pluripotent stem cells that were derived from human embryos created for research purposes; 4) research in which human pluripotent stem cells are derived using somatic cell nuclear transfer; 5) research utilizing human pluripotent stem cells that were derived using somatic cell nuclear transfer; 6) research in which human pluripotent stem cells are combined with an animal embryo; and

7) research in which human pluripotent stem cells are derived using somatic cell nuclear transfer for the purposes of reproductive cloning of a human.

Requirements for Investigators Applying for Funds

With the issuing of the *Guidelines*, the moratorium was lifted, and NIH is accepting applications for this class of research. A request for NIH funds for research using these cells must include a signed assurance that the cells were derived from human embryos or fetal tissue in accordance with the *Guidelines* and that the institution will maintain documentation in support of the assurance.

This assurance must also affirm that the human pluripotent stem cells to be used in the research were obtained through a donation or through a payment that does not exceed the reasonable costs associated with the quality control, processing, transportation, preservation, and storage of the stem cells. It must state that the proposed research is not a class of research that is ineligible for NIH funding.

Investigators must also submit a sample informed consent document, with patient identifier information removed, and a description of the informed consent process along with documentation of IRB approval of the derivation protocol. They must also provide an abstract of the scientific protocol used to derive human pluripotent stem cells along with a title of the research proposal that proposes the use of human pluripotent stem cells.

Ensuring Compliance with the Guidelines

Investigators requesting NIH funds for research using pluripotent stem cells will need to provide documentation that they are in compliance with the *Guidelines* prior to receiving NIH funds for this class of research. Submitted documentation will be reviewed by a newly-created NIH working group called the Human Pluripotent Stem Cell Review Group (HPSCRG).

Members of the working group will review documentation of compliance with the *Guidelines* for funding requests that propose the use of human pluripotent stem cells. They will also advise the NIH Center for Scientific Review Advisory Committee (CSRAC) of the outcome of their review, which, if appropriate, will be approved by the CSRAC. This decision will be forwarded to the funding Institute or Center. The HPSCRG will also hold public meetings when a request proposes the use of a line of human pluripotent stem cells that has not been previously reviewed by the HPSCRG.

The NIH is in the process of finalizing the members of the HPSCRG in preparation for a March deadline for the receipt of requests for NIH funding for human pluripotent stem cell research. The Agency has not, to date, received any applications for this class of research. In no event will NIH fund research or allow existing funds to be used for research using human pluripotent stem cells derived from human embryos or human fetal tissue until the derivation protocol has received HPSCRG review and CSRAC approval. Continued compliance with the *Guidelines* is a term and condition of the NIH award.

Page citation: NIH Fact Sheet on Human Pluripotent Stem Cell Research Guidelines. In *Stem Cell Information*. Bethesda, MD: National Institutes of Health, U.S. Department of Health and Human Services, 2009 Available at
http://stemcells.nih.gov/news/newsArchives/pages/stemfactsheet.aspx

U.S. Department of Health & Human Services (HHS) | USA.gov - Government Made Easy National Institutes of Health (NIH), 9000 Rockville Pike, Bethesda, Maryland 20892 *NIH…Turning Discovery*

Cell Signaling Synopsis from Cancer Principles and Practice of Oncology Vincent T. DeVita, MD

Molecules that initiate signaling cascades include proteins, amino acids, and lipids. Gases and light also initiate signaling cascades.

Most extracellular signals such as growth factors, bind to receptors on the plasma membrane, but others such as cortisol, diffuse into the cell and bind to receptors in the cytoplasm and nucleus. Some signals are continuous, such as those sent by the extracellular matrix, whereas others are episodic, like the secretion of insulin from the pancreatic beta cells in response to increases in blood glucose.

Signaling molecules originate from a variety of sources. Some, such as neurotransmitters, are stored in the cell and are released to provide communication with other cells under specific conditions. Other ligands are stored outside the cell (eg in the extracellular matrix) and become accessible in response to tissue damage or remodeling. Traditionally, signals have been divided based on the cell of origin into those that affect distant cells (endocrine), nearby cells (paracrine), or the same cell (autocrine).

Cells also respond to signals that arise from within. Important examples include the checkpoint pathways that ensure the orderly progression of the cell cycle and the pathways that sense and repair damaged DNA.

Overview of Extracellular Signaling

Communication by extracellular signals usually involves six steps: (1) synthesis and (2) release of the signaling molecule by the signaling cell; (3) transport of the signal to the target cell; (4) detection of the signal by a specific receptor protein; (5) a change in cellular metabolism, function, or development triggered by the receptor-signal complex; and (6) removal of the signal, which often terminates the cellular response.

In many eukaryotic microorganisms (e.g., yeast, slime molds, and protozoans), secreted molecules coordinate the aggregation of free-living cells for sexual mating or differentiation under certain environmental conditions. Chemicals released by one organism that can alter the behavior or gene expression of other organisms of the same species are called pheromones. Yeast mating-type factors discussed later in this chapter are a well-understood example of pheromonemediated cell-to-cell signaling. Some algae and animals also release pheromones, usually dispersing them into the air or water, to attract members of the opposite sex. More important in plants and animals are extracellular signaling molecules that function *within* an organism to control metabolic processes within cells, the growth of tissues, the synthesis and secretion of proteins, and the composition of intracellular and extracellular fluids. This chapter focuses on such cell-to-cell signaling in single-celled eukaryotes and in a variety of higher eukaryotes, particularly mammals.

Signaling Molecules Operate over Various Distances

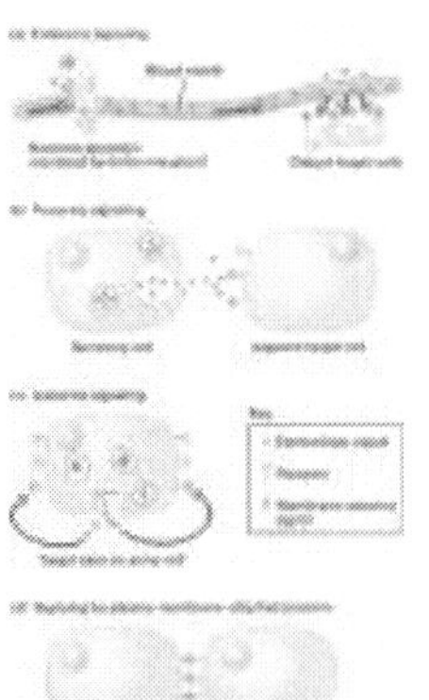

Signaling by extracellular, secreted molecules can be classified into three types — endocrine, paracrine, or autocrine — based on the distance over which the signal acts. In addition, certain membrane-bound proteins on one cell can directly signal an adjacent cell (Figure 1).

Figure 1

General schemes of intercellular signaling in animals. (a – c) Cell-to-cell signaling by extracellular chemicals occurs over distances only a few micrometers in autocrine.

In Endocrine Signaling, signaling molecules, called hormones, act on target cells distant from their site of synthesis by cells of endocrine organs. In animals, an endocrine hormone usually is carried by the blood from its site of release to its target.

In Paracrine Signaling, the signaling molecules released by a cell only affect target cells in close proximity to it. The conduction of an electric impulse from one nerve cell to another or from a nerve cell to a muscle cell (inducing or inhibiting muscle contraction) occurs via paracrine signaling. This type of signaling, mediated by neurotransmitters, is used in transmitting nerve impulses. Many signaling molecules regulating development in multicellular organisms also act at short range.

In Autocrine Signaling, cells respond to substances that they themselves release. Many growth factors act in this fashion, and cultured cells often secrete growth factors that stimulate their own growth and proliferation. This type of signaling is particularly common in tumor cells, many of which overproduce and release growth factors that stimulate inappropriate, unregulated proliferation of themselves as well as adjacent nontumor cells; this process may lead to formation of tumor mass.

Some compounds can act in two or even three types of cell-to-cell signaling. Certain small amino acid derivatives, such as epinephrine, function both as neurotransmitters (paracrine signaling) and as systemic hormones (endocrine signaling). Some protein hormones, such as epidermal growth factor (EGF), are synthesized as the exoplasmic part of a plasma - membrane protein; membrane-bound EGF can bind to and signal an adjacent cell by direct contact. Cleavage by a protease releases secreted EGF, which acts as an endocrine signal on distant cells.

Receptor Proteins Exhibit Ligand-Binding and Effector Specificity

As noted earlier, the cellular response to a particular extracellular signaling molecule depends on its binding to a specific receptor protein located on the surface of a target cell or in its nucleus or cytosol. The signaling molecule (a hormone, pheromone, or neurotransmitter) acts as a ligand, which binds to, or "fits," a site on the receptor. Binding of a ligand to its receptor causes a conformational change in the receptor that initiates a sequence of reactions leading to a specific cellular response.

The response of a cell or tissue to specific hormones is dictated by the particular hormone receptors it possesses and by the intracellular reactions initiated by the binding of any one hormone to its receptor. Different cell types may have different sets of receptors for the same ligand, each of which induces a different response. Or the same receptor may occur on various cell types, and binding of the same ligand may trigger a different response in each type of cell. Clearly, different cells respond in a variety of ways to the same ligand. For instance, acetylcholine receptors are found on the surface of striated muscle cells, heart muscle cells, and pancreatic acinar cells. Release of acetylcholine from a neuron adjacent to a striated muscle cell triggers contraction, whereas release adjacent to a heart muscle slows the rate of contraction. Release adjacent to a pancreatic acinar cell triggers exocytosis of secretory granules that contain digestive enzymes. On the other hand, different receptor -ligand complexes can induce the same cellular response in some cell types. In liver cells, for example, the binding of either glucagon to its receptors or of epinephrine to its receptors can induce degradation of glycogen and release of glucose into the blood.

These examples show that a receptor protein is characterized by binding specificity for a particular ligand, and the resulting hormone-ligand complex exhibits effector specificity (i.e., mediates a specific cellular response) . For instance, activation of either epinephrine or glucagon receptors on liver cells by binding of their respective ligands induces synthesis of cyclic AMP (cAMP), one of several intracellular signaling molecules, termed second messengers, which regulate various metabolic functions; as a result, the effects of both receptors on liver-cell metabolism are the same. Thus, the binding specificity of epinephrine and glucagon receptors differ, but their effector specificity is identical.

In most receptor-ligand systems, the ligand appears to have no function except to bind to the receptor. The ligand is not metabolized to useful products, is not an intermediate in any cellular activity, and has no enzymatic properties. The only function of the ligand appears to be to change the properties of the receptor, which then signals to the cell that a specific product is present in the environment. Target cells often modify or degrade the ligand and, in so doing, can modify or terminate their response or the response of neighboring cells to the signal.

Hormones Can Be Classified Based on Their Solubility and Receptor Location

Most hormones fall into three broad categories: (1) small lipophilic molecules that diffuse across the plasma membrane and interact with intracellular receptors; and (2) hydrophilic or (3) lipophilic molecules that bind to cell-surface receptors (Figure-2) . Recently, nitric oxide, a gas, has been shown to be a key regulator controlling many cellular responses. We discuss this important regulator later in this chapter. Here we briefly describe the three main types of hormones; later we discuss the mechanisms that regulate their synthesis, release, and degradation.

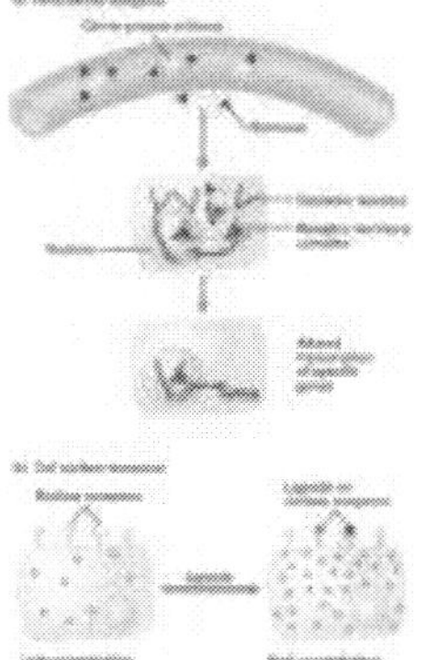

Figure-2

Some hormones bind to intracellular receptors; others, to cell-surface receptors. (a) Steroid hormones, thyroxine, and retinoids, being lipophilic, are transported by carrier proteins.

Lipophilic Hormones with Intracellular Receptors

Many lipid-soluble hormones diffuse across the plasma membrane and interact with receptors in the cytosol or nucleus. The resulting hormone-receptor complexes bind to transcription-control regions in DNA thereby affecting expression of specific genes (see Figure-2a). Hormones of this type include the steroids (e.g., cortisol, progesterone, estradiol, and testosterone), thyroxine, and retinoic acid.

All steroids are synthesized from cholesterol and have similar chemical skeletons. After crossing the plasma membrane, steroid hormones interact with intracellular receptors, forming complexes that can increase or decrease transcription of specific genes. These receptor-steroid complexes also may affect the stability of specific mRNAs. Steroids are effective for hours or days and often influence the growth and differentiation of specific tissues. For example, estrogen and progesterone, the female sex hormones, stimulate the production of egg-white hormones in chickens and cell proliferation in the hen oviduct. In mammals, estrogens stimulate growth of the uterine wall in preparation for embryo implantation. In insects and crustaceans, α-ecdysone (which is chemically related to steroids) triggers the differentiation and maturation of larvae; like estrogens, it induces the expression of specific gene products.

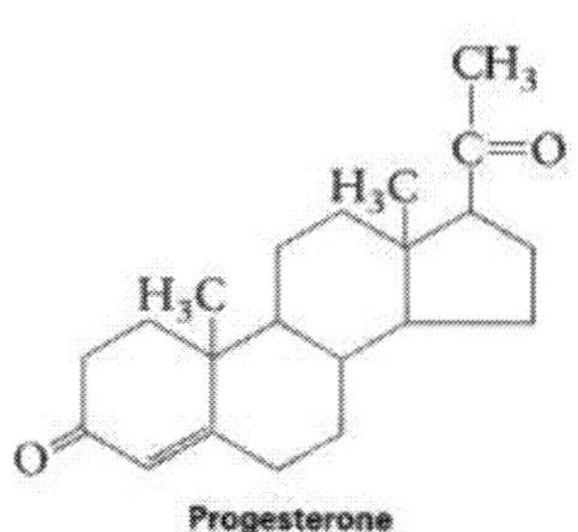

Thyroxine (tetraiodothyronine) and triiodothyronine — the principal iodinated compounds in the body — are formed in the thyroid by intracellular proteolysis of the iodinated protein thyroglobulin and immediately released into the blood.

These two thyroid hormones stimulate increased expression of many

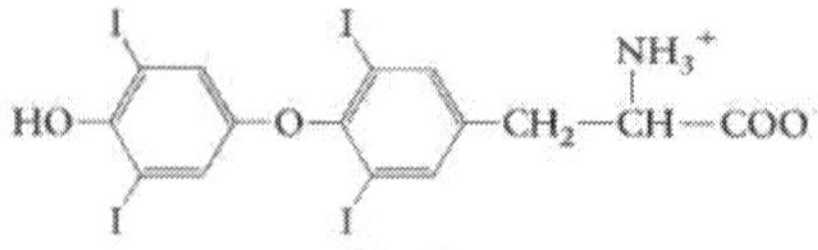

Thyroxine

cytosolic enzymes (e.g., liver hexokinase) that catalyze the catabolism of glucose, fats, and proteins and of mitochondrial enzymes that catalyze oxidative phosphorylation.

Retinoids are polyisoprenoid lipids derived from retinol (vitamin A). They perform multiple regulatory functions in diverse cellular processes. Retinoids regulate cellular proliferation, differentiation, and death, and they have numerous clinical applications. Their diverse effects reflect, at least in part, the multiplicity of retinoid derivatives, the existence of two different classes of receptors that form heterodimers, and differences in their cis-acting regulatory sites on DNA. During development

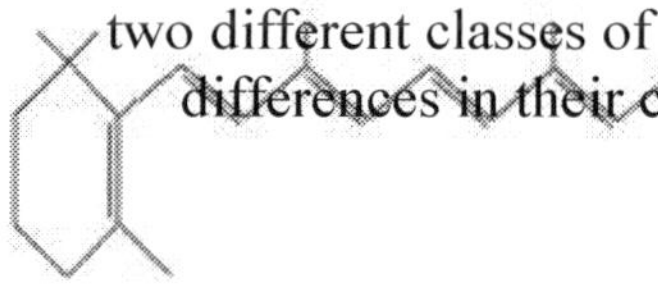

Retinoic acid

retinoids act as local mediators of cell-cell interaction. For instance, during the formation of motor neurons in the chick, one class of motor neurons generates a retinoid signal which regulates the number and type.

Water-Soluble Hormones with Cell-Surface Receptors

Because water-soluble signaling molecules cannot diffuse across the plasma membrane, they all bind to cell-surface receptors. This large class of compounds is composed of two groups: (1) peptide hormones, such as insulin, growth factors, and glucagon, which range in size from a few amino acids to protein-size compounds, and (2) small charged molecules, such as epinephrine and histamine, that are derived from amino acids, and function as hormones and neurotransmitters.

Many water-soluble hormones induce a modification in the activity of one or more enzymes already present in the target cell. In this case, the effects of the surface-bound hormone usually are nearly immediate, but persist for a short period only. These signals also can give rise to changes in gene expression that may persist for hours or days. In yet other cases water-soluble signals may lead to irreversible changes, such as cellular differentiation.

Lipophilic Hormones with Cell-Surface Receptors

The primary lipid-soluble hormones that bind to cell-surface receptors are the prostaglandins. There are at least 16 different prostaglandins in nine different chemical classes, designated PGA – PGI. Prostaglandins are part of an even larger family of 20 carbon–containing hormones called eicosanoid hormones. In addition to prostaglandins, they include prostacyclins, thromboxanes, and leukotrienes. Eicosonoid hormones are synthesized from a common precursor, arachidonic acid. Arachidonic acid is generated from phospholipids and diacylglycerol.

In both vertebrates and invertebrates, prostaglandins are synthesized and secreted continuously by many types of cells and rapidly broken down by enzymes in body fluids.

Many prostaglandins act as local mediators during paracrine and autocrine signaling and are destroyed near the site of their synthesis. They modulate the responses of other hormones and can have profound effects on many cellular processes. Certain prostaglandins cause blood platelets to aggregate and adhere to the walls of blood vessels. Because platelets play a key role in clotting blood and plugging leaks in blood vessels, these prostaglandins can affect the course of vascular disease and wound healing; aspirin inhibits their synthesis by acetylating (and thereby irreversibly inhibiting) prostaglandin H_2 synthase. Other prostaglandins initiate the contraction of smooth muscle cells; they accumulate in the uterus at the time of childbirth and appear to be important in inducing uterine contraction.

Recent studies have shown that a family of plant steroids, called brassinosteroids, regulates many aspects of development. These lipophilic compounds, like prostaglandins, act through cell-surface receptors.

Cell-Surface Receptors Belong to Four Major Classes

The different types of cell-surface receptors that interact with water-soluble ligands are schematically represented in Figure-3. Binding of ligand to some of these receptors induces second-messenger formation, whereas ligand binding to others does not. For convenience, we can sort these receptors into four classes:

G protein – **coupled receptors** (see Figure-3a): Ligand binding activates a G protein, which in turn activates or inhibits an enzyme that generates a specific second messenger or modulates an ion channel, causing a change in membrane potential. The receptors for epinephrine, serotonin, and glucagon are examples.

Ion-channel receptors (see Figure-3b): Ligand binding changes the conformation of the receptor so that specific ions flow through it; the resultant ion movements alter the electric potential across the cell membrane. The acetylcholine receptor at the nerve-muscle junction is an example.

Tyrosine kinase – linked receptors (see Figure-3c): These receptors lack intrinsic catalytic activity, but ligand binding stimulates formation of a dimeric receptor, which then interacts with and activates one or more cytosolic protein-tyrosine kinases. The receptors for many cytokines, the interferons, and human growth factor are of this type. These tyrosine kinase – linked receptors sometimes are referred to as the *cytokine-receptor superfamily.*

Receptors with intrinsic enzymatic activity (see Figure-3d): Several types of receptors have intrinsic catalytic activity, which is activated by binding of ligand. For instance, some activated receptors catalyze conversion of GTP to cGMP; others act as protein phosphatases, removing phosphate groups from phosphotyrosine residues in substrate proteins, thereby modifying their activity. The receptors for insulin and many growth factors are ligand-triggered protein kinases; in most cases, the ligand binds as a dimer, leading to dimerization of the receptor and activation of its kinase activity. These receptors — often referred to as receptor serine/threonine kinases or receptor tyrosine kinases — autophosphorylate residues in their own cytosolic domain and also can phosphorylate various substrate proteins.

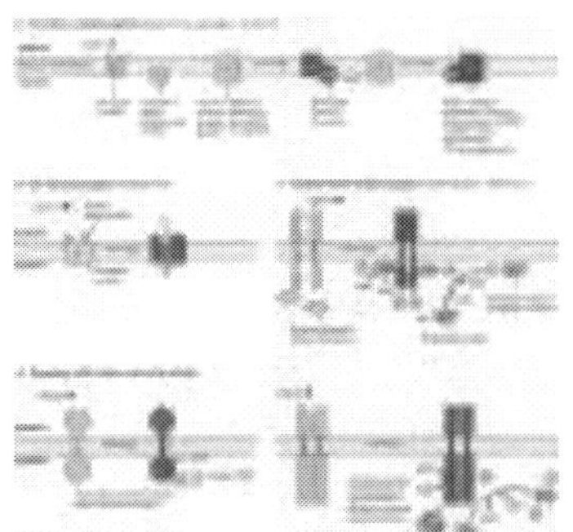

Figure-3

Four classes of ligand-triggered cell-surface receptors. Common ligands for each receptor type are listed in parentheses.

The discussion in this chapter focuses primarily on signaling pathways initiated by G protein – coupled receptors (GPCRs) and receptor tyrosine kinases (RTKs). The general structure and mechanism of action of the intracellular receptors for steroid hormones; ion channels; and certain receptor serine/threonine kinases as well as other developmentally relevant cell-surface receptors are depicted.

Effects of Many Hormones Are Mediated by Second Messengers

The binding of ligands to many cell -surface receptors leads to a short-lived increase (or decrease) in the concentration of the intracellular signaling molecules termed second messengers. These low-molecular-weight signaling molecules include 3′,5′-cyclic AMP (cAMP); 3′,5′-cyclic GMP (cGMP); 1,2-diacylglycerol (DAG); inositol 1,4,5-trisphosphate (IP3); various inositol phospholipids (phosphoinositides); and Ca2+ (Figure-4).

Figure-4

Structural formulas of four common intracellular second messengers. Their abbreviations are indicated. The calcium ion (Ca2+) and several membrane-bound inositol.

The elevated intracellular concentration of one or more second messengers following hormone binding triggers a rapid alteration in the activity of one or more enzymes or nonenzymatic proteins. The metabolic functions controlled by hormone-induced second messengers include uptake and utilization of glucose, storage and mobilization of fat, and secretion of cellular products. These intracellular molecules also control proliferation, differentiation, and survival of cells, in part by regulating the transcription of specific genes. The mode of action of cAMP and other second messengers is discussed in a later section. Removal (or degradation) of a ligand or second messenger, or inactivation of the ligand-binding receptor, can terminate the cellular response to an extracellular signal.

Other Conserved Proteins Function in Signal Transduction

In addition to cell-surface receptors and second messengers, several types of conserved proteins function in signal transduction pathways stimulated by extracellular signals. Here we introduce the three main classes of these intracellular signaling proteins.

GTPase Switch Proteins

A large group of *GTP-binding* proteins act as molecular switches in signal-transduction pathways. These proteins are turned “on” when bound to GTP and turned “off” when bound to GDP (Figure-5a). In the

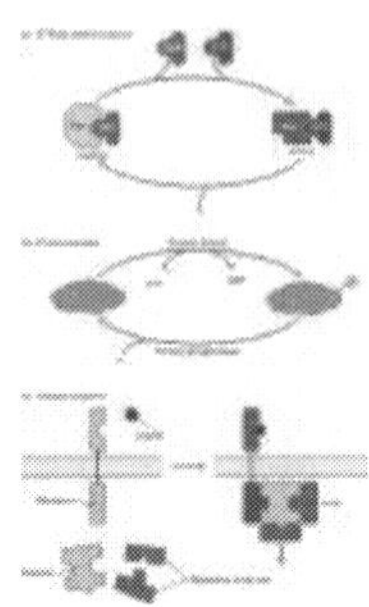

absence of a signal, the protein is bound to GDP. Signals activate the release of GDP, and the subsequent binding to GTP over GDP is favored by the higher concentrations of GTP in the cell. The intrinsic GTPase activity of these GTP-binding proteins hydrolyzes the bound GTP to GDP and Pi, thus converting the active form back to the inactive form. The kinetics of hydrolysis regulates the length of time the switch is "on."

Figure-5

Common intracellular signaling proteins. (a) GTP-binding proteins with GTPase activity function as molecular switches. When bound to GTP they are active; when bound to GDP, they are inactive.

There are two classes of GTPase switch proteins: trimeric G proteins, which as noted already are directly coupled to certain receptors, and monomeric Ras and Ras-like proteins. Both classes contain regions that promote the activity of specific effector proteins by direct protein-protein interactions. These regions are in their active conformation only when the switch protein is bound to GTP. G proteins are coupled directly to activated receptors, whereas Ras is linked only indirectly via other proteins. The two classes of GTPbinding proteins also are regulated in very different ways.

Protein Kinases

Activation of all cell-surface receptors leads to changes in protein phosphorylation through the activation of protein kinases (Figure-5b) . In some cases kinases are part of the receptor itself, and in others they are found in the cytosol or associated with the plasma membrane. Animal cells contain two types of protein kinases: those directed toward tyrosine and those directed toward either serine or threonine. The structures of the catalytic core of both types are very similar. In general, protein kinases become active in response to the stimulation of signaling pathways. The catalytic activities of kinases are modulated by phosphorylation, by direct binding to other proteins, and by changes in the levels of various second messengers. The activity of protein kinases is opposed by the activity of *protein phosphatases,* which remove phosphate groups from specific substrate proteins. The activities of kinases and phosphatases during cell cycle control the function of the cell.

Adapter Proteins

Many signal-transduction pathways contain large multiprotein signaling complexes, which often are held together by *adapter proteins* (Figure-5c). Adapter proteins do not have catalytic activity, nor do they directly activate effector proteins. Rather, they contain different combinations of domains, which function as docking sites for other proteins. For instance, different domains bind to phosphotyrosine residues (SH2 and PTB domains), proline-rich sequences (SH3 and WW domains), phosphoinositides (PH domains), and unique C-terminal sequences with a C-terminal hydrophobic residue (PDZ domains) . In some cases adapter proteins contain arrays of a single binding domain or different combinations of domains. In addition, these binding domains can be found alone or in various combinations in proteins containing catalytic domains. These combinations provide enormous potential for complex interplay and cross-talk between different signaling pathways.

Common Signaling Pathways Are Initiated by Different Receptors in a Class

In general, different members of a particular class of receptors transduce signals by highly conserved pathways. Moreover, analogies are found in the signaling pathways associated with different receptor classes. Figure-6 illustrates the main components of the key signaling pathways downstream from G protein – coupled receptors (GPCRs) and receptor tyrosine kinases (RTKs), the two receptor classes that we consider in detail in this chapter. Although a GTPase switch protein occurs in both types of pathways, its position in the pathway relative to the receptor differs. Second messengers are critical components of most GPCR pathways and some RTK pathways. Adapter proteins function in all RTK pathways but not in the main GPCR pathways. Protein kinases, however, play a key role in all signaling pathways; ultimately an activated protein kinase phosphorylates one or more substrate proteins. The nature of the substrate proteins, which include enzymes, microtubules, histones, and transcription factors, plays an important role in determining the cellular response to a particular signal in a particular cell.

Figure-6

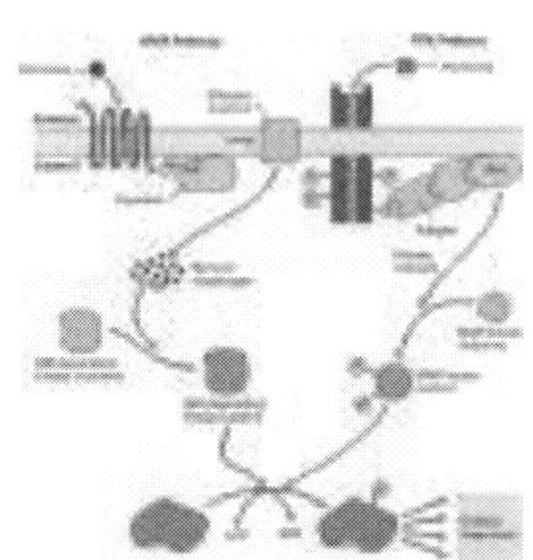

Schematic overview of common signaling pathways downstream from G protein – coupled receptors (GPCRs) and receptor tyrosine kinases (RTKs).

The Synthesis, Release, and Degradation of Hormones Are Regulated

Because of their potent effects, hormones and neurotransmitters must be carefully regulated. The release and degradation of some signaling compounds are regulated to produce rapid, short-term effects; others to produce slower-acting but longer-lasting effects (Table-1). In some cases, complex regulatory networks coordinate the levels of hormones whose effects are interconnected.

Table-1

Characteristic Properties of Principal Types of Mammalian Hormones.

Peptide Hormones and Catecholamines

Organisms must be able to respond instantly to many changes in their internal or external environment. Such rapid responses are mediated primarily by peptide hormones and the catecholamines, epinephrine, norepinephrine, and dopamine. The cells that produce these signaling molecules store them in secretory vesicles just under the plasma membrane. The supply of stored, preformed signaling molecules is sufficient for 1 day in the case of peptide hormones and for several days in the case of catecholamines. All peptide hormones, including insulin and adrenocorticotropic hormone (ACTH), are synthesized as part of a longer propolypeptide, which is cleaved by specific proteases to generate the active molecule just after it is transported to a secretory vesicle.

Stimulation of signaling cells causes immediate exocytosis of the stored peptide hormone or catecholamine into the surrounding medium or the blood. Secreting cells also are stimulated to synthesize the signaling molecule and replenish the cell's supply. Released peptide hormones persist in the blood for only seconds or minutes before being degraded by blood and tissue proteases. Released catecholamines, are rapidly inactivated by different enzymes or taken up by specific cells. The initial actions of these signaling molecules on target cells (the activation or inhibition of specific enzymes), last only seconds or minutes. Thus the catecholamines and some peptide hormones can mediate short responses that are terminated by their own degradation.

Steroid Hormones, Thyroxine, and Retinoic Acid

The pathways for synthesizing steroid hormones from cholesterol involve 10 or more enzymes. Steroid-producing cells, like those in the adrenal cortex, store a small supply of hormone precursor, but none of the mature, active hormone. When stimulated, the cells convert the precursor to the active hormone, which then diffuses across the plasma membrane into the blood. Likewise, thyroglobulin, the iodinated precursor of thyroxine is stored in thyroid follicles. When cells lining these follicles are exposed to thyroid-stimulating hormone (TSH), they take up thyroglobulin; controlled proteolysis of this glycoprotein by lysosomal enzymes yields thyroxine, which is released into the blood.

Because the signaling cells that produce thyroxine and steroid hormones store little of the active hormone, release of these hormones takes from hours to days (see Table-1). These hormones, which are poorly soluble in aqueous solution, are transported in the blood by carrier proteins; the tightly bound active hormones are not rapidly degraded. Thus, cellular responses to thyroxine and steroid hormones take awhile to occur but persist from hours to days.

Retinol is stored in the liver and is found in high concentrations in blood in a complex with serum binding protein. Due to its lipophilic nature, retinol diffuses through the plasma membrane and forms a complex with a cytosolic retinol-binding protein called CRBP. Retinol is converted to retinal through the activity of retinol dehydrogenase, and retinal, in turn, is converted to retinoic acid by retinal dehydrogenase. Retinoic acid can act as a signal in the cell in which it is produced, or it can diffuse through the plasma membrane to influence the development of neighboring cells. Retinoic acid can also be further modified enzymatically to alter its signaling specificity.

Feedback Control of Hormone Levels

The synthesis and/or release of many hormones are regulated by *positive* or *negative feedback.* This type of regulation is particularly important in coordinating the action of multiple hormones on various cell types during growth and differentiation. Often, the levels of several hormones are interconnected by feedback circuits, in which changes in the level of one hormone affect the levels of other hormones. One example is the regulation of estrogen and progesterone, steroid hormones that stimulate the growth and differentiation of cells in the endometrium, the tissue lining the interior of the uterus. Changes in the endometrium prepare the organ to receive and nourish an embryo. The levels of both hormones are regulated by complex feedback circuits involving several other hormones.

SUMMARY of Extracellular Signaling

Extracellular signaling molecules regulate interaction between unicellular organisms and are critical regulators of physiology and development in multicellular organisms.

There are many different types of signals, including membrane-anchored and secreted proteins and peptides, small lipophilic molecules (e.g., steroid hormones, thyroxine), small hydrophilic molecules derived from amino acids (e.g., catecholamines), and gases. Signals can act at short range, long range, or both (see Figure-1).

The multitude of cell-surface receptors fall into four main classes: G protein – coupled receptors, ion-channel receptors, receptors linked to cytosolic tyrosine kinases, and receptors with intrinsic catalytic activity (see Figure-3).

Binding of extracellular signaling molecules to cell-surface receptors trigger intracellular pathways that ultimately modulate cellular metabolism, function, or development.

The level of second messengers, such as Ca^{2+} , cAMP, and IP_3 are modulated in response to binding of a ligand to cell-surface receptors. These intracellular signaling molecules, in turn, regulate the activities of enzymes and nonenzymatic proteins.

Conserved proteins that act in many signal-transduction pathways include GTPase switch proteins (trimeric G proteins and monomeric

Ras-like proteins), protein kinases, and adapter proteins (see Figure-6). The latter coordinate the formation of multicomponent signaling complexes.

Extracellular signals are often integrated into complex regulatory networks in which the synthesis, release, and degradation of hormones are precisely regulated.

Nutrigenetics

Nutrigenetics studies the interaction between diet, genetics, and health. Enzyme deficiency caused by a mutation in the enzyme phenylalanine hydroxylase causes the inability to metabolize foods containing the amino acid phenylalanine. In Diabetes Mellitus two missense mutations in exons coding the aminoterminal transcriptional activating domain of SREBP-1c were found in individuals displaying insulin resistance. This Sterol Response Element Binding Protein -1c or SREBP1-c is a membrane bound transcription factor that directly activates the expression of several genes involved in the synthesis and uptake of cholesterol, fatty acids, triglycerides, and phospholipids.

The E4 Allele in the Apolipoprotein E Gene is a mutation that is associated with hyperlipidemia, atherosclerosis, and coronary artery disease.

Historical Facts on the Genetic Disorders Sickle Cell Disease and Cystic Fibrosis

Sickle Cell Disease

Although James Bryan Herrick, MD 1881-1954 is credited with the description of sickle cell disease and myocardial infarction it was not until Linus Pauling published Sickle Cell Anemia a Molecular Disease in 1949 in the journal Science that the genetic mutation was elucidated as the 6th amino acid Glutamine - substituted for Valine E6V.

Valine's hydrophobic property causes the hemoglobin to collapse on itself causing a sickled shape, preventing malarial infection. Note: a Clinical Trial at the University of California, Los Angeles, Donald Kohn, MD is recruiting participants (Adults with Severe Sickle Cell Disease) to potentially undergo Autologous Transplant of lentivirus vector modified bone marrow, as of 2014.

Cystic Fibrosis

Cystic Fibrosis, also called mucoviscidosis, is caused by inheriting both copies of the CFTR gene, causing an error in the transmembrane conductance regulator which is involved in the production of mucus, digestive fluids, and sweat. Clinical Trials on Animals are being conducted now.

Few Facts on the Genetics of Addiction and Drug Metabolism

It is estimated that 40-60% of the vulnerability to addiction is attributable to genetic factors. Certain genetic alterations of alcohol dehydrogenase are protective against alcoholism.

Polymorphisms in ADH genotypes coding for highly active isoenzyme and aldehyde dehydrogenase (ALDH), coding for less active isoenzyme, may lead to accumulation of acetaldehyde after alcohol consumption, but the significance of this to alcoholic liver disease remains unclear. Other gene polymorphisms (c2 promoter polymorphism of CYP2E1, TNFa, interleukin IL-10 have been associated with alcoholic liver disease.

Certain genetic alterations of cytochrome P450 2A6 are protective against nicotine addiction Acetylcholine (Ach) is our natural (endogenous) nAChR agonist. Nicotine is the exogenous agonist at nAChR. Both cause the release of neurotransmitters, like epinephrine, acetylcholine, serotonin, GABA (gaba aminobutyric acid), glutamate, beta-endorphin, and dopamine.

Acetaldehyde and biogenic amines reduce the activity of the enzymes, monoamine oxidase A and B inhibiting the metabolism of catecholamines including dopamine.

Certain genetic alterations of cytochrome P450 2D6 are protective against codeine abuse The CHRNA5/A3/B4 gene cluster is associated with nicotine dependence Certain genetic alterations of the GABA type A predispose to alcoholism and genetic alterations of the D2-receptor are linked to a higher vulnerability to drug addiction

Public Cord Blood Banking and Donation

Donating your child's cord blood to a public bank is highly beneficial in that once donated, it is available to anyone who needs it, and it costs nothing to donate. Donated cord blood can be used to treat any compatible patient, and may also be used for medical research.

So why don't more women donate their child's cord blood? The answer is simple. Women and their health care providers — including many physicians — are often unaware of the right to donate cord blood. Also, the decision needs to be made early in the pregnancy; donating to a public bank requires foresight. The paperwork must be completed before the birth of your child, as most hospitals are not readily equipped to collect cord blood. The expectant mother has the responsibility to bring the cord blood donation kit supplied by the public bank to the birthing site. Paperwork should be filled out before the 34th week of pregnancy to ensure that you qualify and receive the collection kit in time for your child's birth.

For a closer look at some of the requirements for donation to a public bank, you can download samples here:

See eligibility requirements Sample medical questionnaire

Donating cord blood to a public bank is particularly important for minority populations. Patients in need of a transplant are more likely to find a match from a donor of the same descent. Because statistically there are fewer racial minorities in the national registries, finding a match can be especially difficult.

One of the goals of the Save the Cord Foundation is to simplify the donation process, increase the number of donations, in particular minority donations, and make donating cord blood a standard of care available to every expectant mother.

If you would like more information about donating your child's cord blood, there are two public banks in the U.S. that accept cord blood donations from anywhere in the country:

Cryobanks International 1-800-869-8608

LifeBankUSA 1-877-543-3226

A full list of public banks can be found here.

Please be aware that public banks have budget limitations that sometimes restrict the number of donations that are accepted.

Public Cord Blood Banking & Donation Directory

Arizona Public Cord Blood Banking
Program Phoenix, AZ
(602) 406-7021

Ashley Ross Cord Blood Program of the San Diego Blood Bank
400 Upas St.
San Diego, CA 92103
619-296-6393 ext. 8327

Bonfils Cord Blood Services Belle Bonfils Memorial Blood
Center Denver, CO
303-341-4000

CariCord
12635 E. Montview Blvd, Suite
300 Aurora, CO 80045-7116
844-227-4267

Carolinas Cord Blood Bank
Duke Hospital,Durham Regional Hospital, UNC Hospital, Rex Hospital, Western wake Hospital and Womens Hospital of Greensboro
Duke University Medical Center
Durham, NC 27710
(919) 668-1119919-668-1116

Celebrations Stem Cell
Center 3495 S. Mercy Road
Gilbert, AZ 85297
877.522.2355480-722-9963

Chaim Sheba Medical Center
Tel-Hashomer 972-3-530-3030

Children's Hospital of Orange County Cord Blood
Bank 455 S. Main St.
Orange, CA 92868
714-516-4335714-516-4335

CORD:USE

Collects from Florida, Michigan and North
Carolina 1991 Summit Park Drive, suite 2000
Orlando, FL 32810
1 888 CORDUSE (267 3873)407-667-4844

Core23 BioBank
909 East Republic Road, Suite B104
Springfield, MO 65807
877.286.2323417.222.0323

Cryo-Save India Private Limited
Gayatri Tech Park, Road 1 B, EPIP, KIADB, Whitefield
Bangalore0091-80-42430180
Dan Berger Cord Blood Program
412-209-7479

Gift of Life
800 Yamato Road
Boca Raton, FL 33431
561 988-0100

Hawai'i Cord Blood Bank
1319 Punahou Street
Honolulu, HI 96826
1-855-583-3085(808) 983-2265

Ireland Cancer Center at Case Western Reserve University and
University Hospitals of Cleveland Umbilical Cord Blood Program
10900 Euclid Avenue, WRB 2-125
Cleveland, OH 44106-7284
216-844-4723

ITxM – The Institute for Transfusion Medicine
Clinical Services, Cord Blood 501 Martindale Street 1st floor
Pittsburgh, PA 15212
(412) 209-7479(412) 209-7479

ITxM: The Institute for Transfusion Medicine Blood Services
5505 Pearl Street
Chicago, IL 60018
877-448-2673

J.P. McCarthy Cord Stem Cell Bank
4100 John R St.
Detroit, MI 48201
800-527-6266

KAISER PERMANENTE – LOS ANGELES OB
4867 Sunset Blvd.
Los Angeles, CA 90027
1-323-783-401

LifebankUSA
45 Horsehill Road Cedar
Knolls, NJ 07927
1-877-543-32261-877-543-3226

LifeCord
North Florida Regional Medical Center, Shands at AGH, Shands at UF, Shands at Lakeshore, The Birthing Center, The Patient's Corn
4039 Newberry Rd.
Gainesville, FL 32607
352-224-1600352-224-1600

Lifeforce Cryobanks
270 Northlake Blvd., Suite 1000
Altamonte Springs, FL 32701
1-800-869-86081-407-834-8333

M. D. Anderson Cord Blood Bank
1515 Holcombe Blvd.
Houston, TX 77030
800-392-1611

Michigan Community Blood Centers Cord Blood Bank
Bay Regional(BayCity), Covenant(Saginaw), Hackley(Muskegon), Holland(Holland), Lakeland(St.Joseph and Niles), Mercy General(Musk
1036 Fuller NE
Grand Rapids, MI 49503
866-642-5663616 233-8604

New Jersey Cord Blood Bank
see the web site for list of all participating hospitals

970 Linwood ave W PO Box 39
Paramus, NJ
866-228-1500201-444-3900

New York Blood Center, National Cord Blood Program
Collects from some hospitals in New York, Virginia and Ohio
310 East 67th St.
New York, NY 10021
866-767-6227(212)570-3230

Oklahoma Blood Institute
1001 N. Lincoln Blvd.
Oklahoma City, OK 73104
405-278-3130

Puget Sound Blood Center
921 Terry Ave.
Seattle, WA 98104
206-292-6500

Reliance Life Sciences
R-282, TTC Industrial Area of MIDC, Thane-Belapur Road, Rabale
Navi Mumbai91 (22) 6767 8000

SaneronCCEL
3802 Spectrum Blvd., Suite
145 Tampa, FL 33612
813-977-7664

Sibling Donor Cord Blood Program at Childrens Hospital Oakland
Research Institute
5700 Martin Luther King Jr.
way Oakland, CA 94609-1673
1 866 668-48951 510 450-7605

Singapore Cord Blood Bank
Singapore

South Texas Cord Blood Bank
South Texas Blood and Tissue Center
6211 IH 10 West at First Park Ten
San Antonio, TX 78201

800-292-5534 option 8210 731-5555

St. Louis Cord Blood Bank
Collects cord blood donations from hospitals within 150 mile radius (approx. 30 hospitals)
3662 Park Ave.
St. Louis, MO 63110
888-453-2673314-268-2787

StemCyte
1589 W. Industrial Park
St. Covina, CA 91722
866-783-62981-866-783-6298

StemCyte Tawain Co.
Ltd. 02-26097577
NEW TAIPEI CITY 24402-26097577

Tawain Advance Bio-Pharm, Inc.
12F.-3, No.25, Ln. 169, Kangning St., Xizhi Dist
New Taipei City+886-2-2692-6222#123

The Dan Berger Cord Blood Program, in Western Pennsylvania
412-209-7479412-209-7479 or 412-327-6025

UC Davis Health System
2315 Stockton Blvd.
Sacramento, CA 95817
800-2-UC-DAVIS

University of Arizona Cord Blood Bank
at the University of Arizona Medical Center
1501 N. Campbell Av
Tucson, AZ
520-626-5125520-626-5125

University of Colorado Cord Blood Bank
12635 E. Montview Blvd., Suite 300
Aurora, CO 80045
303-724-1300303-724-1300

University of Iowa's Hematopoietic Stem Cell Bank

125 S. Dubuque St., Suite 371
Iowa City, IA 52242
319-353-3747319-353-3747

Virgin
250 Gunnersbury Avenue, London W4 5QB
0845 620 9665

Virgin Health Bank
P.O. Box 210563
974 4404 9024

Repairing and Replacing Body Parts: What's Next

Researchers are exploring ways to repair, refurbish, or replace human organs.

By **Diane Cole**, for National Geographic News

PUBLISHED April 18, 2013

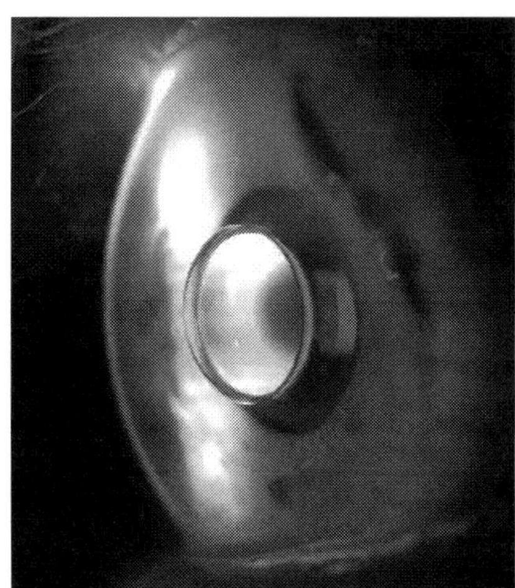

A close-up of a patient's eye with a telescopic implant.

Photograph © U.C. Regents

Advances in medical technology have helped us live longer. Now, researchers are exploring ways to repair, refurbish, or replace human organs that have been damaged by chronic disease, traumatic injury, heart attack, stroke, or just plain aging.

"Medicine is saving people who previously we weren't able to save," says Dr. Doris A. Taylor, director of regenerative medicine research at the Texas Heart Institute in Houston.

Even so, demand for donor organs exceeds the number available. "Each year thousands of people die while waiting for an organ," Taylor says. That gap in supply and demand is one factor that has led researchers to ever more innovative treatments; at times these treatments can sound like science fiction come to life.

Here's a look at what repaired and replacement parts are available to patients now, which treatments are undergoing clinical trials, and what medical scientists are working to achieve in the future.

Eyesight

Here's a new take on magnifying glasses: Surgeons can now implant a tiny telescope within the eye, to help restore some of the vision lost to end-stage age-related macular degeneration (AMD), a disease that affects 1.8 million Americans and is the leading cause of legal blindness for adults age 60 years and older.

Gene Theme Scene Note on Three Gene Based Treatments for Macular Degeneration Patients with

The VEGFR target: Pagaptanib (Pfizer), Ranibizumab (Novartis) and Bevacizumab (Genentech)

The device—which the Food and Drug Administration approved in 2010 and which is becoming more widely available to medical institutions across the country—is implanted via an hour-long outpatient procedure under local anesthesia. It requires about a month of working with an occupational therapist to get used to, says Dr. Mark Mannis, director of the Eye Center at the University of California Davis Health System.

"The reason is that this is not a simple restoration of vision," he says. "It really requires the patient to see in another way, much in the same way that a patient who loses a lower limb and then gets a prosthesis needs to learn how to walk in a new way."

In this case, the patient learns to use one eye—the one with the implant—for detailed vision and the other for peripheral vision.

Regenerative Medicine

As director of the Wake Forest Institute for Regenerative Medicine in Winston-Salem, North Carolina, Dr. Anthony Atala is researching treatments to repair or restore—or "regenerate"—damaged or failing tissues and organs by using the patient's own cells and healing abilities.

That can mean "boosting" healing by injecting stem cells, or by implanting tissues or organs that have been artificially bio-engineered in the lab starting with stem cells usually harvested from the patient, a strategy that minimizes the risk of the tissues or organs being rejected.

"We're working on about 30 different tissues and organs," Atala says. Already a number of implants have been tried successfully in humans, including knee cartilage, skin, blood vessels, urethras, windpipes (trachea), and bladders. Clinical trials are underway to treat urinary incontinence by implanting cells to help boost functionality of the urinary valve and thus keep patients dry.

Says Atala: "The future is focused on making sure that these technologies can get to as many patients and as many conditions as possible."

"Printing" Body Parts

At Atala's lab and other regenerative medicine research centers, 3-D printing is another experimental strategy being used to build bio-artificial body parts and organs.

"We've printed ear lobes and nose parts and miniature kidneys and skin," he says.

"You are laying [down] the cells one layer at a time, and placing the cells right where you need them," by customizing the different layers to form the necessary shape, he says. "If you think of your printer and your ink cartridge, instead of using ink you're using cells and a gel."

Wake Forest is also investigating the possibility of "printing" skin cells directly onto burn wounds.

Stem Cells for Stroke Recovery

Neurologist Dr. Lawrence Wechsler of the University of Pittsburgh's Schools of the Health Sciences is in the early stages of exploring whether stem cells, injected directly into the brain, can aid stroke victims in their recovery.

The first step—now being tested in a clinical trial—is establishing that it's safe just to try the technique. If that goes well, Wechsler says, "then we can design a study that will more reasonably look at the issue of efficacy and clinical benefit."

Such therapies wouldn't "unparalyze" patients, he warns. But small improvements in function could yield big improvements in quality of life. "If you can begin to use your hand to grip something and do some small tasks," Wechsler says, "or gain enough strength in your leg to help you move from being in a wheelchair to walking in some way, that change is a huge benefit."

Growing an Artificial Heart

Perhaps the ultimate goal of regenerative medicine researchers is creating and transplanting a functioning bio-artificial heart.

Is it feasible? Building complex solid organs like the heart, liver, lungs, and pancreas is challenging, and a major issue will be "where do you get those hundreds of billions of cells to do this," says Taylor of the Texas Heart Institute.

But she adds, "we're making huge strides," and predicts that a transplant of one kind of bio-artificial solid complex organ will be possible within five years. "And if I have anything to say about it," Taylor says, "I will be there when it happens."

For more, see this month's National Geographic *magazine cover story, "On Beyond 100."*

The Body Shop

4 pages: | 1 | 2 | 3 | 4 |

Just 15 years ago, Bob Langer and his colleague Joseph Vacanti pioneered a remarkable new process- growing human tissues in the lab. Back in 1987, Langer and Vacanti couldn't get their work published; journal editors didn't see any practical applications. Today, the pair are acknowledged as the fathers of the field of tissue engineering. Now, Langer, Vacanti and his brother Charles, as well as teams of researchers around the world, pursue the day when replacement tissues and organs are readily available, custom-made for those who need them.

REPLACEMENT PARTS

Today, tissue engineered skin, the first so-called "neo-organ" approved by the U.S. Food and Drug Administration, comes to the aid of burn victims and patients with severe skin sores or ulcers. In the not-too-distant future, lab-grown cartilage and bone could relieve arthritis sufferers, while blood vessels, cardiac valves and muscle tissue could save thousands of cardiovascular disease patients. Ultimately, custom-made hearts, livers, breasts, corneas, kidneys, bone marrow and bladders could offer elegant solutions to most life-threatening illnesses.

Ultimately, custom-made hearts, livers, breasts, corneas, kidneys, bone marrow and bladders could offer elegant solutions to most life-threatening illnesses.

"We can't say what the timeline will be," says Dr. Joseph Vacanti, Director of the Tissue Engineering and Organ Fabrication Laboratory at Massachusetts General Hospital in Boston. "But there are thirty plus tissues we're experimenting on in our lab."

Imitating Life

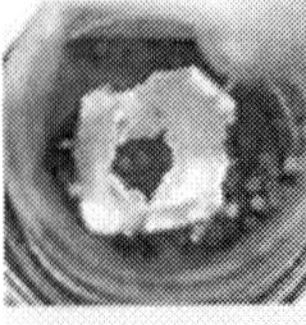

This human heart valve was grown in the lab.

Cultivating tissues in the lab requires closely mimicking the environment in which cells naturally grow. This turns out to be a tall order. Unlocking the biochemical signals that influence growth and development was the first step on the road to tissue engineering. By adding the right combination of compounds, scientists coax cells into growing and proliferating.

But, to produce biologically useful tissues like cartilage and heart valves, tissue engineers must also pay special attention to the physical environment in which cells grow.

In nature, the circulatory system gives each individual cell in a tissue access to nutrients and a means of waste removal. Many of the advances in tissue engineering have been means of replicating this scenario in the lab. One of Langer's major contributions to his filed was his work in biodegradable materials that can serve as scaffolding on which cells can be seeded. Joseph Vacanti deserves credit for the idea of the scaffold itself.

EXTRA ›
Video clip of tissues growing on scaffolding

"The scaffold looks like strands of spaghetti attached together," according to Langer. "The cells are seeded 2 to 3 millimeters apart and the whole apparatus is bathed in a nutritive media."

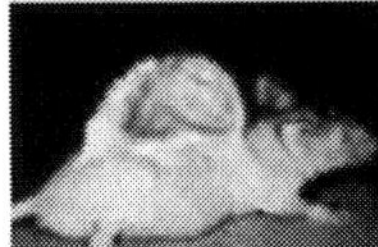

After this human ear is removed, the mouse will remain healthy.

The biodegradable scaffolding provides each cell with better access to nutrients and waste removal. Additionally, since the scaffolding can be molded into any shape or size, the tissue can be custom grown for the intended recipient. For example, to grow an ear like the one on the mouse pictured here, tissue engineers mold the biodegradable scaffold into the proper size and shape. Researchers then "seed" the scaffold with young cartilage cells and surgically implant the mold under the skin. The mouse, hairless and specially bred to lack an immune system that might reject the human tissue, nourishes the ear as the cartilage cells grow.

In the future, bits of scaffolding seeded with young cells could be implanted into ailing organs, where the body's own biochemistry would direct the young cells to grow into a "patch" of healthy tissues.

"Both functions are important," according to Joseph Vacanti. "but, in many circumstances, the shape is less important than the exchange of nutrients. "

....

4 pages: | 1 | 2 | 3 | 4 |

Photos: Charles Vacanti, MD; Advanced Tissue Sciences
Video: Advanced Tissue Sciences

Comment on the Original Replacing Body Parts Video (1/26/11)

Note: in the PBS video January 26 2011 entitled Replacing Body Parts filmed at Massachusetts General Hospital reveals Joseph Vacanti, MD research team member Harold Ott eventually uses a soap like product to wash the original cells of first a rat, later a pig heart that without the original cells was described by Doris Taylor, PhD as appearing "ghost like" until the pig heart scaffold received the recipient's cells, placed in a chamber, given tubes of fluid to create a pulse then an electrical charge until we see a beating heart in a chamber! Dr Taylor also held a pig kidney and remarked that the cell free pig kidney could be covered with the human cells of the recipient and the person would not need anti-rejection medications which was the major point of the presentation!!!!

Pharmacogenomics

Cytochrome P450 2D6 (CYP2D6), CYP2C9, and CYP2C19 determine the status of a normal extensive metabolizer (EM), an intermediate metabolizer (only one variant present), a poor metabolizer (PM) or an ultrarapid metabolizer (UM).The FDA has approved pharmacogenetic testing for the presence or absence of a functional allele to predict how a person will metabolize certain drugs.

Note: Approximately 10% of Caucasians are poor metabolizers based on this CYP2D6 mutation which is an allele on chromosome 22 note: genetically deficient CYP2D6 metabolism provides protection against oral opiate dependence see PubMed (Tyndale RF et al Pharmacogenetics 1997)

CYP2D6 is responsible for the metabolism of approximately 25% of clinically used drugs such as:

Opiods, Anticonvulsants, Cardiac Medications, Psychotropic Drugs as well as Prilosec (Omeprazole).

Grapefruit juice is known to inhibit one of the CYP450 (3A4) inhibiting certain therapeutic effects.

CYP2C9 variations affect the metabolism of Warfarin (Coumadin)

CYP2C19 variations affect the metabolism of Clopidogrel (Plavix)

CYP2D6 variations affect the metabolism of Codeine recent FDA reports caution against potentially lethal effects in children

CYP3A4 variant (22) affects the response to Atorvastin (Lipitor), Simvastatin and Lovastatin

CYP1A2 affects the response to Acetaminophen and Haldol

VKORC1 affects a patient's sensitivity to Warfarin (Coumadin).

COMT Catechol-O-Methytransferase is encoded by the COMT gene needed to metabolize catecholamine drugs such as Dopamine, Methyldopa, and Levodopa

Overview of Chromosomal Information

(NIH - National Library of Medicine)

In the nucleus of each cell, the DNA molecule is packaged into thread-like structures called chromosomes. Each chromosome is made up of DNA tightly coiled many times around proteins called histones that support its structure.

Chromosomes are not visible in the cell's nucleus—not even under a microscope—when the cell is not dividing. However, the DNA that makes up chromosomes becomes more tightly packed during cell division and is then visible under a microscope. Most of what researchers know about chromosomes was learned by observing chromosomes during cell division.

Each chromosome has a constriction point called the centromere, which divides the chromosome into two sections, or "arms." The short arm of the chromosome is labeled the "p arm." The long arm of the chromosome is labeled the "q arm." The location of the centromere on each chromosome gives the chromosome its characteristic shape, and can be used to help describe the location of specific genes.

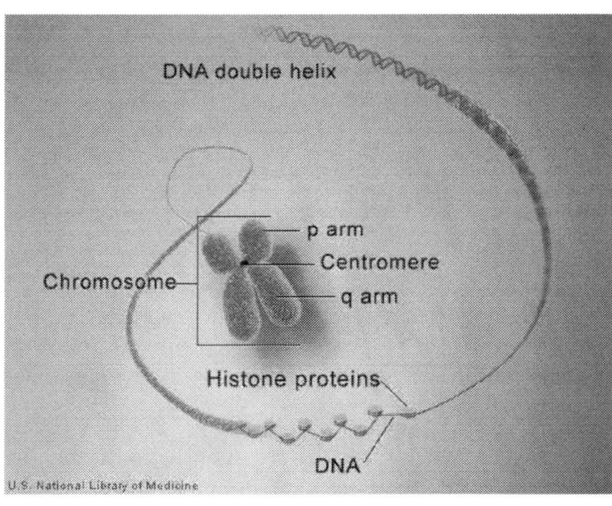

National Center for Biotechnology Information (US). Bethesda (MD): National Center for

Biotechnology Information (US); 1998-.
Chromosome Map

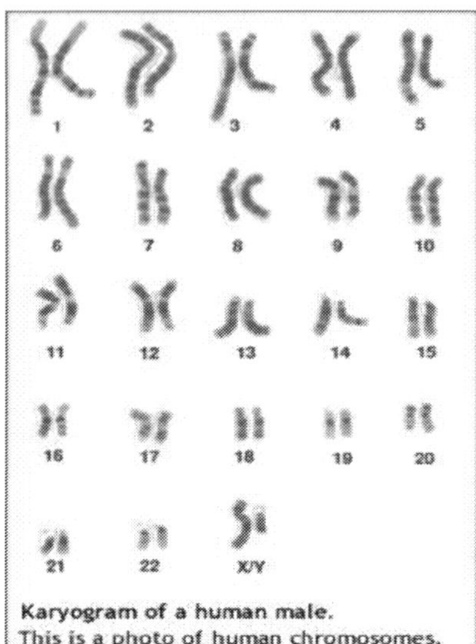

Karyogram of a human male.
This is a photo of human chromosomes. They have been stained and arranged in order of decreasing size. The presence of the Y chromosome in the last pair of chromosomes tells us that these chromosomes are from a man.

Our genetic information is stored in 23 pairs of chromosomes that vary widely in size and shape. Chromosome 1 is the largest and is over three times bigger than chromosome 22. The 23rd pair of chromosomes are two special chromosomes, X and Y, which determine our sex.

Females have a pair of X chromosomes (46, XX), whereas males have one X and one Y chromosomes (46, XY). Chromosomes are made of DNA, and genes are special units of chromosomal DNA. Each chromosome is a very long molecule, so it needs to be wrapped tightly around proteins for efficient packaging.

Near the center of each chromosome is its centromere, a narrow region that divides the chromosome into a long arm (q) and a short arm (p). We can further divide the chromosomes using special stains that produce stripes known as a banding pattern. Each chromosome has a distinct banding pattern, and each band is numbered to help identify a particular region of a chromosome. This method of mapping a gene to a particular band of the chromosome is called cytogenetic mapping. For example, the hemoglobin beta gene (*HBB*) is found on chromosome 11p15.4. This means that the *HBB* gene lies on the short arm (p) of chromosome 11 and is found at the band labeled 15.4.

With the advent of new techniques in DNA analysis, we are able to look at the chromosome in much greater detail. Whereas cytogenetic mapping gives a bird's eye view of the chromosome, more modern methods show DNA at a much higher resolution. The Human Genome Project aims to identify and sequence the ~30,000 genes in human DNA.

What are gene families?

A gene family is a group of genes that share important characteristics. In many cases, genes in a family share a similar sequence of DNA building blocks (nucleotides). These genes provide instructions for making products (such as proteins) that have a similar structure or function. In other cases, dissimilar genes are grouped together in a family because proteins produced from these genes work together as a unit or participate in the same process.

Classifying individual genes into families helps researchers describe how genes are related to each other. Researchers can use gene families to predict the function of newly identified genes based on their similarity to known genes. Similarities among genes in a family can also be used to predict where and when a specific gene is active (expressed). Additionally, gene families may provide clues for identifying genes that are involved in particular diseases.

Sometimes not enough is known about a gene to assign it to an established family. In other cases, genes may fit into more than one family. No formal guidelines define the criteria for grouping genes together. Classification systems for genes continue to evolve as scientists learn more about the structure and function of genes and the relationships between them.

For more information about gene families

Genetics Home Reference provides information about gene families including a brief description of each gene family and a list of the genes included in the family.

The HUGO Gene Nomenclature Committee (HGNC) has classified many human genes into families. Each grouping is given a name and symbol, and contains a table of the genes in that family.

The textbook Human Molecular Genetics (second edition, 1999) provides background information on human gene families.

The Gene Ontology database lists the protein products of genes by their location within the cell (cellular component), biological process, and molecular function.

The Reactome database classifies the protein products of genes based on their participation in specific biological pathways. For example, this resource provides tables of genes involved in controlled cell death (apoptosis), cell division, and DNA repair. What are gene families?

A gene family is a group of genes that share important characteristics. In many cases, genes in a family share a similar sequence of DNA building blocks (nucleotides). These genes provide instructions for making products (such as proteins) that have a similar structure or function. In other cases, dissimilar genes are grouped together in a family because proteins produced from these genes work together as a unit or participate in the same process.

Classifying individual genes into families helps researchers describe how genes are related to each other. Researchers can use gene families to predict the function of newly identified genes based on their similarity to known genes. Similarities among genes in a family can also be used to predict where and when a specific gene is active (expressed). Additionally, gene families may provide clues for identifying genes that are involved in particular diseases.

Sometimes not enough is known about a gene to assign it to an established family. In other cases, genes may fit into more than one family. No formal guidelines define the criteria for grouping genes together. Classification systems for genes continue to evolve as scientists learn more about the structure and function of genes and the relationships between them.

For more information about gene families

Genetics Home Reference provides information about gene families including a brief description of each gene family and a list of the genes included in the family.

The HUGO Gene Nomenclature Committee (HGNC) has classified many human genes into families. Each grouping is given a name and symbol, and contains a table of the genes in that family.

The textbook Human Molecular Genetics (second edition, 1999) provides background information on human gene families.

The Gene Ontology database lists the protein products of genes by their location within the cell (cellular component), biological process, and molecular function.

The Reactome database classifies the protein products of genes based on their participation in specific biological pathways.

For example, this resource provides tables of genes involved in controlled cell death (apoptosis), cell division, and DNA repair.

Gene Families

A gene family is a group of genes that share important characteristics. Classifying individual genes into families helps researchers describe how genes are related to each other. For more information, see what are gene families? In the Handbook.

The following families, defined by the HUGO Gene Nomenclature Committee, are included in Genetics Home Reference.

aaRS (aminoacyl tRNA synthetases) ABC
(ATP-binding cassette transporters)
ABHD (abhydrolase domain containing
genes) ACS (acyl-CoA synthetase family)
ADAMTS (ADAMTS metallopeptidase with thrombospondin type 1 motif)
ALDH (aldehyde dehydrogenases)
ALOX (arachidonate lipoxygenases)
ANKRD (ankyrin repeat domain
containing) ANO (anoctamins)
ARHGAP (Rho GTPase activating proteins)
ARHGEF (Rho guanine nucleotide exchange
factors) ATP (ATPases)
B3GT (beta 3-glycosyltransferases)
bHLH (basic helix-loop-helix)
BIRC (baculoviral IAP repeat-containing
genes) blood group (blood group antigens)
bZIP (basic leucine zipper proteins)
CA (carbonic anhydrases)
CACN (calcium channels)
CATSPER (CatSper
channels) CD (CD molecules)
CDH (cadherins)
CDK (cyclin-dependent kinases)
CHMP (charged multivesicular body proteins) chromatin-
modifying enzymes (chromatin-modifying enzymes) CLCN
(chloride channels, voltage-sensitive)
CNG (cyclic nucleotide-regulated channels)
COL (collagens)
COLEC (collectins)
collagen proteoglycans (collagen proteoglycans)
complement (complement system)
CTS (cathepsins)

CYB (cytochrome b) CYP
(cytochrome P450s) DDX
(DEAD-boxes) DN
(axonemal dyneins)
DNAJ (heat shock proteins, DNAJ (HSP40))
dolichyl D-mannosyl phosphate dependent mannosyltransferases (dolichyl D-mannosyl phosphate dependent mannosyltransferases)
DYN (cytoplasmic dyneins)
EF-hand domain containing (EF-hand domain containing)
endogenous ligands (endogenous ligands)
FANC (Fanconi anemia, complementation groups)
fibronectin type III domain containing (fibronectin type III domain containing)
FOX (forkhead box genes)
GJ (gap junction proteins (connexins))
glycosyltransferase family 8 domain containing (glycosyltransferase family 8 domain containing)
glycosyltransferase group 1 domain containing (glycosyltransferase group 1 domain containing)
GPC (glypicans (GPI-anchored HSPG))
GPCR (G protein-coupled receptors)
GPCRF (class F GPCRs, frizzled-type)
GR (glutamate receptors)
HLA (histocompatibility complex
genes) homeobox (homeoboxes)
IFT (intraflagellar transport homologs) IL
(interleukins and interleukin receptors)
immunoglobulin superfamily, C1-set domain containing (immunoglobulin superfamily, C1-set domain containing)
immunoglobulin superfamily, immunoglobulin-like domain containing (immunoglobulin superfamily, immunoglobulin-like domain containing)
immunoglobulin superfamily, I-set domain containing (immunoglobulin superfamily, I-set domain containing)
intermediate filaments type V, lamins (intermediate filaments type V, lamins)
ITG (integrins)
KCN (potassium
channels) KIF (kinesins)
KLK (kallikreins)
KRT (keratins)

LAM (laminins)
ligand-gated ion channels (ligand-gated ion channels) mitochondrial respiratory chain complex (mitochondrial respiratory chain complex)
mitochondrial respiratory chain complex assembly factors (mitochondrial respiratory chain complex assembly factors)
MUC (mucins)
MYBP (myosin binding proteins) myosins (myosins)
NLR (nucleotide-binding domain and leucine rich repeat containing family)
NR (nuclear hormone receptors)
PAR (pseudoautosomal regions)
PARK (Parkinson disease)
PAX (paired boxes)
PDE (phosphodiesterases)
PDI (protein disulfide isomerases)
PHF (PHD-type zinc fingers)
PIG (phosphatidylinositol glycan anchor biosynthesis)
POLR (RNA polymerase subunits)
proteoglycans (proteoglycans)
PRRT (proline-rich transmembrane proteins) PRSS (serine peptidases)
PSM (proteasome (prosome, macropain) subunits) PTP (protein tyrosine phosphatases)
RAB (RAB, member RAS oncogene family) RNASE (ribonucleases, RNase A)
RNF (RING-type zinc fingers)
RPL (L ribosomal proteins)
RPS (S ribosomal proteins) SC (sodium channels)
SDR (short chain dehydrogenase/reductase superfamily)
serine/threonine phosphatases (serine/threonine phosphatases)
SERPIN (serine (or cysteine) peptidase inhibitors)
SH2 domain containing (SH2 domain containing)
SKOR (SKI transcriptional corepressors)
SLC (solute carriers)
SMAD (SMAD, mothers against DPP homologs) small leucine-rich repeats (small leucine-rich repeats)
small miscellaneous ncRNAs (small miscellaneous ncRNAs) SMC (structural maintenance of chromosomes proteins) SOX (SRY (sex determining region Y)-boxes)

SPINK (serine peptidase inhibitors, Kazal type)
ST3G (sialyltransferases)
sulfotransferases, membrane-bound (sulfotransferases, membrane-bound)
TBX (T-boxes)
TGM (transglutaminases)
THAP domain containing (THAP (C2CH-type zinc finger) domain containing)
TNFRSF (tumor necrosis factor receptor superfamily)
TNFSF (tumor necrosis factor (ligand) superfamily)
TRIM (tripartite motif-containing)
TRNA (transfer RNAs)
TRP (transient receptor potential cation channels)
TTC (tetratricopeptide (TTC) repeat domain
containing) TUB (tubulins)
UBE1 (ubiquitin-like modifier activating enzymes)
UGT (UDP glucuronosyltransferases)
USP (ubiquitin-specific peptidases)
WDR (WD repeat domain containing)
WNT (wingless-type MMTV integration sites)
ZC2HC (zinc fingers, C2HC-type)
ZCCHC (zinc fingers, CCHC domain containing)
ZFYVE (zinc fingers, FYVE-type)
ZMYM (zinc fingers, MYM type)
ZMYND (zinc fingers, MYND -
type) ZNF (zinc fingers, C2H2-type)

Published: *November 17, 2014*

The State of Alabama
Chromosome One

University of Alabama School of Medicine (Birmingham) (Prostate Cancer)

University of South Alabama College of Medicine (Mobile) (Glaucoma) (Porphyria Cutanea Tarda)

Alabama College of Osteopathic Medicine (Dothan) (Alzheimer Disease) (Gaucher disease)

Chromosome 1

Contains over 3000 genes
Contains over 240 million base pairs, of which ~90% have been determined
See the diseases associated with chromosome 1 in the MapViewer.

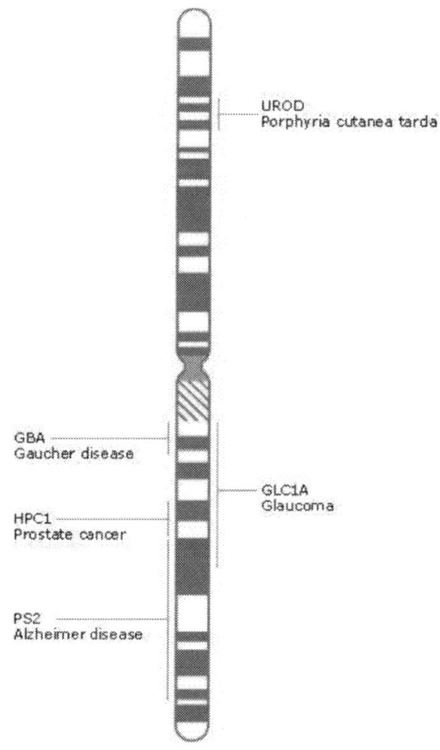

State of Arizona
Chromosome Two

University of Arizona College of Medicine (Tuscan) (Essential Tremor)

Midwestern University Arizona College of Osteopathic Medicine (Glendale) (Colon Cancer)

AT Still University School of Osteopathic Medicine in Arizona (Mesa) (Waardenberg Syndrome)

Chromosome 2

Contains over 2500 genes
Contains over 240 million base pairs, of which ~95% have been determined
See the diseases associated with chromosome 2 in the MapViewer.

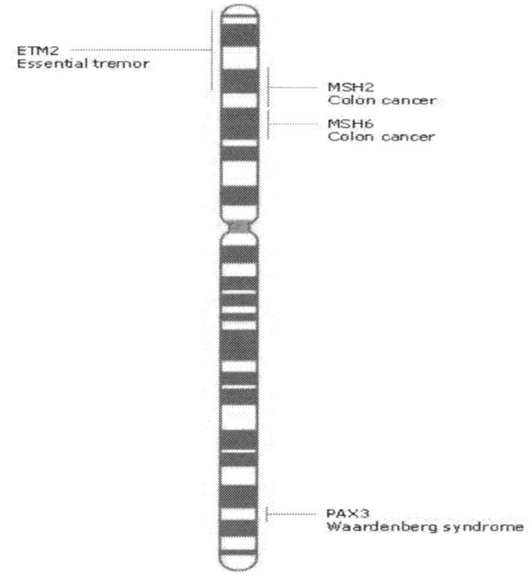

State of Arkansas
Chromosome Three

University of Arkansas for Medical Sciences (Little Rock)
(Lung Cancer) (Von Hippel-Lindau)

Chromosome 3

Contains approximately 1900 genes
Contains approximately 200 million base pairs, of which ~95% have been determined
See the diseases associated with chromosome 3 in the MapViewer.

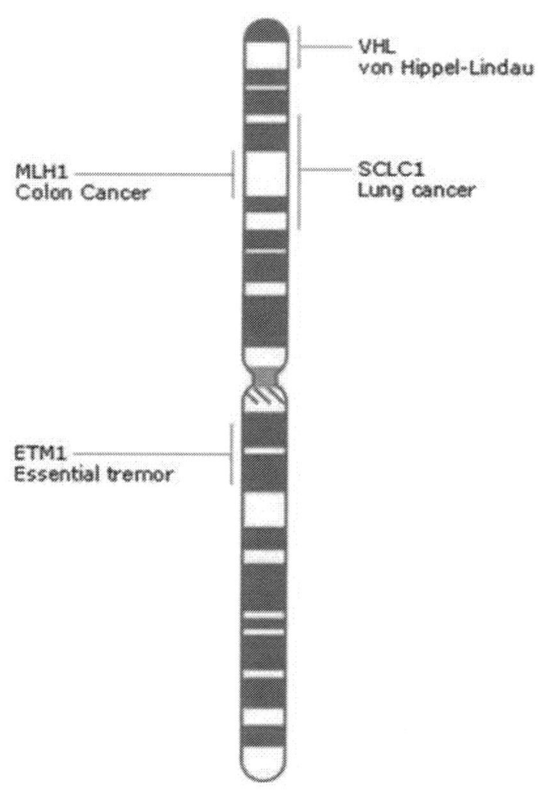

State of California
Chromosome Four

UC Davis School of Medicine (Sacramento) (Huntington Disease)

University of California Irvine School of Medicine (Irvine) (Parkinson Disease)

Keck School of Medicine Southern California Los Angeles) (Fibrodysplasia Ossificans Progressiva)

Loma Linda University School of Medicine (Loma Linda) (Achondroplasia)

Stanford University School of Medicine (Palo Alto) (Narcolepsy)

David Geffen School of Medicine at UCLA (Los Angeles) (Ellis Van Crevald)

UC Riverside School of Medicine (Riverside) (AD Polycystic Kidney Disease)

University of San Diego School of Medicine (San Diego) (Chronic Lymphoid Leukemia)

University of California San Francisco School of Medicine (San Francisco) (Bladder Cancer)

Western University of Health Sciences College of Osteopathic Medicine (Pomona)

Chromosome 4

Contains approximately 1600 genes
Contains approximately 190 million base pairs, of which ~95% have been determined
See the diseases associated with chromosome 4 in the MapViewer

State of California - Chromosome Four

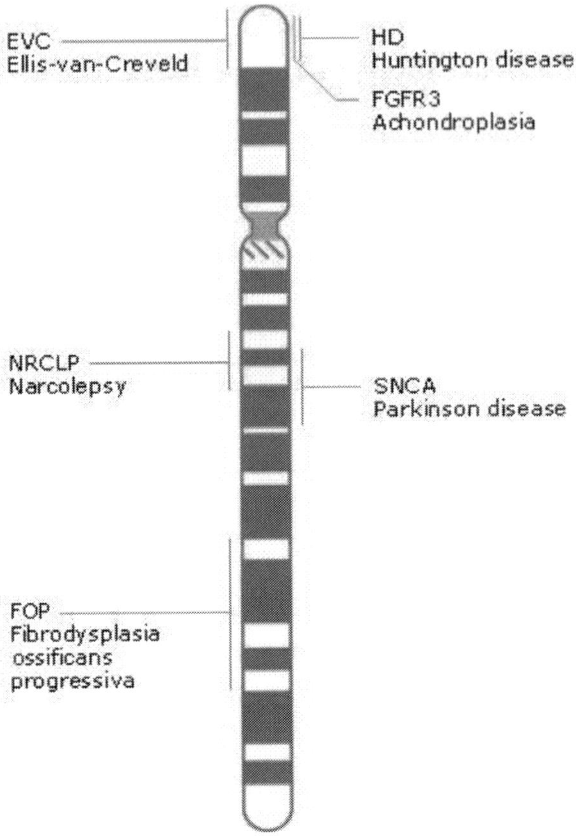

State of Colorado
Chromosome Five

University of Colorado School of Medicine (Aurora) (Spinal Muscular Atrophy)

Rocky Vista University College of Osteopathic Medicine (Parker) (Asthma)

Chromosome 5

Contains approximately 1700 genes
Contains approximately 180 million base pairs, of which over 95% have been determined
See the diseases associated with chromosome 5 in the MapViewer.

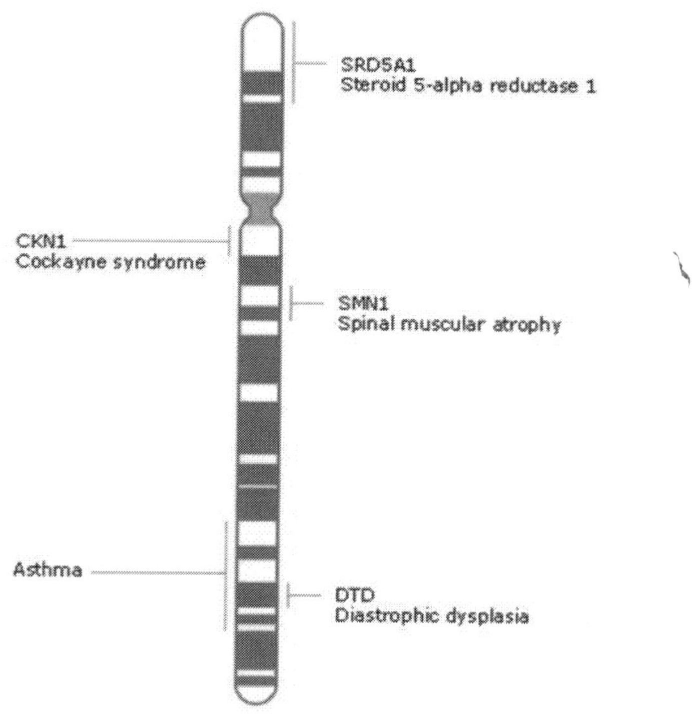

State of Connecticut
Chromosome Six

Yale School of Medicine (New Haven) (Epilepsy)

University of Connecticut School of Medicine (Farmington) (Diabetes)

Frank H. Netter School of Medicine Quinnipiac University (North Haven) (Spinocerebellar Ataxia)

Chromosome 6

Contains approximately 1900 genes
Contains approximately 170 million base pairs, of which over 95% have been determined

See the diseases associated with chromosome 6 in the MapViewer

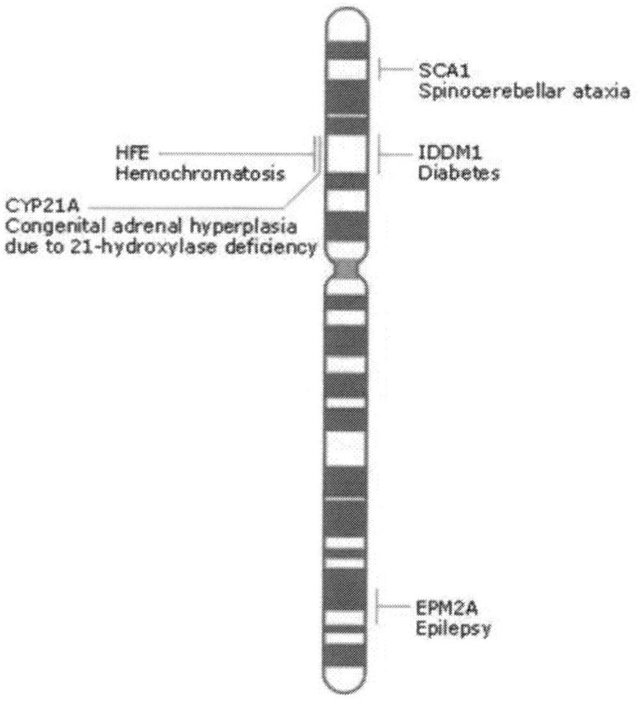

District of Columbia
Chromosome Seven

Georgetown University School of Medicine (Washington, D.C.) (Diabetes) (Pendred Syndrome)

Howard University College of Medicine (Washington, D.C.) (Obesity)

George Washington University Medical School (Washington, D.C.) (Cystic Fibrosis) (Williams Syndrome)

Chromosome 7

Contains approximately 1800 genes
Contains over 150 million base pairs, of which over 95% have been determined

See the diseases associated with chromosome 7 in the MapViewer.

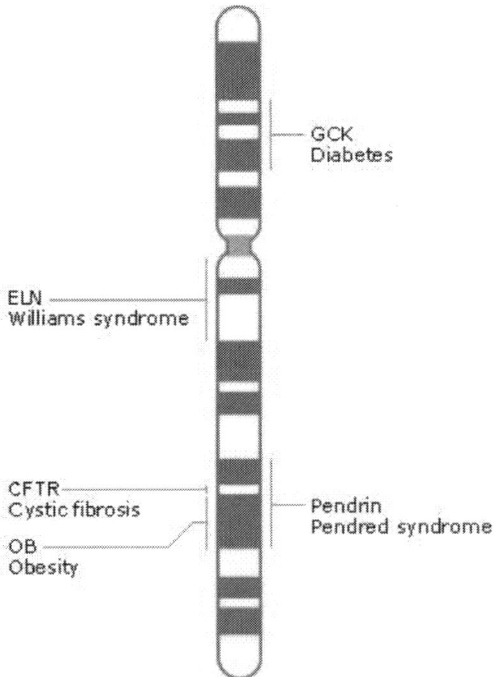

State of Florida
Chromosome Eight

Florida State University College of Medicine (Tallahasse) (Familiar Hyperaldosteronism)

University of South Florida College of Medicine (Tampa) (Hereditary Exostosis)

University of Florida College of Medicine (Gainsville) (Mucopolysaccharidos Type 111)

University of Central Florida College of Medicine (Orlando) (Idiopathic Pulmonary Fibrosis)

Chromosome 8

Contains over 1400 genes
Contains over 140 million base pairs, of which over 95% have been determined

See the diseases associated with chromosome 8 in the MapViewer.

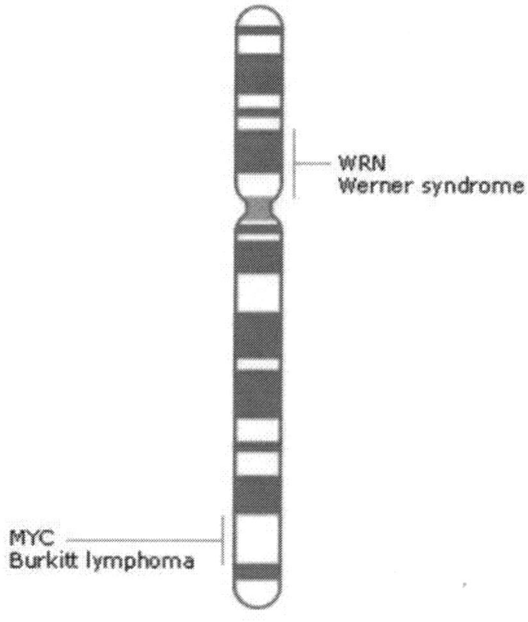

State of Georgia
Chromosome Nine

Emory University School of Medicine (Atlanta) (Malignant Melanoma)

Georgia Regents University Medical College of Georgia (Augusta) (Chronic Myeloid Leukemia)

Mercer University School of Medicine (Macon) (Freidrich's Ataxia)

Morehouse School of Medicine (Atlanta) (Tuberous Sclerous)

Philadelphia College of Osteopathic Medicine (Atlanta) (Tangier Disease)

Chromosome 9

Contains over 1400 genes
Contains over 130 million base pairs, of which over 85% have been determined
See the diseases associated with chromosome 9 in the MapViewer.

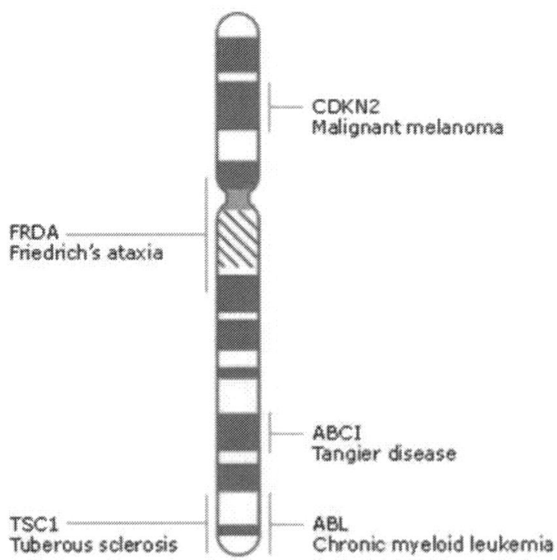

State of Hawaii
Chromosome Ten

Univerity of Hawaii at Manoa John A Burns School of Medicine (Honolulu) (Gyrate Atrophy) (Refsum Disease)

Chromosome 10

Contains over 1400 genes
Contains over 130 million base pairs, of which over 95% have been determined
See the diseases associated with chromosome 10 in the MapViewer.

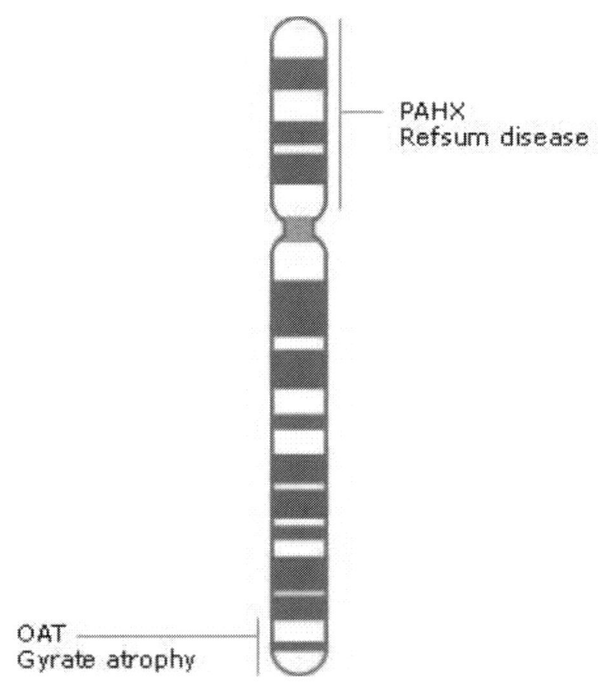

State of Illinois
Chromosome Eleven

Northwestern University Feinberg School of Medicine (Chicago) (Harvey Ras Oncogene)

Rush Medical College (Chicago) (Multiple Endocrine Neoplasia)

Rosalind Franklin University Chicago Medical School (North Chicago) (Best Disease)

Southern Illinois University School of Medicine (Springfield) (Insulin Dependent Diabetes Mellitus 2)

University of Chicago Pritzker School of Medicine (Chicago) (Ataxia Telangiectasia)

University of Illinois College of Medicine (Chicago) (Long QT Syndrome)

Chromosome 11

Contains approximately 2000 genes

Contains over 130 million base pairs, of which over 95% have been determined

See the diseases associated with chromosome 11 in the MapViewer.

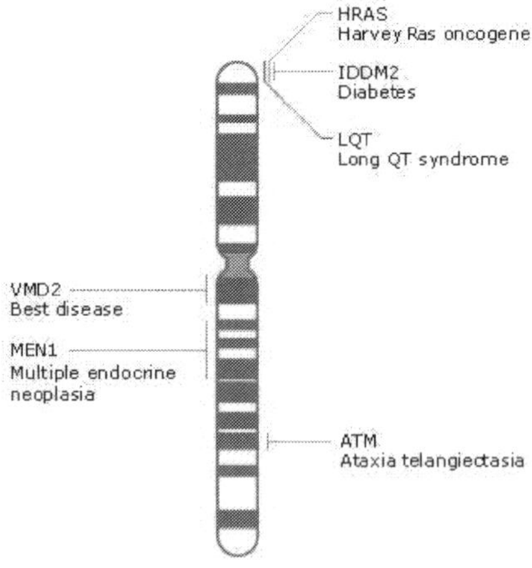

State of Indiana
Chromosome Twelve

Indiana University School of Medicine (Indianapolis) (Zellweger Syndrome)

Marian University College of Osteopathic Medicine (Indianapolis) (Phenylketonuria)

Chromosome 12

Contains over 1600 genes
Contains over 130 million base pairs, of which over 95% have been determined
See the diseases associated with chromosome 12 in the MapViewer.

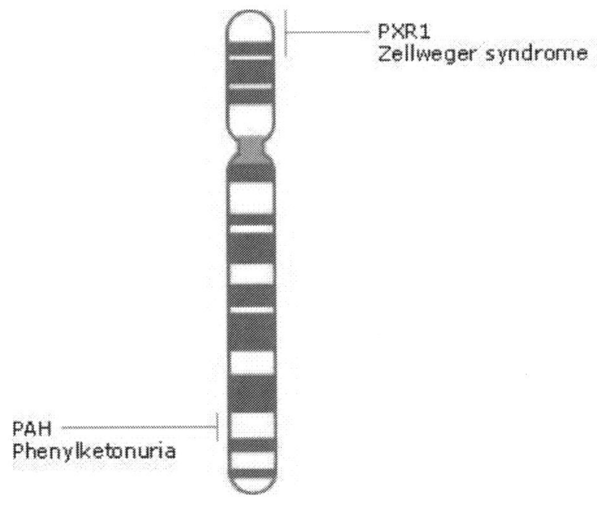

State of Iowa
Chromosome Thirteen

University of Iowa Roy and Lucille Carver College of Medicine (Iowa City) (Breast Cancer) (Autosomal Recessive Neurosensory Deafness)

Des Moines University College of Osteopathic Medicine (Des Moines) (Retinoblastoma) (Wilson Disease)

Chromosome 13

Contains approximately 800 genes
Contains over 110 million base pairs, of which over 80% have been determined
See the diseases associated with chromosome 13 in the MapViewer.

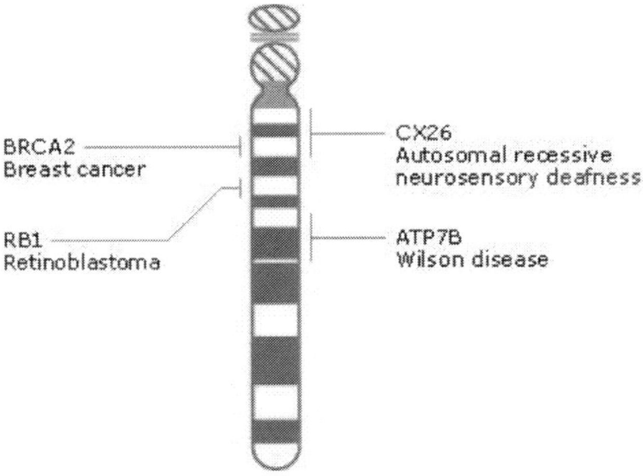

State of Kansas
Chromosome Fourteen

University of Kansas School of Medicine (Kansas City) (Alzheimer Disease) (Alpha 1 Antitrypsin Deficiency)

Chromosome 14

Contains approximately 1200 genes
Contains over 100 million base pairs, of which over 80% have been determined
See the diseases associated with chromosome 14 in the MapViewer.

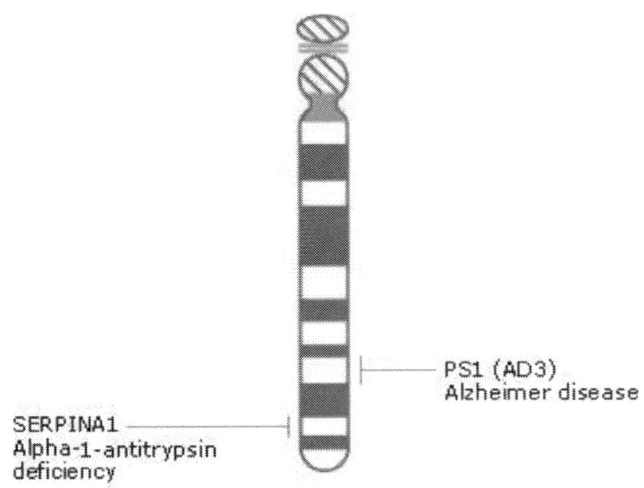

State of Kentucky
Chromosome Fifteen

University of Louisville School of Medicine (Louisville) (Marfan Syndrome) (Angelman Syndrome)

University of Pikeville Kentucky College of Osteopathic Medicine (Pikeville) (Tay - Sachs disease) (Prader-Willi Syndrome)

Chromosome 15

Contains approximately 1200 genes
Contains approximately 100 million base pairs, of which over 80% have been determined
See the diseases associated with chro mosome 15 in the MapViewer.

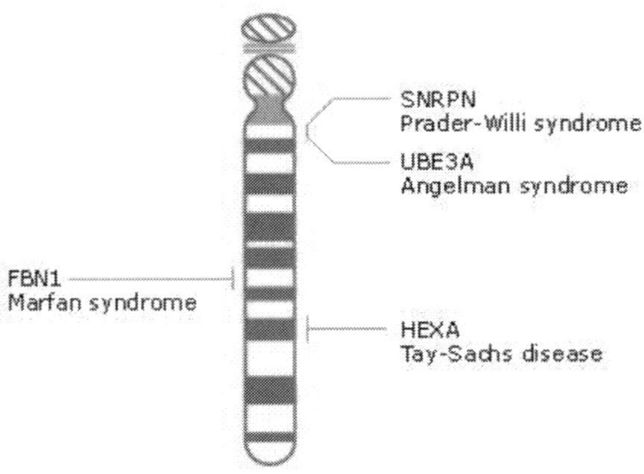

State of Louisiana
Chromosome Sixteen

Tulane University School of Medicine (New Orleans) (Familial Mediterranean fever)

Louisiana State University School of Medicine in New Orleans (New Orleans) (Crohn's Disease)

Louisiana State University School of Medicine in Shreveport (Shreveport) (Alpha Thalassemia)

Chromosome 16

Contains approximately 1300 genes
Contains approximately 90 million base pairs, of which over 85% have been determined

See the diseases associated with chromosome 16 in the MapViewer.

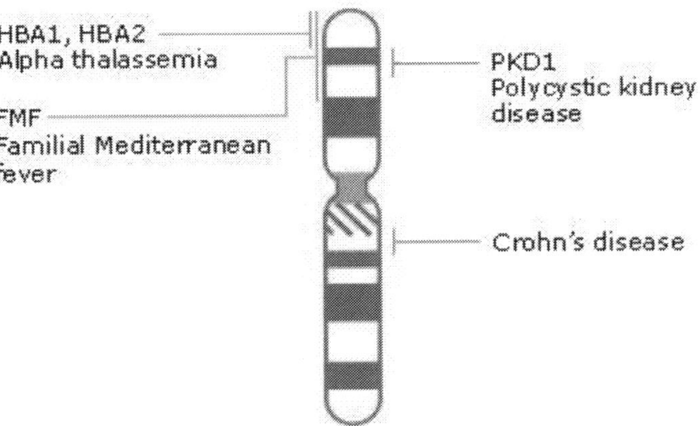

State of Maine
Chromosome Seventeen

University of New England College of Osteopathic Medicine (Biddeford) (BRCA1 Breast Cancer) (P53Tumor Suppressor Protein) (Charcot-Marie-Tooth Syndrome) (Prostate Cancer SPOP Gene)

Chromosome 17

Contains over 1600 genes
Contains approximately 80 million base pairs, of which over 95% have been determined
See the diseases associated with chromosome 17 in the MapViewer.

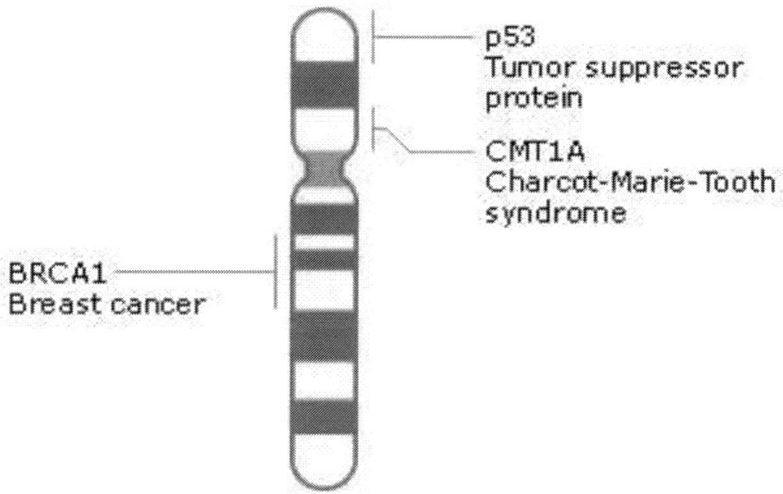

State of Maryland
Chromosome Eighteen

John Hopkins University School of Medicine (Baltimore) (Pancreatic Cancer)

University of Maryland School of Medicine (Baltimore) (Porphyria) (Newman-Pick Disease)

Uniformed Services University of the Health Sciences F Edward Herbert School of Medicine (Bethesda) (Hereditary Hemorrhagic Telangiectasia)

Chromosome 18

Contains over 600 genes
Contains over 70 million base pairs, of which over 95% have been determined
See the diseases associated with chromosome 18 in the MapViewer.

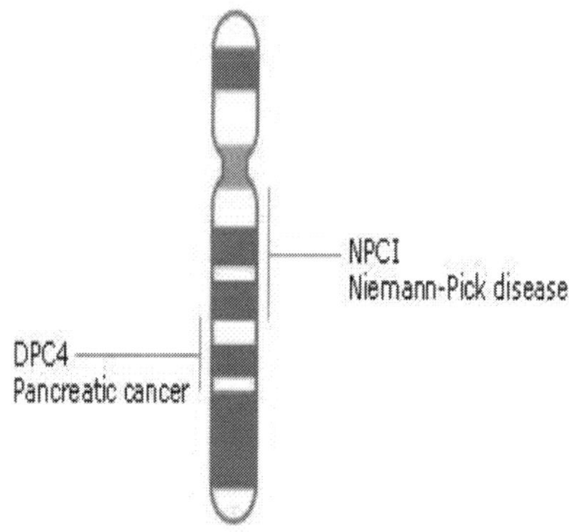

State of Massachusetts
Chromosome Nineteen

Tufts University School of Medicine (Boston) (Myotonic Dystrophy) Harvard Medical School (Boston) (Atherosclerosis)

Boston University School of Medicine (Boston) (Severe Combined Immunodeficiency)

Chromosome 19

Contains over 1700 genes

Contains over 60 million base pairs, of which over 85% have been determined

See the diseases associated with chromosome 19 in the MapViewer.

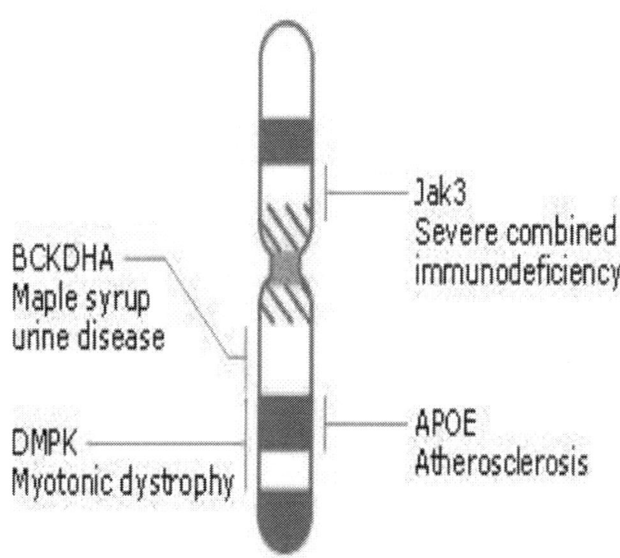

State of Michigan
Chromosome Twenty

Western Michigan University Homer Stryker M.D. School of Medicine (Kalamazoo) (Celiac Disease)

University of Michigann Medical School (Ann Harbor) (Maturity onset Diabetes of the Young Type 1)

Wayne State University School of Medicine (Detroit) (Albright Hereditary Osteodystrophy)

Michigan State University College of Osteopathic Medicine (East Lansing) (Arterial Tortuosity Syndrome)

Michigan State University College of Human Medicine (East Lansing) (Pantothenate Kinase Associated Neurodegeneration)

Central Michigan University College of Medicine (Mount Pleasant)

Oakland University William Beaumont School of Medicine (Rochester)

Chromosome 20

Contains over 900 genes
Contains over 60 million base pairs, of which over 90% have been determined
See the diseases associated with chromosome 20 in the MapViewer.

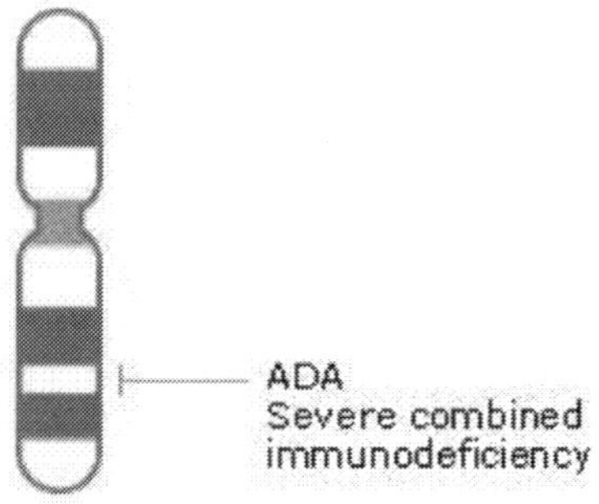

State of Minnesota
Chromosome Twenty One

University of Minnesota Medical School (Minneapolis) (Autoimmune Polyglandular Syndrome)

Mayo Clinic College of Medicine (Rochester) (Amyotrophic Lateral Sclerosis)

Chromosome 21

Contains over 400 genes
Contains over 40 million base pairs, of which over 70% have been determined
See the diseases associated with chromosome 21 in the MapViewer.

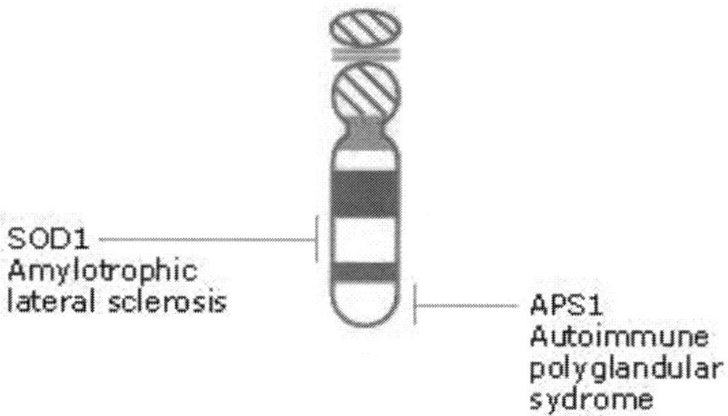

State of Mississippi
Chromosome Twenty Two

University of Mississippi School of Medicine (Jackson) (Chronic Myeloid Leukemia) (Glucose Galactase Malabsorption)
William Carey University College of Osteopathic Medicine (Hattiesburg) (Neurofibromatosis) (DiGeorge Syndrome)

Chromosome 22

Contains over 800 genes
Contains over 40 million base pairs, of which approximately 70% have been determined
See the diseases associated with chromosome 22 in the MapViewer.

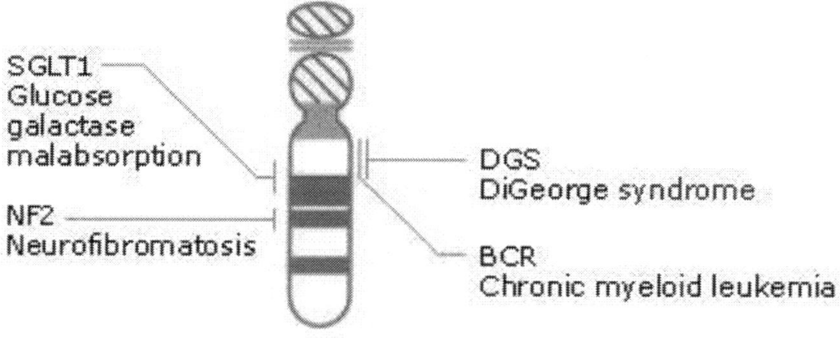

State of Missouri
Chromosome X

AT Still University Kirksville (Kirksville) (Alport Syndrome)

Saint Louis University School of Medicine (St. Louis) (Adrenoleukodystrophy)

Kansas City University of Medicine and Biosciences College of Osteopathic Medicine) (Kansas City) (Rett Syndrome)

University of Missouri-Columbia School of Medicine (Columbia) (Fragile X Syndrome)

University of Missouri Kansas City School of Medicine (Kansas City) (Lesch Nyham Syndrome)

Washington University School of Medicine (St Louis) (Hemophilia A)

Chromosome X

Contains over 1400 genes
Contains over 150 million base pairs, of which approximately 95% have been determined
See the diseases associated with chromosome X in the

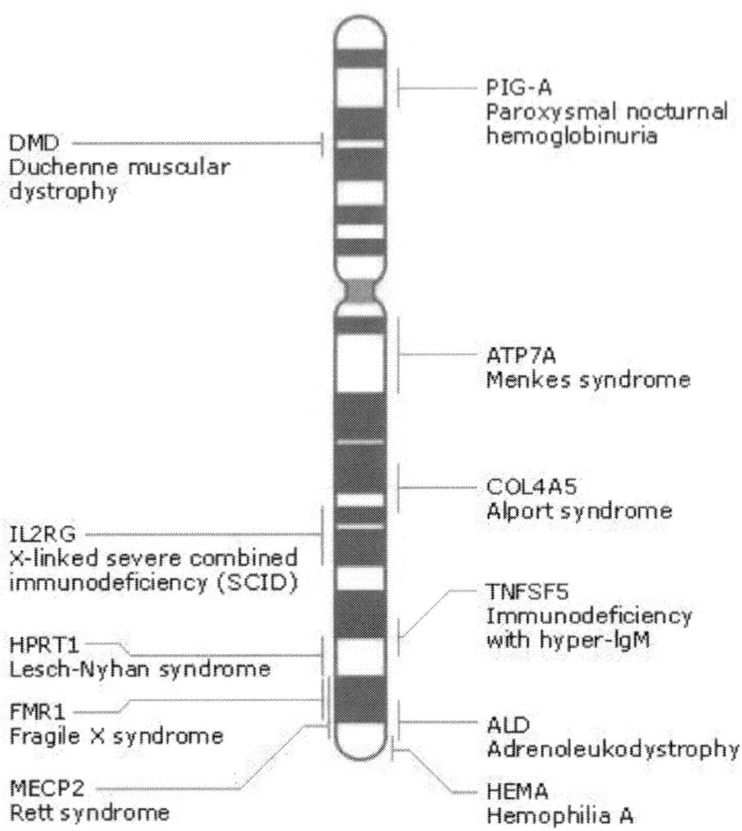
PIG-A
Paroxysmal nocturnal hemoglobinuria
DMD
Duchenne muscular dystrophy
ATP7A
Menkes syndrome
COL4A5
Alport syndrome
IL2RG
X-linked severe combined immunodeficiency (SCID)
TNFSF5
Immunodeficiency with hyper-IgM
HPRT1
Lesch-Nyhan syndrome
FMR1
Fragile X syndrome
ALD
Adrenoleukodystrophy
MECP2
Rett syndrome
HEMA
Hemophilia A

State of Nebraska
Chromosome Y

University of Nebraska College of Medicine (Omaha) (Kleinefelter Syndrome)

Creighton University School of Medicine (Omaha) (Testes Determining Factor)

Chromosome Y

Contains over 200 genes

Contains over 50 million base pairs, of which approximately 50% have been determined

See the diseases associated with chromosome Y in the MapViewer.

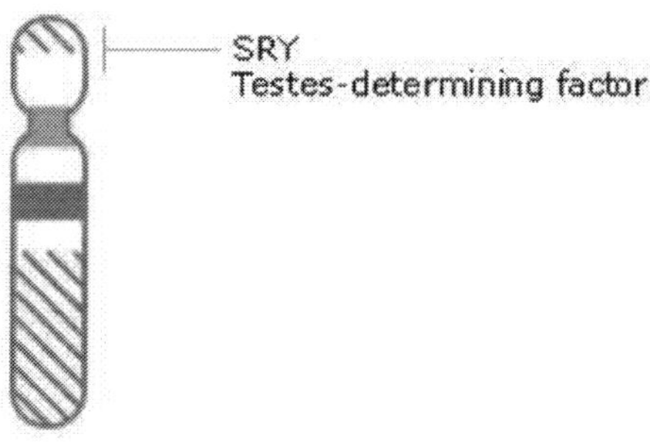

http://www.ncbi.nlm.nih.gov/books/NBK22266/ - _A273

State of Nevada
Chromosome One

University of Nevada School of Medicine (Reno) (Hypercholesterolemia)

Touro University Nevada College of Osteopathic Medicine (Henderson) (Ankylosing Spondolysis)

What is chromosome 1?

Humans normally have 46 chromosomes in each cell, divided into 23 pairs. Two copies of chromosome 1, one copy inherited from each parent, form one of the pairs. Chromosome 1 is the largest human chromosome, spanning about 249 million DNA building blocks (base pairs) and representing approximately 8 percent of the total DNA in cells.

Identifying genes on each chromosome is an active area of genetic research. Because researchers use different approaches to predict the number of genes on each chromosome, the estimated number of genes varies. Chromosome 1 likely contains 2,000 to 2,100 genes that provide instructions for making proteins. These proteins perform a variety of different roles in the body.

Genes on chromosome 1 are among the estimated 20,000 to 25,000 total genes in the human genome.

How are changes in chromosome 1 related to health conditions?

Many genetic conditions are related to changes in particular genes on chromosome 1. This list of disorders associated with genes on chromosome 1 provides links to additional information.

Changes in the structure or number of copies of a chromosome can also cause problems with health and development. The following chromosomal conditions are associated with such changes in chromosome 1.

1p36 deletion syndrome

1p36 deletion syndrome is caused by a deletion of genetic material from a specific region in the short (p) arm of chromosome 1. The signs and symptoms of this disorder, which include intellectual disability, distinctive facial features, and structural abnormalities in several body systems, are probably related to the loss of multiple genes in this region. The size of the deletion varies among affected individuals.

1q21.1 microdeletion

1q21.1 microdeletion is a chromosomal change in which a small piece of the long (q) arm of chromosome 1 is deleted in each cell. Specifically, affected individuals are missing about 1.35 million DNA building blocks (base pairs), also written as 1.35 megabases (Mb), in the q21.1 region. The exact size of the deleted region varies, but it typically contains at least nine genes. The loss of several of these genes probably contributes to the various signs and symptoms that can be associated with a 1q21.1 microdeletion. Related features can include delayed development, intellectual disability, physical abnormalities, and neurological and psychiatric problems; however, some individuals with a 1q21.1 microdeletion have no obvious signs or symptoms.

Neuroblastoma

Deletions within region 1p36 have also been associated with another condition called neuroblastoma. Neuroblastoma is a type of cancerous tumor composed of immature nerve cells (neuroblasts). These deletions are somatic mutations, which means they occur during a person's lifetime and are present only in the cells that become cancerous. About 25 percent of people with neuroblastoma have a deletion of 1p36.1-1p36.3, which is associated with a more severe form of neuroblastoma. Researchers believe the deleted region could contain a gene that keeps cells from growing and dividing too quickly or in an uncontrolled way, called a tumor suppressor gene. When tumor suppressor genes are deleted, cancer can occur. Researchers have identified several possible tumor suppressor genes in the deleted region of chromosome 1, and more research is needed to understand what role these genes play in neuroblastoma development.

Thrombocytopenia-absent radius syndrome

A deletion in the 1q21.1 region of chromosome 1 is involved in most cases of thrombocytopenia-absent radius (TAR) syndrome. TAR syndrome is characterized by the absence of a bone called the radius in each forearm and a shortage (deficiency) of blood cells involved in clotting (platelets).

The deletion in chromosome 1 involved in TAR syndrome eliminates at least 200,000 DNA building blocks (200 kilobases, or 200 kb) from the long (q) arm of the chromosome, including a gene called *RBM8A*.

Most people with TAR syndrome have the deletion in one copy of chromosome 1, which removes one copy of the *RBM8A* gene, and a mutation in the other copy of the *RBM8A* gene in each cell. The *RBM8A* gene provides instructions for making a protein called RNA-binding motif protein 8A. This protein is believed to be involved in a number of important cellular functions involving the production of other proteins.

RBM8A gene mutations that cause TAR syndrome reduce the amount of RNA-binding motif protein 8A in cells. The deletion on chromosome 1 eliminates one copy of the *RBM8A* gene in each cell and the RNA-binding motif protein 8A that would have been produced from it. The reduced total amount of RNA-binding motif protein 8A is thought to cause problems in the development of certain tissues, but it is unknown how it causes the specific signs and symptoms of TAR syndrome. No cases have been reported in which individuals have deletions on both copies of chromosome 1 that include both copies of the *RBM8A* gene; studies indicate that the complete loss of RNA-binding motif protein 8A is not compatible with life.

Researchers sometimes refer to the deletion in chromosome 1 associated with TAR syndrome as the 200-kb deletion to distinguish it from another chromosomal abnormality called a 1q21.1 microdeletion. People with a 1q21.1 microdeletion are missing a different, larger DNA segment in the chromosome 1q21.1 region near the area where the 200-kb deletion occurs. The chromosomal change related to 1q21.1 microdeletion is often called the recurrent distal 1.35-Mb deletion.

Other cancers

Changes in the structure of chromosome 1 are associated with other forms of cancer and conditions related to cancer. These changes are typically somatic, which means they are acquired during a person's lifetime and are present only in tumor cells.

Deletions in the short (p) arm of the chromosome have been identified in tumors of the brain and kidney. Duplications in the long (q) arm of the chromosome have been reported in a disorder called myelodysplastic syndrome, which is a disease of the blood and bone marrow. People with this condition have a low number of red blood cells (anemia) and an increased risk of developing leukemia.

Other chromosomal conditions

Other changes in the number or structure of chromosome 1 can have a variety of effects, including delayed growth and development, distinctive facial features, birth defects, and other health problems. Changes to chromosome 1 may include an extra segment of the short (p) or long (q) arm of the chromosome in each cell (partial trisomy 1p or 1q), a missing segment of the short or long arm of the chromosome in each cell (partial monosomy 1p or 1q), or a circular structure called ring chromosome 1. Ring chromosomes occur when a chromosome breaks in two places and the ends of the chromosome arms fuse together to form a circular structure.

Is there a standard way to diagram chromosome 1?

Geneticists use diagrams called ideograms as a standard representation for chromosomes. Ideograms show a chromosome's relative size and its banding pattern. A banding pattern is the characteristic pattern of dark and light bands that appears when a chromosome is stained with a chemical solution and then viewed under a microscope. These bands are used to describe the location of genes on each chromosome.

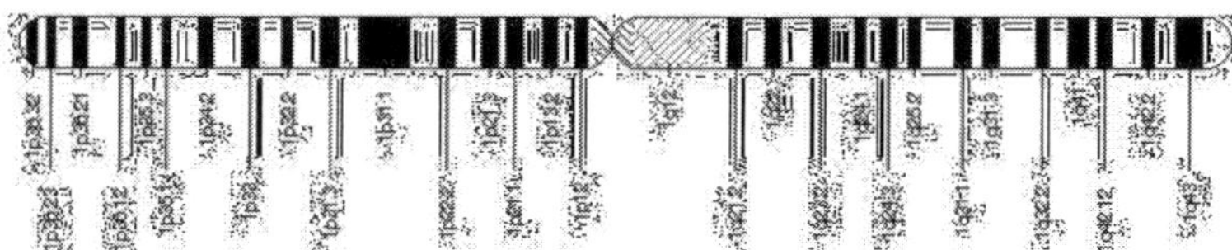

Conditions related to genes on chromosome 1

Genetics Home Reference includes these conditions related to genes on chromosome 1:

3- beta-hydroxysteroid dehydrogenase deficiency
3-hydroxy-3-methylglutaryl-CoA lyase deficiency
actin-accumulation myopathy
activated PI3K-delta syndrome
adenosine monophosphate deaminase
deficiency age-related macular degeneration
Alagille syndrome
alternating hemiplegia of
childhood Alzheimer disease
amyotrophic lateral
sclerosis anencephaly

ankylosing spondylitis
arrhythmogenic right ventricular cardiomyopathy
atypical hemolytic-uremic syndrome auriculo-
condylar syndrome
autoimmune Addison disease
autoimmune lymphoproliferative syndrome
autosomal dominant nocturnal frontal lobe epilepsy
autosomal recessive primary microcephaly
Bartter syndrome
breast cancer cap
myopathy
carnitine palmitoyltransferase II deficiency
catecholaminergic polymorphic ventricular tachycardia
Charcot-Marie-Tooth disease
Chediak-Higashi syndrome
chronic granulomatous disease
Coffin-Siris syndrome
color vision deficiency
congenital disorder of glycosylation type Ic
congenital fiber-type disproportion
congenital hypothyroidism
congenital insensitivity to pain with anhidrosis
core binding factor acute myeloid leukemia
Cowden syndrome
Crohn disease
cytogenetically normal acute myeloid leukemia
dense deposit disease
desmosterolosis Diamond-
Blackfan anemia
dihydropyrimidine dehydrogenase deficiency
early-onset glaucoma
Ehlers-Danlos syndrome Emery-
Dreifuss muscular dystrophy
erythrokeratodermia variabilis et progressiva
essential thrombocythemia
factor V deficiency
factor V Leiden thrombophilia
familial adenomatous polyposis
familial cold autoinflammatory syndrome
familial dilated cardiomyopathy
familial erythrocytosis
familial hemiplegic migraine

familial hypertrophic cardiomyopathy
familial hypobetalipoproteinemia
familial isolated hyperparathyroidism
familial restrictive cardiomyopathy
frontonasal dysplasia
Fuchs endothelial dystrophy
fucosidosis
fumarase deficiency
galactosemia
gastrointestinal stromal tumor
Gaucher disease
Gitelman syndrome
GLUT1 deficiency syndrome
glycogen storage disease type III
Graves disease
Greenberg dysplasia
hemochromatosis
hereditary antithrombin deficiency
hereditary leiomyomatosis and renal cell cancer
hereditary paraganglioma-pheochromocytoma
hereditary sensory and autonomic neuropathy type V
hidradenitis suppurativa
homocystinuria
Hutchinson-Gilford progeria syndrome
hypercholesterolemia
hypermanganesemia with dystonia, polycythemia, and
cirrhosis hyperparathyroidism-jaw tumor syndrome
hyperprolinemia
hypohidrotic ectodermal
dysplasia hypokalemic periodic
paralysis hypophosphatasia
idiopathic inflammatory myopathy
infantile neuronal ceroid lipofuscinosis
intranuclear rod myopathy
junctional epidermolysis bullosa
juvenile Batten disease juvenile
idiopathic arthritis Kufs disease
late-infantile neuronal ceroid
lipofuscinosis Leber congenital amaurosis
leptin receptor deficiency

leukoencephalopathy with brainstem and spinal cord involvement and lactate elevation
leukoencephalopathy with vanishing white matter
limb-girdle muscular dystrophy
Loeys-Dietz syndrome
Mabry syndrome
malignant hyperthermia
mandibuloacral dysplasia
maple syrup urine disease
medium-chain acyl-CoA dehydrogenase deficiency
medullary cystic kidney disease type 1 Meier-Gorlin syndrome
Muckle-Wells syndrome
Müllerian aplasia and hyperandrogenism
multiminicore disease
multiple epiphyseal dysplasia
nemaline myopathy
neonatal onset multisystem inflammatory disease
neuroblastoma
nonsyndromic deafness
nonsyndromic
paraganglioma Noonan
syndrome osteogenesis
imperfecta Parkinson disease
phosphoglycerate dehydrogenase deficiency
popliteal pterygium syndrome
porphyria
primary myelofibrosis
psoriatic arthritis pyruvate
kinase deficiency renal
tubular dysgenesis
REN-related kidney disease
retinitis pigmentosa
rheumatoid arthritis
rhizomelic chondrodysplasia punctata
severe congenital neutropenia
Shprintzen-Goldberg syndrome spina
bifida
sporadic hemiplegic migraine
Stargardt macular degeneration
Stickler syndrome
systemic lupus erythematosus

systemic scleroderma
thiamine-responsive megaloblastic anemia
syndrome thrombocytopenia -absent radius
syndrome trimethylaminuria
type 1 diabetes
Usher syndrome
van der Woude
syndrome vitiligo
Vohwinkel syndrome

Reviewed: *October 2012*

Published: *November 17, 2014*

Genes on chromosome 1

Genetics Home Reference includes these genes on chromosome 1:

ABCA4
ACADM
ACTA1
AGL AGT
ALDH4A1
ALG6
ALPL
ALX3
AMPD1
ARID1A
ASPM
ATP1A2
BSND
CACNA1S
CASQ2
CDC73
CFH
CFHR5
CHRNB2
CLCNKA
CLCNKB
COL8A2
COL9A2

COL11A1
CPT2
CRB1
DARS2
DBT
DHCR24
DIRAS3
DPYD
EDARADD
EGLN1
EIF2B3
ESPN
F5
FH
FMO3
FUCA1
GALE
GBA
GJB3
GJB4
GNAI3
GNAT2
GNPAT
HAX1
HFE2
HMGCL
HSD3B2
IL23R
IRF6
KCNQ4
KIF1B
LAMB3
LAMC2
LBR
LDLRAP1
LEPR
LEPRE1
LMNA
LOR
LYST
MFN2
MPL

MPZ
MTHFR
MTR
MUC1
MUTYH
MYOC
NCF2
NCSTN
NGF
NLRP3
NOTCH2
NRAS
NTRK1
ORC1
PARK7
PCSK9
PHGDH
PIGV
PIK3CD
PINK1
PKLR
PLOD1
PPOX
PPT1
PSEN2
PTPN22
RBM8A
REN
RPE65
RPL5
RPL11
RYR2
SDHB
SDHC
SEPN1
SERPINC1
SKI
SLC2A1
SLC19A2
SLC30A10
TARDBP
TGFB2

TNNT2
TPM3
TSHB
UROD
USH2A
WNT4
YARS
ZMPSTE24

Reviewed: *October 2012*

Published: *November 17, 2014*

State of New Hampshire Chromosome Two

State of New Hampshire Chromosome Two

Geisel School of Medicine at Dartmouth (Hanover) (Ehlers Danlos Syndrome)

Chromosome 2

What is chromosome 2?

Humans normally have 46 chromosomes in each cell, divided into 23 pairs. Two copies of chromosome 2, one copy inherited from each parent, form one of the pairs. Chromosome 2 is the second largest human chromosome, spanning about 243 million building blocks of DNA (base pairs) and representing almost 8 percent of the total DNA in cells.

Identifying genes on each chromosome is an active area of genetic research. Because researchers use different approaches to predict the number of genes on each chromosome, the estimated number of genes varies. Chromosome 2 likely contains 1,300 to 1,400 genes that provide instructions for making proteins. These proteins perform a variety of different roles in the body.

Genes on chromosome 2 are among the estimated 20,000 to 25,000 total genes in the human genome.

How are changes in chromosome 2 related to health conditions?

Many genetic conditions are related to changes in particular genes on chromosome 2. This list of disorders associated with genes on chromosome 2 provides links to additional information.

Changes in the structure or number of copies of a chromosome can also cause problems with health and development. The following chromosomal conditions are associated with such changes in chromosome 2.

2q37 deletion syndrome

2q37 deletion syndrome is caused by a deletion of genetic material from a specific region in the long (q) arm of chromosome 2. The deletion occurs near the end of the chromosome at a location designated 2q37. The size of the deletion varies among affected individuals. The signs and symptoms of this disorder, which may include intellectual disability, autism, short stature, obesity, and characteristic facial features, are probably related to the loss of multiple genes in this region.

Cancers

Changes in chromosome 2 have been identified in several types of cancer. These genetic changes are somatic, which means they are acquired during a person's lifetime and are present only in certain cells. For example, a rearrangement (translocation) of genetic material between chromosomes 2 and 3 has been associated with cancers of a certain type of blood cell originating in the bone marrow (myeloid malignancies).

Trisomy 2, in which cells have three copies of chromosome 2 instead of the usual two copies, has been found in myelodysplastic syndrome. This disease affects the blood and bone marrow. People with myelodysplastic syndrome have a low number of red blood cells (anemia) and an increased risk of developing a form of blood cancer known as acute myeloid leukemia.

Other chromosomal conditions

Another chromosome 2 abnormality is known as a ring chromosome 2. A ring chromosome is formed when breaks occur at both ends of the chromosome and the broken ends join together to form a circular structure. Individuals with this chromosome abnormality often have developmental delay, small head size (microcephaly), and slow growth before and after birth, heart defects, and distinctive facial features. The severity of symptoms typically depends on how many and which types of cells contain the ring chromosome 2.

Other changes involving the number or structure of chromosome 2 include an extra piece of the chromosome in each cell (partial trisomy 2) or a missing segment of the chromosome in each cell (partial monosomy 2). These changes can have a variety of effects on health and development, including intellectual disability, slow growth, characteristic facial features, weak muscle tone (hypotonia), and abnormalities of the fingers and toes.

Is there a standard way to diagram chromosome 2?

Geneticists use diagrams called ideograms as a standard representation for chromosomes. Ideograms show a chromosome's relative size and its banding pattern. A banding pattern is the characteristic pattern of dark and light bands that appears when a chromosome is stained with a chemical solution and then viewed under a microscope. These bands are used to describe the location of genes on each chromosome.

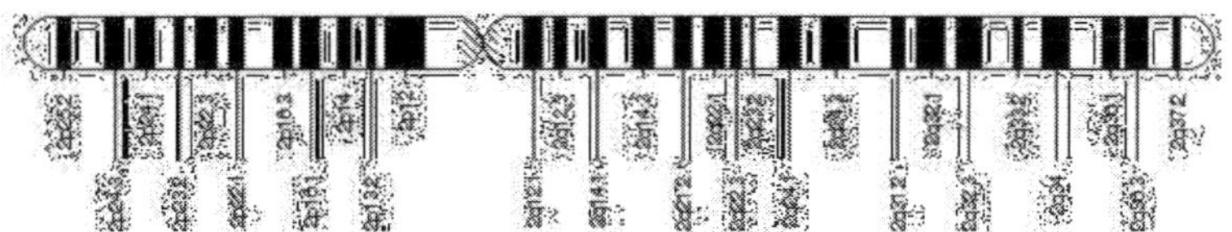

Conditions related to genes on chromosome 2

Genetics Home Reference includes these conditions related to genes on chromosome 2:

2- hydroxyglutaric aciduria
5-alpha reductase deficiency
African iron overload Alport
syndrome
Alström syndrome
amyotrophic lateral sclerosis
ankylosing spondylitis
benign recurrent intrahepatic cholestasis
Bethlem myopathy
biotin-thiamine-responsive basal ganglia disease
Björnstad syndrome
breast cancer
carbamoyl phosphate synthetase I deficiency
centronuclear myopathy
cerebrotendinous xanthomatosis
color vision deficiency
congenital hypothyroidism
congenital insensitivity to pain
cranioectodermal dysplasia
craniofacial-deafness-hand syndrome
Crigler-Najjar syndrome
Crohn disease
cystinuria
cytogenetically normal acute myeloid leukemia
deoxyguanosine kinase deficiency

Diamond-Blackfan anemia
Donnai-Barrow syndrome
dopa-responsive dystonia
early-onset glaucoma
Ehlers-Danlos syndrome
epidermolysis bullosa with pyloric atresia
episodic ataxia
erythromelalgia
familial dilated cardiomyopathy
familial erythrocytosis
familial hemiplegic migraine familial
hypertrophic cardiomyopathy familial
hypobetalipoproteinemia
familial male-limited precocious puberty
familial paroxysmal nonkinesigenic dyskinesia
Feingold syndrome
fibrodysplasia ossificans progressiva
Gilbert syndrome
GM3 synthase deficiency
GRACILE syndrome
Griscelli syndrome
harlequin ichthyosis
hemochromatosis
hepatic veno-occlusive disease with
immunodeficiency hereditary myopathy with early
respiratory failure homocystinuria
hypercholesterolemia
hyperphosphatemic familial tumoral
calcinosis hypohidrotic ectodermal dysplasia
idiopathic inflammatory myopathy
infantile-onset ascending hereditary spastic
paralysis intrahepatic cholestasis of pregnancy
isolated Duane retraction syndrome
Joubert syndrome
juvenile idiopathic arthritis
juvenile myoclonic epilepsy
juvenile primary lateral sclerosis
lactose intolerance
lamellar ichthyosis
leukoencephalopathy with vanishing white
matter Leydig cell hypoplasia
limb-girdle muscular dystrophy

long-chain 3-hydroxyacyl-CoA dehydrogenase deficiency
Lynch syndrome
malignant migrating partial seizures of infancy
Meier-Gorlin syndrome
methylmalonic acidemia microcephaly-
capillary malformation syndrome
mitochondrial complex III deficiency
mitochondrial trifunctional protein deficiency
Miyoshi myopathy
Mowat-Wilson syndrome
MPV17-related hepatocerebral mitochondrial DNA depletion
syndrome
multiple epiphyseal dysplasia
multiple pterygium syndrome
myofibrillar myopathy myostatin-
related muscle hypertrophy nemaline
myopathy nephronophthisis

neuroblastoma
nonsyndromic deafness
nonsyndromic holoprosencephaly
nonsyndromic paraganglioma
Noonan syndrome
paroxysmal extreme pain disorder
Perry syndrome
Peters anomaly
primary hyperoxaluria
primary myelofibrosis
progressive familial intrahepatic cholestasis
proopiomelanocortin deficiency
protein C deficiency pulmonary
arterial hypertension
renal tubular acidosis with deafness
rheumatoid arthritis
rhizomelic chondrodysplasia punctata
Salih myopathy
Schimke immuno-osseous dysplasia
Senior-Løken syndrome sepiapterin
reductase deficiency sitosterolemia

small fiber neuropathy
spastic paraplegia type 4

succinate-CoA ligase deficiency
surfactant dysfunction systemic
lupus erythematosus systemic
scleroderma
tibial muscular dystrophy
trichothiodystrophy
Ullrich congenital muscular dystrophy
Waardenburg syndrome
xeroderma pigmentosum
ZAP70-related severe combined immunodeficiency

Reviewed: *April 2009*

Published: *November 17, 2014*

Genes on chromosome 2

Genetics Home Reference includes these genes on chromosome 2:

ABCA12
ABCB11
ABCG5
ABCG8
ACVR1
AGPS
AGXT
ALK
ALMS1
ALS2
APOB
ATG16L1
ATP6V1B1
BARD1
BCS1L
BIN1
BMPR2
CACNB4
CHN1
CHRNG
CNGA3
COL3A1
COL4A3

COL4A4
COL5A2
COL6A3
CPS1
CYP1B1
CYP27A1
D2HGDH
DCTN1
DES
DFNB59
DGUOK
DNMT3A
DYSF
EDAR
EIF2B4
EPAS1
EPCAM
ERCC3
GALNT3
HADHA
HADHB
HS1BP3
IDH1
IL1A
ITGA6
LCT
LHCGR
LRP2
MATN3
MCEE
MCM6
MLPH
MMADHC
MPV17
MSH2
MSH6
MSTN
MYCN
NEB
NPHP1
NR4A2
ORC4

OTOF
PAX3
PAX8
PNKD
POMC
PROC
RPS7
SCN1A
SCN9A
SFTPB
SIX3
SLC3A1
SLC19A3
SLC40A1
SMARCAL1
SOS1
SP110
SPAST
SPR
SRD5A2
ST3GAL5
STAMBP
STAT4
SUCLG1
TMEM127
TPO
TTN
UGT1A1
WDR35
ZAP70
ZEB2

State of New Jersey
Chromosome Three

Rutgers New Jersey Medical School (Newark) (Autosomal Dominant Congenital Night Blindness)

Cooper Medical School of Rowan University (Camden) (Familial Isolated Hyperparathyroidism)

Rowan University School of Osteopathic Medicine (Stratford) (Alkaptonuria)

Rutgers Robert Wood Johnson Medical School (New Brunswick) (Dystrophic Epidermolysis Bullosa)

Chromosome 3

What is chromosome 3?

Humans normally have 46 chromosomes in each cell, divided into 23 pairs. Two copies of chromosome 3, one copy inherited from each parent, form one of the pairs. Chromosome 3 spans about 198 million base pairs (the building blocks of DNA) and represents approximately 6.5 percent of the total DNA in cells.

Identifying genes on each chromosome is an active area of genetic research. Because researchers use different approaches to predict the number of genes on each chromosome, the estimated number of genes varies. Chromosome 3 likely contains 1,000 to 1,100 genes that provide instructions for making proteins. These proteins perform a variety of different roles in the body.

Genes on chromosome 3 are among the estimated 20,000 to 25,000 total genes in the human genome.

How are changes in chromosome 3 related to health conditions?

Many genetic conditions are related to changes in particular genes on chromosome 3. This list of disorders associated with genes on chromosome 3 provides links to additional information.

Is there a standard way to diagram chromosome 3?

Geneticists use diagrams called ideograms as a standard representation for chromosomes. Ideograms show a chromosome's relative size and its banding pattern. A banding pattern is the characteristic pattern of dark and light bands that appears when a chromosome is stained with a chemical solution and then viewed under a microscope. These bands are used to describe the location of genes on each chromosome.

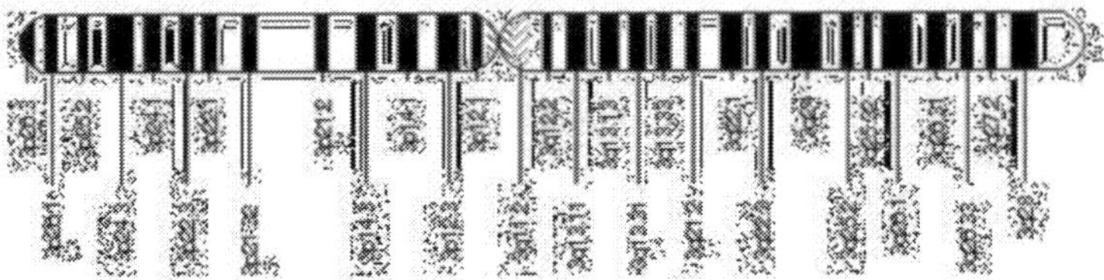

Genetics Home Reference includes these conditions related to genes on chromosome 3:

Conditions related to genes on chromosome 3

3- methylcrotonyl-CoA carboxylase
deficiency aceruloplasminemia
Adams-Oliver syndrome
adult polyglucosan body disease
Aicardi-Goutieres syndrome
alkaptonuria
aminoacylase 1 deficiency
ankyloblepharon-ectodermal defects-cleft lip/palate
syndrome asphyxiating thoracic dystrophy
atelosteogenesis type 1
atelosteogenesis type 3
autosomal dominant congenital stationary night blindness
autosomal recessive hypotrichosis
benign chronic pemphigus
biotinidase deficiency
blepharophimosis, ptosis, and epicanthus inversus
syndrome boomerang dysplasia
Brugada syndrome
carnitine-acylcarnitine translocase deficiency

CAV3-related distal myopathy
cerebral cavernous malformation
Chanarin-Dorfman syndrome
Charcot-Marie-Tooth disease
CHMP2B-related frontotemporal dementia
combined pituitary hormone deficiency
congenital myasthenic syndrome
congenital sucrase-isomaltase deficiency
cranioectodermal dysplasia
desmoid tumor Diamond-
Blackfan anemia
dilated cardiomyopathy with ataxia syndrome
dyskeratosis congenita
dystrophic epidermolysis bullosa
essential thrombocythemia
familial dilated cardiomyopathy
familial encephalopathy with neuroserpin inclusion bodies
familial erythrocytosis
familial isolated hyperparathyroidism
familial thoracic aortic aneurysm and dissection
glycine encephalopathy
glycogen storage disease type IV
GM1 gangliosidosis
gray platelet syndrome
Hermansky-Pudlak syndrome
idiopathic pulmonary fibrosis
isolated hyperCKemia
Kallmann syndrome
Larsen syndrome
Leigh syndrome
leukoencephalopathy with vanishing white matter
limb-girdle muscular dystrophy
Loeys-Dietz syndrome
Lynch syndrome
megalencephaly-capillary malformation syndrome
microphthalmia
mucopolysaccharidosis type IV
multiple lentigines syndrome
multiple sulfatase deficiency
myotonic dystrophy
nonsyndromic deafness
nonsyndromic paraganglioma

Noonan syndrome optic
atrophy type 1
osteogenesis imperfecta
pilomatricoma
pontocerebellar hypoplasia
porphyria
propionic acidemia
protein S deficiency
pseudocholinesterase deficiency
pyruvate dehydrogenase deficiency
renal tubular dysgenesis
retinitis pigmentosa
rippling muscle disease
Romano-Ward syndrome
septo-optic dysplasia sick
sinus syndrome small
fiber neuropathy
SOX2 anophthalmia syndrome
spondylocarpotarsal synostosis syndrome
systemic lupus erythematosus
Usher syndrome
von Hippel-Lindau syndrome
Waardenburg syndrome
xeroderma pigmentosum

Published*: November 17, 2014*

Genes on chromosome 3

Genetics Home Reference includes these genes on chromosome 3

ABHD5
ACY1
AGTR1
ALAS1
AMT
ARHGAP31
ATP2B2
ATP2C1
BCHE BTD

CASR
CAV3
CHMP2B
CLRN1
CNBP
COL7A1
COLQ
CP
CPOX
CRTAP
CTNNB1
DNAJC19
DRD3
EIF2B5
EOGT
FLNB
FOXL2
GBE1
GLB1
GNAT1
HESX1
HGD
HPS3
IFT80
IFT122
LIPH
MCCC1
MITF
MLH1
NBEAL2
OPA1
PCCB
PDCD10
PDHB
PIK3CA
PROK2
PROS1
RAB7A
RAF1
RHO
RPL35A
SCN5A

SCN10A
SERPINI1
SI
SLC25A20
SOX2
SUMF1
TERC
TGFBR2
THPO
TMIE
TP63
TREX1
TSEN2
VHL
XPC

State of New Mexico
Chromosome Four

University of New Mexico School of Medicine (Albuquerque)
(Gastrointestinal Stromal Tumor)

Chromosome 4

What is chromosome 4?

Humans normally have 46 chromosomes in each cell, divided into 23 pairs. Two copies of chromosome 4, one copy inherited from each parent, form one of the pairs. Chromosome 4 spans about 191 million DNA building blocks (base pairs) and represents more than 6 percent of the total DNA in cells.

Identifying genes on each chromosome is an active area of genetic research. Because researchers use different approaches to predict the number of genes on each chromosome, the estimated number of genes varies. Chromosome 4 likely contains 1,000 to 1,100 genes that provide instructions for making proteins. These proteins perform a variety of different roles in the body.

Genes on chromosome 4 are among the estimated 20,000 to 25,000 total genes in the human genome.

How are changes in chromosome 4 related to health conditions?

Many genetic conditions are related to changes in particular genes on chromosome 4. This list of disorders associated with genes on chromosome 4 provides links to additional information.

Changes in the structure or number of copies of a chromosome can also cause problems with health and development. The following chromosomal conditions are associated with such changes in chromosome 4.

Cancers

Changes in chromosome 4 have been identified in several types of human cancer. These genetic changes are somatic, which means they are acquired during a person's lifetime and are present only in certain cells. For example, rearrangements (translocations) of genetic material between chromosome 4 and several other chromosomes have been associated with leukemias, which are cancers of blood-forming cells.

A specific translocation involving chromosome 4 and chromosome 14 is commonly found in multiple myeloma, which is a cancer that starts in cells of the bone marrow. The translocation, which is written as t(4;14)(p16;q32), abnormally fuses the *WHSC1* gene on chromosome 4 with part of another gene on chromosome 14. The fusion of these genes overactivates *WHSC1*, which appears to promote the uncontrolled growth and division of cancer cells.

Facioscapulohumeral muscular dystrophy

Facioscapulohumeral muscular dystrophy is caused by genetic changes involving the long (q) arm of chromosome 4. This condition is characterized by muscle weakness and wasting (atrophy) that worsens slowly over time. It results from changes in a region of DNA known as D4Z4, located near the end of the chromosome at a position described as 4q35. The D4Z4 region consists of 11 to more than 100 repeated segments, each of which is about 3,300 DNA base pairs (3.3 kb) long. The entire D4Z4 region is normally hypermethylated, which means that it has a large number of methyl groups (consisting of one carbon atom and three hydrogen atoms) attached to the DNA. Facioscapulohumeral muscular dystrophy results when the region is hypomethylated, with too few methyl groups attached. In facioscapulohumeral muscular dystrophy type 1 (FSHD1), hypomethylation occurs because the D4Z4 region is abnormally shortened (contracted), containing between 1 and 10 repeats instead of the usual 11 to 100 repeats. In facioscapulohumeral muscular dystrophy type 2 (FSHD2), hypomethylation most often results from mutations in a gene called *SMCHD1*, which normally hypermethylates the D4Z4 region.

The segment of the D4Z4 region closest to the end of chromosome 4 contains a gene called *DUX4*. Hypermethylation of the D4Z4 region normally keeps the *DUX4* gene turned off (silenced) in most adult cells and tissues. In people with facioscapulohumeral muscular dystrophy, hypomethylation of the D4Z4 region prevents the *DUX4* gene from being silenced in cells and tissues where it is usually turned off. Although little is known about the function of the *DUX4* gene when it is turned on (active), researchers believe that it influences the activity of other genes, particularly in muscle cells. It is unknown how abnormal activity of the *DUX4* gene damages or destroys these cells, leading to progressive muscle weakness and atrophy.

The *DUX4* gene is located next to a regulatory region of DNA known as a Plam sequence, which is necessary for the production of the DUX4 protein. Some copies of chromosome 4 have a functional Plam sequence, while others do not. Copies of chromosome 4 with a functional Plam sequence are described as 4Qa or "permissive." Those without a functional Plam sequence are described as 4Qb or "non-permissive." Without a functional Plam sequence, no DUX4 protein is made. Because there are two copies of chromosome 4 in each cell, individuals may have two "permissive" copies of chromosome 4, two "non-permissive" copies, or one of each. Facioscapulohumeral muscular dystrophy can only occur in people who have at least one "permissive" copy of chromosome 4. Whether an affected individual has a contracted D4Z4 region or a *SMCHD1* gene mutation, the disease results only if a functional Plam sequence is also present to allow DUX4 protein to be produced.

PDGFRA-associated chronic eosinophilic leukemia

PDGFRA-associated chronic eosinophilic leukemia is caused by genetic abnormalities that involve the *PDGFRA* gene, a gene found on chromosome 4. This condition is a type of blood cell cancer characterized by an increased number of eosinophils, a type of white blood cell involved in allergic reactions.

The *PDGFRA* gene abnormalities are somatic mutations, which are mutations acquired during a person's lifetime that are present only in certain cells. The most common of these abnormalities is a deletion of genetic material from chromosome 4 that removes approximately 800 DNA building blocks (nucleotides) and brings together parts of two genes, *FIP1L1* and *PDGFRA*, creating the *FIP1L1-PDGFRA* fusion gene. Occasionally, through mechanisms other than deletion, genes other than *FIP1L1* are fused with the *PDGFRA* gene. Rarely, mutations that change single DNA building blocks in the *PDGFRA* gene (point mutations) cause this condition.

The protein produced from the *FIP1L1-PDGFRA* fusion gene (as well as other *PDGFRA* fusion genes) has the function of the PDGFRA protein, which stimulates signaling pathways inside the cell that control many important cellular processes, such as cell growth and division (proliferation) and cell survival. Unlike the normal PDGFRA protein, however, the fusion protein is constantly turned on (constitutively activated), which means the cells are always receiving signals to proliferate. Similarly, point mutations in the *PDGFRA* gene can result in a constitutively activated PDGFRA protein. When the *FIP1L1-PDGFRA* fusion gene or point mutations in the *PDGFRA* gene occur in blood cell precursors, the growth of eosinophils (and occasionally other blood cells) is poorly controlled, leading to *PDGFRA*-associated chronic eosinophilic leukemia. It is unclear why eosinophils are preferentially affected by this genetic change.

Wolf-Hirschhorn syndrome

Wolf-Hirschhorn syndrome is caused by a deletion of genetic material near the end of the short (p) arm of chromosome 4 at a position described as 4p16.3. The signs and symptoms of this condition are related to the loss of multiple genes from this part of the chromosome. The size of the deletion varies among affected individuals; studies suggest that larger deletions tend to result in more severe intellectual disability and physical abnormalities than smaller deletions.

The region of chromosome 4 that is deleted most often in people with Wolf-Hirschhorn syndrome is known as Wolf-Hirschhorn syndrome critical region 2 (WHSCR-2). This region contains several genes, some of which are known to play important roles in early development. A loss of these genes leads to developmental delay, a distinctive facial appearance, and other characteristic features of the condition. Scientists are working to identify additional genes at the end of the short arm of chromosome 4 that contribute to the characteristic features of Wolf-Hirschhorn syndrome.

Other chromosomal conditions

Some deletions of genetic material from the short (p) arm of chromosome 4 do not involve the critical region WHSCR-2. These deletions cause signs and symptoms that are distinct from those of Wolf-Hirschhorn syndrome, including mild intellectual disability and, in some cases, rapid (accelerated) growth. People with this type of deletion usually do not have seizures.

Trisomy 4 occurs when cells have three copies of chromosome 4 instead of the usual two copies. Full trisomy 4, which occurs when all of the body's cells contain an extra copy of chromosome 4, is not compatible with life. A similar but somewhat less severe condition called mosaic trisomy 4 occurs when only some of the body's cells have an extra copy of 'chromosome 4. The signs and symptoms of mosaic trisomy 4 vary widely and can include heart defects, abnormalities of the fingers and toes, and other birth defects. Mosaic trisomy 4 is very rare; only a few cases have been reported.

Other changes in the number or structure of chromosome 4 can have a variety of effects including delayed growth and development, intellectual disability, distinctive facial features, heart defects, and other medical problems. Changes involving chromosome 4 include an extra piece of the chromosome in each cell (partial trisomy 4), a missing segment of the chromosome in each cell (partial monosomy 4), and a circular structure called a ring chromosome 4. Ring chromosomes occur when a chromosome breaks in two places and the ends of the chromosome arms fuse together to form a circular structure.

Is there a standard way to diagram chromosome 4?

Geneticists use diagrams called ideograms as a standard representation for chromosomes. Ideograms show a chromosome's relative size and its banding pattern. A banding pattern is the characteristic pattern of dark and light bands that appears when a chromosome is stained with a chemical solution and then viewed under a microscope. These bands are used to describe the location of genes on each chromosome.

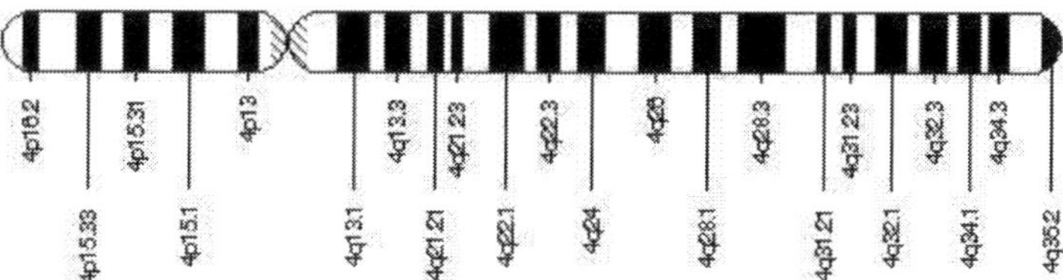

Conditions related to genes on chromosome 4

Genetics Home Reference includes these conditions related to genes on chromosome 4

3-hydroxyacyl-CoA dehydrogenase deficiency
abetalipoproteinemia
achondroplasia Adams-
Oliver syndrome
age-related macular degeneration
amelogenesis imperfecta
aspartylglucosaminuria
atypical hemolytic-uremic syndrome
autosomal dominant congenital stationary night blindness

Axenfeld-Rieger syndrome
benign essential blepharospasm
beta-mannosidosis
Bietti crystalline dystrophy
bladder cancer
cherubism
complement factor I deficiency
congenital afibrinogenemia
congenital central hypoventilation
syndrome congenital hyperinsulinism
congenital myasthenic syndrome
core binding factor acute myeloid leukemia
cranioectodermal dysplasia
Crouzonodermoskeletal syndrome
dentinogenesis imperfecta
Ellis-van Creveld
syndrome epidermal nevus
essential thrombocythemia
facioscapulohumeral muscular dystrophy
Fraser syndrome
gastrointestinal stromal tumor
glutaric acidemia type II
Hennekam syndrome
Huntington disease
hypochondroplasia
infantile systemic hyalinosis
juvenile hyaline fibromatosis
lacrimo-auriculo-dento-digital syndrome
late-infantile neuronal ceroid lipofuscinosis
limb-girdle muscular dystrophy
methylmalonic acidemia
mucopolysaccharidosis type I
Muenke syndrome
multiple system atrophy
nephronophthisis
neuroblastoma
nonsyndromic deafness
Parkinson disease
PDGFRA-associated chronic eosinophilic leukemia
Peters anomaly
piebaldism
polycystic kidney disease

polycythemia vera
prekallikrein deficiency
primary myelofibrosis
progressive external ophthalmoplegia
pseudohypoaldosteronism type 1
retinitis pigmentosa
rheumatoid arthritis
Romano-Ward syndrome
SADDAN Senior-Løken
syndrome
tetrahydrobiopterin deficiency
thanatophoric dysplasia UV-
sensitive syndrome Weyers
acrofacial dysostosis Wolf-
Hirschhorn syndrome
Wolfram syndrome

Reviewed: *February 2012*

Published: *November 17, 2014*

State of New York
Chromosome Five

Albert Einstein College of Medicine at Yeshiva University (Bronx) (Familial Adenomatous Polyposis)

Albany Medical College (Albany) (Spinal Muscular Atrophy) (Nicotine Dependency)

Columbia University College of Physicians and Surgeons (Manhattan) (Homocystinuria)

Hofstra University North Shore LIJ School of Medicine (Hempstead) (Myelodysplastic Syndrome)

Icahn School of Medicine at Mount Sinai (Manhattan) (Recessive Multiple Epiphyseal Dysplasia)

New York Institute of Technology College of Osteopathic Medicine (Old Westbury) (Diastrophic Dysplasia)

New York Medical College (Yalhalla) (Granular Corneal Dystrophy Type 1 and 2)

New York University School of Medicine (Manhattan) (Acute Myeloid Leukemia)

Stony Brook University School of Medicine (Stony Brook)(Achondrogenesis Type 1 B)

State University of New York Upstate Medical University (Syracuse) (Gangliosidosis, AB Variant)

State University of New York Downstate Medical Center College of Medicine (Brooklyn) (Cri du Chat)

Touro College of Osteopathic Medicine (Manhattan) (Motor Neuron Spinal Muscular Atrophy)

University at Buffalo School of Medicine and Biomedical Sciences (Buffalo) (Atelosteogenesis, Type 11)

University of Rochester School of Medicine (Rochester) (Corneal Dystrophy Type 1 and 11)

Weil Cornell Medical College (Manhattan) (Cockayne Syndrome)

What is chromosome 5?

Humans normally have 46 chromosomes in each cell, divided into 23 pairs. Two copies of chromosome 5, one copy inherited from each parent, form one of the pairs. Chromosome 5 spans about 181 million DNA building blocks (base pairs) and represents almost 6 percent of the total DNA in cells.

Identifying genes on each chromosome is an active area of genetic research. Because researchers use different approaches to predict the number of genes on each chromosome, the estimated number of genes varies. Chromosome 5 likely contains about 900 genes that provide instructions for making proteins. These proteins perform a variety of different roles in the body.

Genes on chromosome 5 are among the estimated 20,000 to 25,000 total genes in the human genome.

How are changes in chromosome 5 related to health conditions?

Many genetic conditions are related to changes in particular genes on chromosome 5. This list of disorders associated with genes on chromosome 5 provides links to additional information.

Changes in the structure or number of copies of a chromosome can also cause problems with health and development. The following chromosomal conditions are associated with such changes in chromosome 5.

Cancers

Changes in the structure of chromosome 5 are associated with certain forms of cancer and conditions related to cancer. These changes are typically somatic, which means they are acquired during a person's lifetime and are present only in tumor cells. Deletions in the long (q) arm of the chromosome have been identified in a form of blood cancer known as acute myeloid leukemia (AML). These deletions also frequently occur in a disorder called myelodysplastic syndrome, which is a disease of the blood and bone marrow. People with this condition have a low number of red blood cells (anemia) and an increased risk of developing AML. When MDS is associated with a specific deletion in the long arm of chromosome 5, it is known as 5q- (5q minus) syndrome.

Studies suggest that some genes on chromosome 5 play critical roles in the growth and division of cells. When segments of the chromosome are deleted, as in some cases of AML and MDS, these important genes are missing. Without these genes, cells can grow and divide too quickly and in an uncontrolled way. Researchers are working to identify the specific genes on chromosome 5 that are related to AML and MDS.

Cri-du-chat syndrome

Cri-du-chat (cat's cry) syndrome is caused by a deletion of the end of the short (p) arm of chromosome 5. This chromosomal change is written as 5p- (5p minus). The signs and symptoms of cri-du -chat syndrome are probably related to the loss of multiple genes in this region. Researchers are working to determine how the loss of these genes leads to the features of the disorder. They have discovered that in people with cri-du-chat syndrome, larger deletions tend to result in more severe intellectual disability and developmental delays than smaller deletions. Researchers have also defined regions of the short arm of chromosome 5 that are associated with particular features of cri-du-chat syndrome. A specific region designated 5p15.3 is associated with a cat-like cry, and a nearby region called 5p15.2 is associated with intellectual disability, small head size (microcephaly), and distinctive facial features.

Crohn disease

Several regions of chromosome 5 have been associated with the risk of developing Crohn disease. For example, a combination of genetic variations in a region of DNA on the long (q) arm of the chromosome (5q31) has been shown to increase a person's chance of developing Crohn disease. Together, these variations are known as the inflammatory bowel disease 5 (IBD5) locus. This region of chromosome 5 contains several related genes that may influence Crohn disease risk, including *SLC22A4* and *SLC22A5*.

Variations in a region of the short (p) arm of chromosome 5 designated 5p13.1 are also associated with Crohn disease risk. Researchers refer to this part of chromosome 5 as a "gene desert" because it contains no known genes; however, it may contain stretches of DNA that help regulate nearby genes such as *PTGER4*. Research studies are under way to examine a possible connection between the *PTGER4* gene and Crohn disease.

PDGFRB-associated chronic eosinophilic leukemia

Translocations involving chromosome 5 are involved in a type of blood cell cancer called *PDGFRB*-associated chronic eosinophilic leukemia.

This condition is characterized by an increased number of eosinophils, a type of white blood cell. The most common translocation that causes this condition fuses part of the *PDGFRB* gene from chromosome 5 with part of the *ETV6* gene from chromosome 12, written as t(5;12)(q31-33;p13). Translocations fusing the *PDGFRB* gene with one of more than 20 other genes have also been found to cause *PDGFRB*-associated chronic eosinophilic leukemia, but these other genetic changes are relatively uncommon. These translocations are acquired during a person's lifetime and are present only in cancer cells. This type of genetic change, called a somatic mutation, is not inherited.

The protein produced from the *ETV6-PDGFRB* fusion gene, called ETV6/PDGFRβ, functions differently than the proteins normally produced from the individual genes. The ETV6 protein normally turns off (represses) gene activity and the PDGFRβ protein plays a role in turning on (activating) signaling pathways.

The ETV6/PDGFRβ protein is always turned on, activating signaling pathways and gene activity. When the *ETV6-PDGFRB* fusion gene mutation occurs in cells that develop into blood cells, the growth of eosinophils (and occasionally other white blood cells, such as neutrophils and mast cells) is poorly controlled, leading to *PDGFRB*-associated chronic eosinophilic leukemia. It is unclear why eosinophils are preferentially affected by this genetic change.

Periventricular heterotopia

In a few cases, abnormalities in chromosome 5 have been associated with periventricular heterotopia, a disorder characterized by abnormal clumps of nerve cells (neurons) around fluid-filled cavities (ventricles) near the center of the brain. In each case, the affected individual had extra genetic material caused by an abnormal duplication of part of this chromosome. It is not known how this duplicated genetic material results in the signs and symptoms of periventricular heterotopia.

Other chromosomal conditions

Other changes in the number or structure of chromosome 5 can have a variety of effects, including delayed growth and development, distinctive facial features, birth defects, and other health problems. Changes to chromosome 5 include an extra segment of the short (p) or long (q) arm of the chromosome in each cell (partial trisomy 5p or 5q), a missing segment of the long arm of the chromosome in each cell (partial monosomy 5q), and a circular structure called ring chromosome 5.

Ring chromosomes occur when a chromosome breaks in two places and the ends of the chromosome arms fuse together to form a circular structure.

Is there a standard way to diagram chromosome 5?

Geneticists use diagrams called ideograms as a standard representation for chromosomes. Ideograms show a chromosome's relative size and its banding pattern. A banding pattern is the characteristic pattern of dark and light bands that appears when a chromosome is stained with a chemical solution and then viewed under a microscope. These bands are used to describe the location of genes on each chromosome.

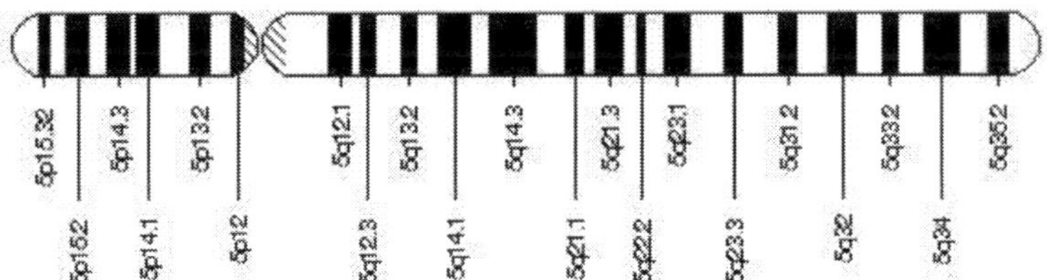

Conditions related to genes on chromosome 5

Genetics Home Reference includes these conditions related to genes on chromosome 5:

3-methylcrotonyl-CoA carboxylase deficiency
achondrogenesis
acute promyelocytic leukemia alpha-
methylacyl-CoA racemase deficiency
amyotrophic lateral sclerosis
ankylosing spondylitis
atelosteogenesis type 2
autosomal recessive axonal neuropathy with
neuromyotonia autosomal recessive congenital stationary
night blindness breast cancer
capillary malformation-arteriovenous malformation
syndrome Charcot-Marie-Tooth disease
chylomicron retention disease
Cockayne syndrome
combined pituitary hormone deficiency
congenital contractural arachnodactyly
Cornelia de Lange syndrome
craniometaphyseal dysplasia cri-du-
chat syndrome

Crohn disease
cytogenetically normal acute myeloid leukemia
D-bifunctional protein deficiency
desmoid tumor
diastrophic dysplasia
distal myopathy 2
dyskeratosis congenita
Ehlers-Danlos syndrome
enlarged parietal foramina
episodic ataxia
familial adenomatous polyposis
familial dilated cardiomyopathy
familial idiopathic basal ganglia calcification
gastrointestinal stromal tumor GM2-
gangliosidosis, AB variant
hereditary angioedema
hereditary diffuse leukoencephalopathy with spheroids
hereditary hyperekplexia
hereditary sensory and autonomic neuropathy type II
homocystinuria
idiopathic pulmonary fibrosis
juvenile myoclonic epilepsy
lacrimo-auriculo-dento-digital syndrome
lattice corneal dystrophy type I
Leigh syndrome
limb-girdle muscular dystrophy
Marinesco-Sjögren syndrome
Milroy disease
molybdenum cofactor deficiency
mucopolysaccharidosis type VI
multiple epiphyseal dysplasia
multiple sclerosis
myofibrillar myopathy
Netherton syndrome
nonsyndromic paraganglioma
oculocutaneous albinism Paget
disease of bone
Parkes Weber syndrome
PDGFRB-associated chronic eosinophilic leukemia
primary carnitine deficiency
primary ciliary dyskinesia
pyridoxine-dependent epilepsy

Sandhoff disease
short stature, hyperextensibility, hernia, ocular depression, Rieger anomaly, and teething delay
Sotos syndrome
spinal muscular atrophy
STING-associated vasculopathy with onset in infancy
succinyl-CoA:3-ketoacid CoA transferase deficiency
Treacher Collins syndrome
trichohepatoenteric
syndrome Usher syndrome
UV-sensitive syndrome
Wagner syndrome
Weaver syndrome

Reviewed: *February 2013*

Published: *November 17, 2014*

Genes on chromosome 5 genes on chromosome 5:

ADAMTS2
ALDH7A1
AMACR
ANKH
APC
ARSB
CSF1R
CTNND2
DNAH5
ERAP1
ERCC8
F12
FAM134B
FBN2
FGF10
FGFR4
FLT4
GABRA1
GLRA1
GM2A
GPR98
GRM6
HEXB

HINT1
HSD17B4
IL7R
IRGM
MATR3
MCCC2
MOCS2
MSX2
MTRR
MYOT
NIPBL
NPM1
NSD1
OXCT1
PDGFRB
PIK3R1
PROP1
RAD50
RASA1
SAR1B
SDHA
SGCD
SH3TC2
SIL1
SLC1A3
SLC22A5
SLC26A2
SLC45A2
SMN1
SMN2
SNCAIP
SPINK5
SQSTM1
TCOF1
TERT
TGFBI
TMEM173
TTC37
VCAN

State of North Carolina
Chromomosome Six

Duke University School of Medicine (Durham) (Polycytic Kidney Disease)

University of North Carolina School of Medicine (Chapel Hill) (Systemic Lupus Erythematosis)

Wake Forest School of Medicine (Winston Salem) (Parkinson Disease)

Campbell University School of Osteopathic Medicine (Buies Creek) (Rheumatoid Arthritis HLA-DR)

The Brody School of Medicine at East Carolina University (Greenville) (Guillain Barre Syndrome)

Chromosome 6

What is chromosome 6?

Humans normally have 46 chromosomes in each cell, divided into 23 pairs. Two copies of chromosome 6, one copy inherited from each parent, form one of the pairs. Chromosome 6 spans about 171 million DNA building blocks (base pairs) and represents between 5.5 and 6 percent of the total DNA in cells.

Identifying genes on each chromosome is an active area of genetic research. Because researchers use different approaches to predict the number of genes on each chromosome, the estimated number of genes varies. Chromosome 6 likely contains 1,000 to 1,100 genes that provide instructions for making proteins. These proteins perform a variety of different roles in the body.

Genes on chromosome 6 are among the estimated 20,000 to 25,000 total genes in the human genome.

How are changes in chromosome 6 related to health conditions?

Many genetic conditions are related to changes in particular genes on chromosome 6. This list of disorders associated with genes on chromosome 6 provides links to additional information.

Changes in the structure or number of copies of a chromosome can also cause problems with health and development. The following chromosomal conditions are associated with such changes in chromosome 6.

6q24-related transient neonatal diabetes mellitus

6q24-related transient neonatal diabetes mellitus, a type of diabetes that occurs in infants, is caused by the overactivity (overexpression) of certain genes in a region of the long (q) arm of chromosome 6 called 6q24. People inherit two copies of their genes, one from their mother and one from their father. Usually both copies of each gene are active, or "turned on," in cells. In some cases, however, only one of the two copies is normally turned on. Which copy is active depends on the parent of origin: some genes are normally active only when they are inherited from a person's father; others are active only when inherited from a person's mother. This phenomenon is known as genomic imprinting.

The 6q24 region includes paternally expressed imprinted genes, which means that normally only the copy of each gene that comes from the father is active. The copy of each gene that comes from the mother is inactivated (silenced) by a mechanism called methylation.

There are three ways that overexpression of paternally expressed imprinted genes in the 6q24 region can occur. About 40 percent of cases of 6q24-related transient neonatal diabetes mellitus are caused by a genetic change known as paternal uniparental disomy (UPD) of chromosome 6. In paternal UPD, people inherit both copies of the affected chromosome from their father instead of one copy from each parent.

Paternal UPD causes people to have two active copies of paternally expressed imprinted genes, rather than one active copy from the father and one inactive copy from the mother.

Another 40 percent of cases of 6q24-related transient neonatal diabetes mellitus occur when the copy of chromosome 6 that comes from the father has a duplication of genetic material including the paternally expressed imprinted genes in the 6q24 region.

The third mechanism by which overexpression of genes in the 6q24 region can occur is by impaired silencing of the maternal copy of the genes (maternal hypomethylation). Approximately 20 percent of cases of 6q24-related transient neonatal diabetes mellitus are caused by maternal hypomethylation. Some people with this disorder have a genetic change in the maternal copy of the 6q24 region that prevents genes in that region from being silenced. Other affected individuals have a more generalized impairment of gene silencing involving many imprinted regions, called hypomethylation of imprinted loci (HIL). Because HIL can cause overexpression of many genes, this mechanism may account for the additional health problems that occur in some people with 6q24-related transient neonatal diabetes mellitus.

It is not well understood how overexpression of genes in the 6q24 region causes 6q24-related transient neonatal diabetes mellitus and why the condition improves after infancy. This form of diabetes is characterized by high blood sugar levels (hyperglycemia) resulting from a shortage of the hormone insulin.

Insulin controls how much glucose (a type of sugar) is passed from the blood into cells for conversion to energy.

The protein produced from one gene in the 6q24 region may help control insulin secretion by beta cells in the pancreas. In addition, overexpression of this protein has been shown to stop the cycle of cell division and lead to the self-destruction of cells (apoptosis). Researchers suggest that overexpression of this gene may reduce the number of insulin-secreting beta cells or impair their function in affected individuals.

Lack of sufficient insulin results in the signs and symptoms of diabetes mellitus. In individuals with 6q24-related transient neonatal diabetes mellitus, these signs and symptoms are most likely to occur during times of physiologic stress, including the rapid growth of infancy, childhood illnesses, and pregnancy. Because insulin acts as a growth promoter during early development, a shortage of this hormone may account for the slow growth before birth (intrauterine growth retardation) seen in 6q24-related transient neonatal diabetes mellitus.

Cancers

Duplications of genetic material in the short (p) arm of chromosome 6 have been associated with the growth and spread of several types of cancer. These duplications are somatic, which means they are acquired during a person's lifetime and are present only in certain cells. Researchers believe that some of the genes in the duplicated region on chromosome 6p are oncogenes.

Oncogenes play roles in several critical cell functions, including cell division, the maturation of cells to carry out specific functions (cell differentiation), and the self-destruction of cells (apoptosis) . When mutated, oncogenes have the potential to cause normal cells to become cancerous. The presence of extra copies of the oncogenes may allow cells to grow and divide in an uncontrolled way, leading to the progression and spread of cancer.

Other chromosomal conditions

Other changes in the number or structure of chromosome 6 can have a variety of effects, including delayed growth and development, intellectual disability, distinctive facial features, birth defects, and other health problems. Changes to chromosome 6 may include deletions or duplications of genetic material in the short (p) or long (q) arm of the chromosome in each cell, or a circular structure called ring chromosome 6. Ring chromosomes occur when a chromosome breaks in two places and the ends of the chromosome arms fuse together to form a circular structure.

Is there a standard way to diagram chromosome 6?

Geneticists use diagrams called ideograms as a standard representation for chromosomes. Ideograms show a chromosome's relative size and its banding pattern. A banding pattern is the characteristic pattern of dark and light bands that appears when a chromosome is stained with a chemical solution and then viewed under a microscope. These bands are used to describe the location of genes on each chromosome.

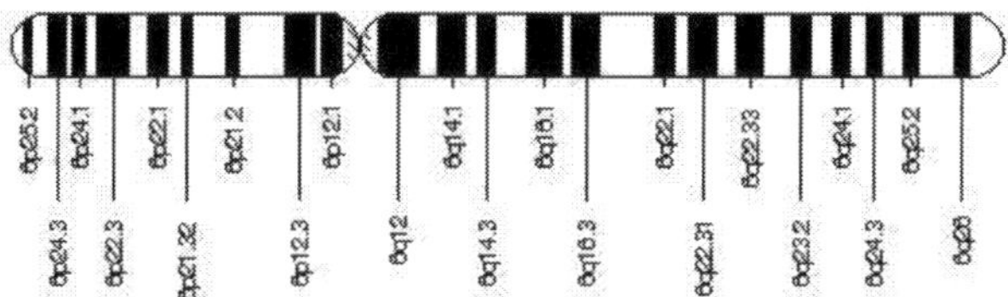

Conditions related to genes on chromosome 6

Genetics Home Reference includes these conditions related to genes on chromosome 6:

3-M syndrome
6q24-related transient neonatal diabetes mellitus
21-hydroxylase deficiency age-
related macular degeneration
amyotrophic lateral sclerosis

ankylosing spondylitis
arginase deficiency
autoimmune Addison disease
autosomal recessive cerebellar ataxia type
1 Axenfeld-Rieger syndrome
Behçet disease branchio-oculo-
facial syndrome Carpenter
syndrome
celiac disease Charcot-
Marie-Tooth disease Char
syndrome
chordoma
cleidocranial dysplasia
Coffin-Siris syndrome
complement component 2 deficiency
congenital plasminogen deficiency
critical congenital heart disease
Dandy-Walker syndrome Diamond-
Blackfan anemia Ehlers-Danlos
syndrome
familial dilated cardiomyopathy
granulomatosis with polyangiitis
Graves disease
Hashimoto thyroiditis
hemochromatosis
Huntington disease-like syndrome
hypermethioninemia
idiopathic inflammatory myopathy
juvenile idiopathic arthritis juvenile
myoclonic epilepsy
Lafora progressive myoclonus
epilepsy LAMA2-related muscular
dystrophy maple syrup urine disease
MEGDEL syndrome
methylmalonic acidemia
molybdenum cofactor deficiency
multiple epiphyseal dysplasia
multiple sclerosis Nakajo-
Nishimura syndrome narcolepsy

nonsyndromic deafness
oculodentodigital dysplasia

otospondylomegaepiphyseal dysplasia
Parkinson disease
Peters anomaly
polycystic kidney disease
polycystic lipomembranous osteodysplasia with sclerosing leukoencephalopathy
pontocerebellar hypoplasia
porphyria
progressive pseudorheumatoid dysplasia
psoriatic arthritis
recurrent hydatidiform mole
Refsum disease
retinitis pigmentosa
rheumatoid arthritis
rhizomelic chondrodysplasia punctata
sialic acid storage disease
sialidosis
spinocerebellar ataxia type 1
Stargardt macular degeneration
Stickler syndrome
succinic semialdehyde dehydrogenase deficiency
Treacher Collins syndrome
trichohepatoenteric syndrome
trichothiodystrophy
type 1 diabetes
vitelliform macular dystrophy
Weissenbacher-Zweymüller syndrome
xeroderma pigmentosum
X-linked sideroblastic anemia

Genes on chromosome 6

Genetics Home Reference includes these genes on chromosome 6:

ALDH5A1
ARG1
ARID1B
ATXN1
BCKDHB
C2
COL9A1

COL11A2
CUL7
CYP21A2
EFHC1
ELOVL4
EPM2A
EYA4
FIG4
FOXC1
GJA1
GNMT
GTF2H5
HFE
HLA-B
HLA-DPB1
HLA-DQA1
HLA-DQB1
HLA-DRB1
KHDC3L
LAMA2
LHFPL5
MOCS1
MUT
MYO6
NEU1
NHLRC1
PARK2
PEX7
PKHD1
PLAGL1
PLG
POLH
POLR1C
PRPH2
PSMB8
RAB23
RARS2
RPS10
RUNX2
SERAC1
SKIV2L
SLC17A5

SYNE1
T
TBP
TFAP2A
TFAP2B
TNXB
TREM2
WISP3
ZFP57

State of North Dakota
Chromosome Seven

University of North Carolina School of Medicine and Health Sciences (Grand Forks) (Myelodysplastic Syndrome)

Chromosome 7

What is chromosome 7?

Humans normally have 46 chromosomes in each cell, divided into 23 pairs. Two copies of chromosome 7, one copy inherited from each parent, form one of the pairs. Chromosome 7 spans about 159 million DNA building blocks (base pairs) and represents more than 5 percent of the total DNA in cells.

Identifying genes on each chromosome is an active area of genetic research. Because researchers use different approaches to predict the number of genes on each chromosome, the estimated number of genes varies. Chromosome 7 likely contains 900 to 1,000 genes that provide instructions for making proteins. These proteins perform a variety of different roles in the body.

Genes on chromosome 7 are among the estimated 20,000 to 25,000 total genes in the human genome.

How are changes in chromosome 7 related to health conditions?

Many genetic conditions are related to changes in particular genes on chromosome 7. This list of disorders associated with genes on chromosome 7 provides links to additional information.

Changes in the structure or number of copies of a chromosome can also cause problems with health and development. The following chromosomal conditions are associated with such changes in chromosome 7.

Cancers

Changes in the number or structure of chromosome 7 occur frequently in human cancers. These changes are typically somatic, which means they are acquired during a person's lifetime and are present only in tumor cells. Many forms of cancer are associated with damage to chromosome 7. In particular, changes in this chromosome have been identified in cancers of blood-forming tissue (leukemias) and cancers of immune system cells (lymphomas). A loss of part or all of one copy of chromosome 7 is common in myelodysplastic syndrome, which is a disease of the blood and bone marrow. People with this disorder have an increased risk of developing leukemia.

Studies suggest that some genes on chromosome 7 may play critical roles in controlling the growth and division of cells. Without these genes, cells could grow and divide too quickly or in an uncontrolled way, resulting in a cancerous tumor. Researchers are working to identify the genes on chromosome 7 that are involved in the development and progression of cancer.

Greig cephalopolysyndactyly syndrome

Abnormalities of chromosome 7 are responsible for some cases of Greig cephalopolysyndactyly syndrome. These chromosomal changes involve a region of the short (p) arm of chromosome 7 that contains the *GLI3* gene. This gene plays an important role in the development of many tissues and organs before birth.

In some cases, Greig cephalopolysyndactyly syndrome results from a rearrangement (translocation) of genetic material between chromosome 7 and another chromosome. Other cases are caused by the deletion of several genes, including *GLI3*, from the short arm of chromosome 7. The loss of multiple genes can cause a more severe form of this disorder called Greig cephalopolysyndactyly contiguous gene deletion syndrome. People with this form of the disorder have characteristic developmental problems involving the limbs, head, and face along with seizures, developmental delay, and intellectual disability.

Russell-Silver syndrome

People normally inherit one copy of each chromosome from their mother and one copy from their father. For most genes, both copies are expressed, or "turned on," in cells. For some genes, however, only the copy inherited from a person's father (the paternal copy) is expressed. For other genes, only the copy inherited from a person's mother (the maternal copy) is expressed. These parent -specific differences in gene expression are caused by a phenomenon called genomic imprinting. Chromosome 7 contains a group of genes that normally undergo genomic imprinting.

Abnormalities involving these genes appear to be responsible for many cases of Russell-Silver syndrome.

In 7 percent to 10 percent of cases of Russell-Silver syndrome, people inherit both copies of chromosome 7 from their mother instead of one copy from each parent. This phenomenon is called maternal uniparental disomy (UPD). Maternal UPD causes people to have two active copies of maternally expressed imprinted genes rather than one active copy from the mother and one inactive copy from the father. These individuals do not have a paternal copy of chromosome 7 and therefore do not have any copies of genes that are active only on the paternal copy. In cases of Russell-Silver syndrome caused by maternal UPD, an imbalance in active paternal and maternal genes on chromosome 7 underlies the signs and symptoms of the disorder.

Saethre-Chotzen syndrome

Some cases of Saethre -Chotzen syndrome result from abnormalities of chromosome 7. These chromosomal changes involve a region of the short (p) arm of chromosome 7 that contains the *TWIST1* gene. This gene plays an important role in early development of the head, face, and limbs.

The chromosome abnormalities responsible for Saethre-Chotzen syndrome include translocations of genetic material between chromosome 7 and another chromosome, a rearrangement of genetic material within chromosome 7 (an inversion), or the formation of an abnormal circular structure called a ring chromosome 7. Each of these chromosomal changes alters or deletes the *TWIST1* gene and may also affect nearby genes.

When Saethre-Chotzen syndrome is caused by a chromosomal deletion instead of a mutation within the *TWIST1* gene, affected children are much more likely to have intellectual disability, developmental delay, and learning difficulties. These features are typically not seen in classic cases of Saethre-Chotzen syndrome. Researchers believe that a loss of other genes on the short arm of chromosome 7 may be responsible for these additional features.

Williams syndrome

Williams syndrome is caused by the deletion of genetic material from a portion of the long (q) arm of chromosome 7. The deleted region, which is located at position 11.23 (written as 7q11.23), is designated the Williams-Beuren region. This region includes more than 25 genes, and researchers believe that the characteristic features of Williams syndrome are probably related to the loss of several of these genes.

While a few of the specific genes related to Williams syndrome have been identified, the relationship between most of the genes in the deleted region and the signs and symptoms of Williams syndrome is unknown.

Other chromosomal conditions

Whereas Williams syndrome is caused by a deletion of genes in the Williams-Beuren region of chromosome 7, another syndrome is caused by the abnormal duplication (copying) of genes in this region. This duplication appears to be associated with delayed expressive language skills (vocabulary and the production of speech) and delayed development. Very few people with a duplication of the Williams-Beuren region have been identified.

Other changes in the number or structure of chromosome 7 can cause delayed growth and development, intellectual disability, distinctive facial features, skeletal abnormalities, delayed speech, and other medical problems. Changes in chromosome 7 include an extra copy of some genetic material from this chromosome in each cell (partial trisomy 7) or a missing segment of the chromosome in each cell (partial monosomy 7). In some cases, several DNA building blocks (nucleotides) are abnormally deleted or duplicated in part of chromosome 7. A circular structure called ring chromosome 7 is also possible. Ring chromosomes occur when a chromosome breaks in two places and the ends of the chromosome arms fuse together to form a circular structure.

Is there a standard way to diagram chromosome 7?

Geneticists use diagrams called ideograms as a standard representation for chromosomes. Ideograms show a chromosome's relative size and its banding pattern. A banding pattern is the characteristic pattern of dark and light bands that appears when a chromosome is stained with a chemical solution and then viewed under a microscope. These bands are used to describe the location of genes on each chromosome.

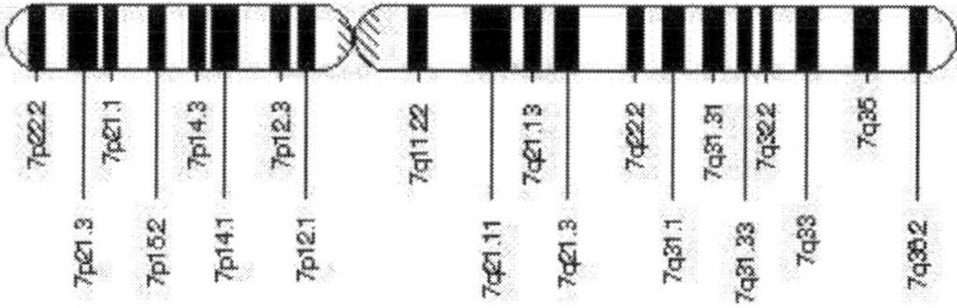

Conditions related to genes on chromosome 7

Genetics Home Reference includes these conditions related to genes on chromosome 7:

argininosuccinic aciduria
aromatic l-amino acid decarboxylase deficiency
Baraitser-Winter syndrome
cardiofaciocutaneous syndrome
cerebral cavernous malformation
Charcot -Marie-Tooth disease
chronic granulomatous disease
citrullinemia
coloboma
color vision deficiency
congenital bilateral absence of the vas deferens
congenital disorder of glycosylation type IIi
congenital leptin deficiency
cutis laxa
cystic fibrosis
cytochrome P450 oxidoreductase deficiency
dihydrolipoamide dehydrogenase deficiency
distal hereditary motor neuropathy, type II
distal hereditary motor neuropathy, type V
Ehlers-Danlos syndrome
familial hypertrophic cardiomyopathy
gastrointestinal stromal tumor Ghosal
hematodiaphyseal dysplasia
Greig cephalopolysyndactyly
syndrome hand-foot-genital syndrome
hemochromatosis
hereditary pancreatitis
hyperlysinemia
hypomyelination and congenital cataract
intrahepatic cholestasis of pregnancy
isolated growth hormone deficiency
Langerhans cell histiocytosis
lissencephaly with cerebellar hypoplasia
Lynch syndrome
microphthalmia
mucopolysaccharidosis type
VII multiple lentigines
syndrome myoclonus-dystonia

myotonia congenita
nonsyndromic deafness
nonsyndromic holoprosencephaly
Noonan syndrome
osteogenesis imperfecta
Pallister-Hall syndrome
Pendred syndrome
phosphoglycerate mutase deficiency
progressive familial intrahepatic cholestasis
renal tubular acidosis with deafness
rheumatoid arthritis
Romano-Ward syndrome
Saethre-Chotzen syndrome short
QT syndrome Shwachman-
Diamond syndrome
spondylocostal dysostosis
supravalvular aortic stenosis
systemic lupus erythematosus
systemic scleroderma
trichothiodystrophy Walker-
Warburg syndrome Weaver
syndrome
Williams syndrome Wolff-
Parkinson-White syndrome
Zellweger spectrum

Reviewed: *April 2008*

Published: *November 17, 2014*

Genes on chromosome 7

Genetics Home Reference includes these genes on chromosome 7:

AASS
ABCB4
ACTB
ASL
ATP6V0A4
BRAF
CCM2
CFTR
CLCN1
CLIP2

COG5
COL1A2
DDC
DFNA5
DLD
ELN
EZH2
FAM126A
GARS
GHRHR
GLI3
GTF2I
GTF2IRD1
GUSB
HOXA13
HSPB1
IRF5
ISPD
KCNH2
KRIT1
LEP
LFNG
LIMK1
MPLKIP
NCF1
OPN1SW
PEX1
PGAM2
PMS2
POR
PRKAG2
PRSS1
RELN
SBDS
SGCE
SHH
SLC25A13
SLC26A4
TBXAS1
TFR2
TWIST1

State of Ohio
Chromosome Eight

Case Western Reserve University School of Medicine (Cleveland) (Burkitt's Lymphoma)

Cleveland Clinic Lerner College of Medicine (Cleveland) (Cleft Lip and Palate)

The Ohio State University College of Medicine (Columbus) (Congenital Hypothyroidism)

Northeast Ohio Medical University (Rootstown) (Hereditary Multiple Exostosis)

Boonshoft School of Medicine at Wright State University (Dayton) (Cohen Syndrome)

University of Cincinnati College of Medicine (Cincinnati) (Schizophrenia 8p21-22locus)

Ohio University Heritage College of Osteopathic Medicine (Athens) (Primary Microcephaly)

University of Toledo College of Medicine (Toledo) (Familial Lipoprotein Lipase Deficiency)

Oklahoma State University Center for Health Sciences College of Osteopathic Medicine (Tulsa) (Langer-Giedion Syndrome)

University of Oklahoma College of Medicine (Oklahoma City) (Werner Syndrome)

Chromosome 8

What is chromosome 8?

Humans normally have 46 chromosomes in each cell, divided into 23 pairs. Two copies of chromosome 8, one copy inherited from each parent, form one of the pairs. Chromosome 8 spans more than 146 million DNA building blocks (base pairs) and represents between 4.5 and 5 percent of the total DNA in cells.

Identifying genes on each chromosome is an active area of genetic research. Because researchers use different approaches to predict the number of genes on each chromosome, the estimated number of genes varies. Chromosome 8 likely contains about 700 genes that provide instructions for making proteins. These proteins perform a variety of different roles in the body.

Genes on chromosome 8 are among the estimated 20,000 to 25,000 total genes in the human genome.

How are changes in chromosome 8 related to health conditions?

Many genetic conditions are related to changes in particular genes on chromosome 8. This list of disorders associated with genes on chromosome 8 provides links to additional information.

Changes in the structure or number of copies of a chromosome can also cause problems with health and development. The following chromosomal conditions are associated with such changes in chromosome 8.

8p11 myeloproliferative syndrome

Translocations of genetic material between chromosome 8 and other chromosomes can cause 8p11 myeloproliferative syndrome. This condition is characterized by an increased number of white blood cells (myeloproliferative disorder) and the development of lymphoma, a blood-related cancer that causes tumor formation in the lymph nodes. The myeloproliferative disorder usually develops into another form of blood cancer called acute myeloid leukemia. The most common translocation involved in this condition, written as t(8;13)(p11;q12), fuses part of the *FGFR1* gene on chromosome 8 with part of the *ZMYM2* gene on chromosome 13. The translocations are found only in cancer cells.

The protein produced from the normal *FGFR1* gene can turn on cellular signaling that helps the cell respond to its environment, for example by stimulating cell growth. The protein produced from the fused gene, regardless of the partner gene involved, leads to constant FGFR1 signaling. The uncontrolled signaling promotes continuous cell growth and division, leading to cancer.

Core binding factor acute myeloid leukemia

A type of blood cancer known as core binding factor acute myeloid leukemia (CBF-AML) is associated with a rearrangement (translocation) of genetic material between chromosomes 8 and 21.

This rearrangement is associated with approximately 7 percent of acute myeloid leukemia cases in adults. The translocation, written as t(8;21), fuses part of the *RUNX1T1* gene (also known as *ETO*) from chromosome 8 with part of the *RUNX1* gene from chromosome 21. This mutation is acquired during a person's lifetime and is present only in certain cells. This type of genetic change, called a somatic mutation, is not inherited.

The fusion protein produced from the t(8;21) translocation, called RUNX1-ETO, retains some function of the two individual proteins. The normal RUNX1 protein, produced from the *RUNX1* gene, is part of a protein complex called core binding factor (CBF) that attaches (binds) to DNA and turns on genes involved in blood cell development. The normal ETO protein, produced from the *RUNX1T1* gene, turns off gene activity. The fusion protein forms CBF and attaches to DNA, but instead of turning on genes that stimulate the development of blood cells, it turns those genes off. This change in gene activity blocks the maturation (differentiation) of blood cells and leads to the production of abnormal, immature white blood cells called myeloid blasts. While t(8;21) is important for leukemia development, one or more additional genetic changes are typically needed for the myeloid blasts to develop into cancerous leukemia cells.

Langer-Giedion syndrome

Langer-Giedion syndrome is caused by a deletion or mutation in several genes on the long (q) arm of chromosome 8 at a position described as 8q24.1. This condition causes bone abnormalities, including noncancerous bone tumors known as exostoses, and distinctive facial features. The signs and symptoms of this condition are related to the deletion or mutation in at least two genes from this part of the chromosome. Researchers have determined that the loss of a functional *EXT1* gene is responsible for the multiple noncancerous (benign) bone tumors called exostoses seen in people with Langer-Giedion syndrome. Loss of a functional *TRPS1* gene may cause the other bone and facial abnormalities. One copy of the *EXT1* gene and the *TRPS1* gene are always missing or mutated in affected individuals; however, neighboring genes may also be involved. The loss of additional genes from this region of chromosome 8 likely contributes to the varied features of Langer-Giedion syndrome.

Recombinant 8 syndrome

A rearrangement of chromosome 8 causes recombinant 8 syndrome, a condition that involves heart and urinary tract abnormalities, moderate to severe intellectual disability, and a distinctive facial appearance. This rearrangement results in a deletion of a piece of the short (p) arm and a duplication of a piece of the long (q) arm. This chromosome abnormality is written rec(8)dup(8q)inv(8)(p23.1q22.1).

The signs and symptoms of recombinant 8 syndrome are related to the loss of genetic material on the short arm of chromosome 8 and the presence of extra genetic material on the long arm of chromosome 8. Researchers are working to determine which genes are involved in the deletion and duplication on chromosome 8.

Other cancers

Translocations between chromosome 8 and other chromosomes have been associated with other types of cancer. For example, Burkitt lymphoma (a cancer of white blood cells called B cells that occurs most often in children and young adults) can be caused by a translocation between chromosomes 8 and 14. This translocation, written t(8;14)(q24;q32), leads to continuous cell division without control or order, which likely contributes to the development of Burkitt lymphoma. Less frequently, Burkitt lymphoma can be caused by translocations between chromosomes 8 and 2 or chromosomes 8 and 22.

Other chromosomal conditions

Trisomy 8 occurs when cells have three copies of chromosome 8 instead of the usual two copies. Full trisomy 8, which occurs when all of the body's cells contain an extra copy of chromosome 8, is not compatible with life. A similar but less severe condition called mosaic trisomy 8 occurs when only some of the body's cells have an extra copy of chromosome 8.

The signs and symptoms of mosaic trisomy 8 vary widely and can include intellectual disability, absence of the tissue connecting the left and right halves of the brain (corpus callosum), skeletal defects, heart problems, kidney and liver malformations, and facial abnormalities. Trisomy 8 mosaicism is also associated with an increased risk of acute myeloid leukemia.

Another chromosomal condition called inversion duplication 8p is caused by a rearrangement of genetic material on the short (p) arm of chromosome 8. This rearrangement results in an abnormal duplication and an inversion of a segment of the chromosome. An inversion involves the breakage of a chromosome in two places; the resulting piece of DNA is reversed and reinserted into the chromosome. People with inversion duplication 8p typically have severe intellectual disability, a thin or absent corpus callosum, weak muscle tone (hypotonia), abnormal curvature of the spine (scoliosis), and minor facial abnormalities. Some individuals with this condition may also have heart defects, underdeveloped kidneys, or eye abnormalities. Older individuals usually develop abnormal muscle stiffness (spasticity). The signs and symptoms of inversion duplication 8p tend to depend on the size and location of the chromosome segment involved. For example, inclusion of chromosome region 8p21 is thought to be associated with more severe symptoms.

Is there a standard way to diagram chromosome 8?

Geneticists use diagrams called ideograms as a standard representation for chromosomes. Ideograms show a chromosome's relative size and its banding pattern. A banding pattern is the characteristic pattern of dark and light bands that appears when a chromosome is stained with a chemical solution and then viewed under a microscope. These bands are used to describe the location of genes on each chromosome.

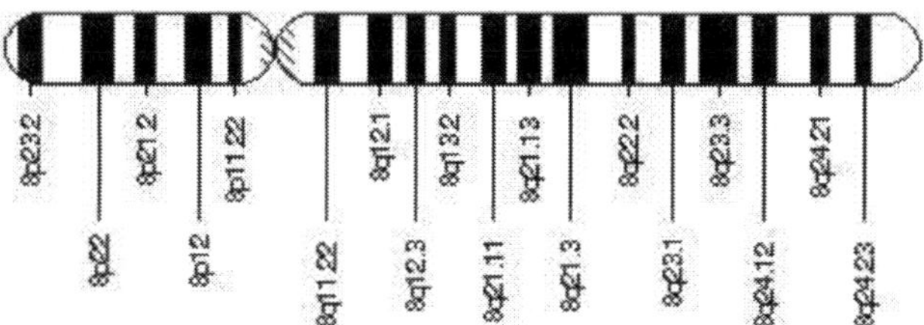

Conditions related to genes on chromosome 8

Genetics Home Reference includes these conditions related to genes on chromosome 8:

8p11 myeloproliferative syndrome
ataxia with vitamin E deficiency
autosomal dominant nocturnal frontal lobe epilepsy
Baller-Gerold syndrome
benign familial neonatal seizures
branchiootorenal syndrome
breast cancer Charcot-Marie-
Tooth disease CHARGE
syndrome
Cohen syndrome
coloboma

color vision deficiency
congenital adrenal hyperplasia due to 11-beta-hydroxylase deficiency
congenital hypothyroidism
congenital myasthenic syndrome
core binding factor acute myeloid leukemia
corticosterone methyloxidase deficiency
dihydropyrimidinase deficiency
dystonia 6
epidermolysis bullosa simplex
epidermolysis bullosa with pyloric atresia
familial hyperaldosteronism
familial idiopathic basal ganglia calcification
familial lipoprotein lipase deficiency
Farber lipogranulomatosis
Graves disease
Hashimoto thyroiditis
hereditary multiple exostoses
hereditary spherocytosis
idiopathic pulmonary
fibrosis juvenile Paget
disease Kallmann syndrome
Klippel-Feil syndrome
Langer-Giedion syndrome
late-infantile neuronal ceroid lipofuscinosis
mal de Meleda
microphthalmia
mucopolysaccharidosis type
III Nijmegen breakage
syndrome Northern epilepsy
osteoglophonic dysplasia
Paget disease of bone
Pfeiffer syndrome
piebaldism
pyruvate dehydrogenase deficiency
RAPADILINO syndrome
Roberts syndrome Rothmund-
Thomson syndrome spastic
paraplegia type 8
spinal muscular atrophy with progressive myoclonic
epilepsy surfactant dysfunction
Waardenburg syndrome

Werner syndrome

Reviewed: *November 2013*

Published: *November 17, 2014*

Genes on chromosome 8

Genetics Home Reference includes these genes on chromosome 8:

ANK1
ASAH1
CHD7
CHRNA2
CLN8
CNGB3
CYP11B1
CYP11B2
DPYS
ESCO2
EXT1
EYA1
FGFR1
GDAP1
GDF6
HGSNAT
KCNQ3
KIAA0196
LPL
NBN
NDRG1
NEFL
NEFM
PDP1
PLEC
RECQL4
RUNX1T1
SFTPC
SLC20A2
SLURP1
SNAI2 TG
THAP1

TNFRSF11B

TRPS1
TTPA
VPS13B
WRN

State of Oregon
Chromosome Nine

Oregon Health and Science University School of Medicine (Portland) (Galactosemia)

College of Osteopathic Medicine of the Pacific Northwest (Lebanon) (Polycythemia Vera)

Chromosome 9

What is chromosome 9?

Humans normally have 46 chromosomes in each cell, divided into 23 pairs. Two copies of chromosome 9, one copy inherited from each parent, form one of the pairs. Chromosome 9 is made up of about 141 million DNA building blocks (base pairs) and represents approximately 4.5 percent of the total DNA in cells.

Identifying genes on each chromosome is an active area of genetic research. Because researchers use different approaches to predict the number of genes on each chromosome, the estimated number of genes varies. Chromosome 9 likely contains 800 to 900 genes that provide instructions for making proteins. These proteins perform a variety of different roles in the body.

Genes on chromosome 9 are among the estimated 20,000 to 25,000 total genes in the human genome.

How are changes in chromosome 9 related to health conditions?

Many genetic conditions are related to changes in particular genes on chromosome 9. This list of disorders associated with genes on chromosome 9 provides links to additional information.

Changes in the structure or number of copies of a chromosome can also cause problems with health and development. The following chromosomal conditions are associated with such changes in chromosome 9.

9q22.3 microdeletion

9q22.3 microdeletion is a chromosomal change in which a small piece of the long (q) arm of chromosome 9 is deleted in each cell. Affected individuals are missing at least 352,000 base pairs, also written as 352 kilobases (kb), in the q22.3 region of chromosome 9. This 352-kb segment is known as the minimum critical region because it is the smallest deletion that has been found to cause the signs and symptoms related to 9q22.3 microdeletions. These signs and symptoms include delayed development, intellectual disability, certain physical abnormalities, and the characteristic features of a genetic condition called Gorlin syndrome (also known as nevoid basal cell carcinoma syndrome). 9q22.3 microdeletions can also be much larger; the largest reported deletion included 20.5 million base pairs (20.5 Mb).

People with a 9q22.3 microdeletion are missing from two to more than 270 genes on chromosome 9. All known 9q22.3 microdeletions include the *PTCH1* gene. Researchers believe that many of the features associated with 9q22.3 microdeletions, particularly the signs and symptoms of Gorlin syndrome, result from a loss of the *PTCH1* gene. Other signs and symptoms related to 9q22.3 microdeletions probably result from the loss of additional genes in the q22.3 region. Researchers are working to determine which missing genes contribute to the other features associated with the deletion.

Bladder cancer

Deletions of part or all of chromosome 9 are commonly found in bladder cancers. These chromosomal changes are seen only in cancer cells and typically occur early in tumor formation. Researchers believe that several genes that play a role in bladder cancer may be located on chromosome 9. They suspect that these genes may be tumor suppressors, which means they normally help prevent cells from growing and dividing in an uncontrolled way. Researchers are working to determine which genes, when altered or missing, are involved in the development and progression of bladder tumors.

Kleefstra syndrome

Most people with Kleefstra syndrome, a disorder with signs and symptoms involving many parts of the body, are missing a sequence of about 1 million DNA building blocks (base pairs) on one copy of chromosome 9 in each cell. The deletion occurs near the end of the long (q) arm of the chromosome at a location designated q34.3, a region containing a gene called *EHMT1*. Some affected individuals have shorter or longer deletions in the same region.

The loss of the *EHMT1* gene from one copy of chromosome 9 in each cell is believed to be responsible for the characteristic features of Kleefstra syndrome in people with the 9q34.3 deletion. However, the loss of other genes in the same region may lead to additional health problems in some affected individuals.

The *EHMT1* gene provides instructions for making an enzyme called euchromatic histone methyltransferase 1. Histone methyltransferases are enzymes that modify proteins called histones. Histones are structural proteins that attach (bind) to DNA and give chromosomes their shape. By adding a molecule called a methyl group to histones, histone methyltransferases can turn off (suppress) the activity of certain genes, which is essential for normal development and function. A lack of euchromatic histone methyltransferase 1 enzyme impairs proper control of the activity of certain genes in many of the body's organs and tissues, resulting in the abnormalities of development and function characteristic of Kleefstra syndrome.

Other cancers

Changes in the structure of chromosome 9 have been found in many types of cancer. These changes, which occur only in cancer cells, usually involve a loss of part of the chromosome or a rearrangement of chromosomal material. For example, a loss of part of the long (q) arm of chromosome 9 has been identified in some types of brain tumor. It is unclear how these chromosomal changes are related to the development and growth of cancers.

A rearrangement (translocation) of genetic material between chromosomes 9 and 22 is associated with several types of blood cancer known as leukemias. This chromosomal abnormality, which is commonly called the Philadelphia chromosome, is found only in cancer cells.

It fuses part of a specific gene from chromosome 22 (the *BCR* gene) with part of another gene from chromosome 9 (the *ABL1* gene). The protein produced from these fused genes signals tumor cells to continue dividing abnormally and prevents them from adequately repairing DNA damage.

The Philadelphia chromosome has been identified in most cases of a slowly progressing form of blood cancer called chronic myeloid leukemia (CML). It also has been found in some cases of more rapidly progressing blood cancers known as acute leukemias. The presence of the Philadelphia chromosome can help predict how a cancer will progress and provides a target for molecular therapies.

Other chromosomal conditions

Other changes in the structure or number of copies of chromosome 9 can have a variety of effects. Intellectual disability, delayed development, distinctive facial features, and an unusual head shape are common features. Changes to chromosome 9 include an extra piece of the chromosome in each cell (partial trisomy), a missing segment of the chromosome in each cell (partial monosomy), and a circular structure called a ring chromosome 9. A ring chromosome occurs when both ends of a broken chromosome are reunited. Rearrangements (translocations) of genetic material between chromosome 9 and other chromosomes can also lead to extra or missing chromosome segments.

Is there a standard way to diagram chromosome 9?

Geneticists use diagrams called ideograms as a standard representation for chromosomes. Ideograms show a chromosome's relative size and its banding pattern. A banding pattern is the characteristic pattern of dark and light bands that appears when a chromosome is stained with a chemical solution and then viewed under a microscope. These bands are used to describe the location of genes on each chromosome.

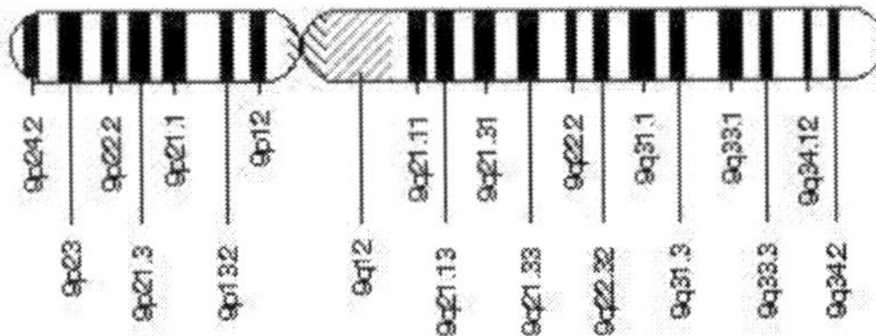

See How do geneticists indicate the location of a gene? in the Handbook.

Conditions related to genes on chromosome 9

Genetics Home Reference includes these conditions related to genes on chromosome 9:

3-methylglutaconyl-CoA hydratase deficiency
9q22.3 microdeletion
17-beta hydroxysteroid dehydrogenase 3 deficiency
amyotrophic lateral sclerosis
ataxia with oculomotor apraxia
autosomal dominant nocturnal frontal lobe epilepsy
benign essential blepharospasm
Berardinelli-Seip congenital
lipodystrophy bladder cancer

cap myopathy cartilage-
hair hypoplasia chorea-
acanthocytosis
citrullinemia
cytochrome c oxidase deficiency
distal arthrogryposis type 1
dopamine beta-hydroxylase deficiency
early-onset primary dystonia Ehlers-
Danlos syndrome
essential thrombocythemia
familial dysautonomia
familial HDL deficiency
familial thoracic aortic aneurysm and dissection
Fanconi anemia
Friedreich ataxia
Fukuyama congenital muscular dystrophy
galactosemia
geleophysic dysplasia
glycine encephalopathy
Gorlin syndrome
hereditary fructose intolerance
hereditary hemorrhagic telangiectasia
hereditary sensory neuropathy type 1
inclusion body myopathy 2
inclusion body myopathy with early-onset Paget disease and
frontotemporal dementia
Kleefstra syndrome
lattice corneal dystrophy type II
Leigh syndrome
limb-girdle muscular dystrophy
Loeys-Dietz syndrome
lymphangioleiomyomatosis
Mabry syndrome
malignant migrating partial seizures of infancy
Manitoba oculotrichoanal syndrome
multiple cutaneous and mucosal venous malformations
nail-patella syndrome
nemaline myopathy
nonsyndromic deafness
nonsyndromic holoprosencephaly
oculocutaneous albinism
polycythemia vera

porphyria
primary ciliary dyskinesia
primary hyperoxaluria
primary myelofibrosis
Robinow syndrome
sialuria
Swyer syndrome
Tangier disease
thrombotic thrombocytopenic purpura
tuberous sclerosis complex VLDLR-
associated cerebellar hypoplasia Walker-
Warburg syndrome
xeroderma pigmentosum

Genes on chromosome 9

Genetics Home Reference includes these genes on chromosome 9:

ABCA1
ADAMTS13
ADAMTSL2
AGPAT2
ALAD
ALDOB
APTX
ASS1
AUH
C9orf72
COL5A1
DBH
DFNB31
DNAI1
EHMT1
ENG
FANCC
FANCG
FKTN
FREM1
FXN
GALT
GLDC
GNE
GRHPR

GSN
HSD17B3
IKBKAP
JAK2
KCNT1
LMX1B
NR5A1
PIGO
POMT1
PTCH1
RMRP
ROR2
SETX
SPTLC1
SURF1
TEK
TGFBR1
TMC1
TOR1A
TPM2
TSC1
TYRP1
VCP
VLDLR
VPS13A
XPA

State of Pennsylvania
Chromosome Ten

Pennsylvania State University College of Medicine (Hershey) (Glioblastoma Multiform)

Dexel University College of Medicine (Philadelphia) (Multiple Endocrine Neoplasia Type 2)

The Commonwealth Medical College (Scranton) (Hirschprung Disease)

Lake Erie College of Osteopathic Medicine (Erie) (Non Syndromic Deafness)

Perelman School of Medicine at the Univerisity of Pennslyvania (Philadelphia) (Usher Syndrome)

Philadelphia College of Osteopathic Medicine (Philadelphia) (Multiple Endocrine Neoplasia)

Reviewed August 2007

What is chromosome 10?

Humans normally have 46 chromosomes in each cell, divided into 23 pairs. Two copies of chromosome 10, one copy inherited from each parent, form one of the pairs. Chromosome 10 spans more than 135 million DNA building blocks (base pairs) and represents between 4 and 4.5 percent of the total DNA in cells.

Identifying genes on each chromosome is an active area of genetic research. Because researchers use different approaches to predict the number of genes on each chromosome, the estimated number of genes varies. Chromosome 10 likely contains 700 to 800 genes that provide instructions for making proteins. These proteins perform a variety of different roles in the body.

Genes on chromosome 10 are among the estimated 20,000 to 25,000 total genes in the human genome.

How are changes in chromosome 10 related to health conditions?

Many genetic conditions are related to changes in particular genes on chromosome 10. This list of disorders associated with genes on chromosome 10 provides links to additional information.

Changes in the structure or number of copies of a chromosome can also cause problems with health and development. The following chromosomal conditions are associated with such changes in chromosome 10.

cancers

Changes in the number and structure of chromosome 10 are associated with several types of cancer. For example, a loss of all or part of chromosome 10 is often found in brain tumors called gliomas, particularly in aggressive, fast-growing gliomas. The association of cancerous tumors with a loss of chromosome 10 suggests that some genes on this chromosome play critical roles in controlling the growth and division of cells. Without these genes, cells could grow and divide too quickly or in an uncontrolled way, resulting in cancer. Researchers are working to identify the specific genes on chromosome 10 that may be involved in the development and progression of gliomas.

A complex rearrangement (translocation) of genetic material between chromosomes 10 and 11 is associated with several types of blood cancer known as leukemias. This chromosomal abnormality is found only in cancer cells. It fuses part of a specific gene from chromosome 11 (the *MLL* gene) with part of another gene from chromosome 10 (the *MLLT10* gene). The abnormal protein produced from this fused gene signals cells to divide without control or order, leading to the development of cancer.

Crohn disease

Variations in a particular region of chromosome 10 have been associated with the risk of developing Crohn disease. These genetic changes are located on the long (q) arm of the chromosome at a position designated 10q21.1. Researchers refer to this part of chromosome 10 as a "gene desert" because it contains no known genes. However, it may contain stretches of DNA that help regulate nearby genes such as *ERG2*. This gene has a potential role in immune system function, and researchers are interested in studying the gene further to determine whether it is associated with Crohn disease risk.

other chromosomal conditions

Other changes in the number or structure of chromosome 10 can have a variety of effects. Intellectual disability, delayed growth and development, distinctive facial features, and heart defects are common features. Changes to chromosome 10 include an extra piece of the chromosome in each cell (partial trisomy), a missing segment of the chromosome in each cell (partial monosomy), and an abnormal structure called a ring chromosome 10. Ring chromosomes occur when a chromosome breaks in two places and the ends of the chromosome arms fuse together to form a circular structure. Rearrangements (translocations) of genetic material between chromosomes can also lead to extra or missing material from chromosome 10.

Is there a standard way to diagram chromosome 10?

Geneticists use diagrams called ideograms as a standard representation for chromosomes. Ideograms show a chromosome's relative size and its banding pattern. A banding pattern is the characteristic pattern of dark and light bands that appears when a chromosome is stained with a chemical solution and then viewed under a microscope. These bands are used to describe the location of genes on each chromosome.

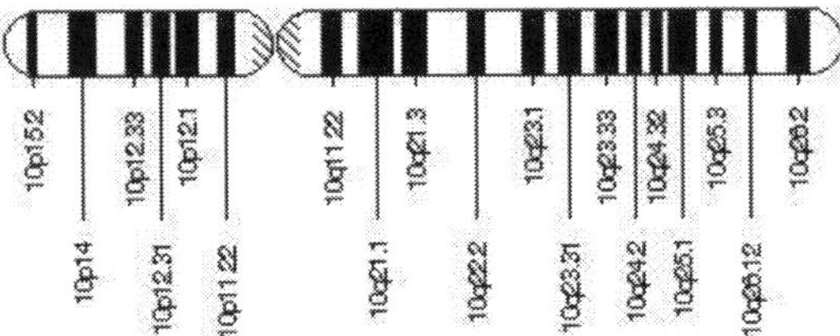

Chromosome 10

Reviewed August 2007

What is chromosome 10?

Humans normally have 46 chromosomes in each cell, divided into 23 pairs. Two copies of chromosome 10, one copy inherited from each parent, form one of the pairs. Chromosome 10 spans more than 135 million DNA building blocks (base pairs) and represents between 4 and 4.5 percent of the total DNA in cells.

Identifying genes on each chromosome is an active area of genetic research. Because researchers use different approaches to predict the number of genes on each chromosome, the estimated number of genes varies. Chromosome 10 likely contains 700 to 800 genes that provide instructions for making proteins. These proteins perform a variety of different roles in the body.

Genes on chromosome 10 are among the estimated 20,000 to 25,000 total genes in the human genome.

How are changes in chromosome 10 related to health conditions?

Many genetic conditions are related to changes in particular genes on chromosome 10. This list of disorders associated with genes on chromosome 10 provides links to additional information.

Changes in the structure or number of copies of a chromosome can also cause problems with health and development. The following chromosomal conditions are associated with such changes in chromosome 10.

cancers

Changes in the number and structure of chromosome 10 are associated with several types of cancer. For example, a loss of all or part of chromosome 10 is often found in brain tumors called gliomas, particularly in aggressive, fast-growing gliomas. The association of cancerous tumors with a loss of chromosome 10 suggests that some genes on this chromosome play critical roles in controlling the growth and division of cells. Without these genes, cells could grow and divide too quickly or in an uncontrolled way, resulting in cancer. Researchers are working to identify the specific genes on chromosome 10 that may be involved in the development and progression of gliomas.

A complex rearrangement (translocation) of genetic material between chromosomes 10 and 11 is associated with several types of blood cancer known as leukemias. This chromosomal abnormality is found only in cancer cells. It fuses part of a specific gene from chromosome 11 (the *MLL* gene) with part of another gene from chromosome 10 (the *MLLT10* gene). The abnormal protein produced from this fused gene signals cells to divide without control or order, leading to the development of cancer.

Crohn disease

Variations in a particular region of chromosome 10 have been associated with the risk of developing Crohn disease. These genetic changes are located on the long (q) arm of the chromosome at a position designated 10q21.1. Researchers refer to this part of chromosome 10 as a "gene desert" because it contains no known genes. However, it may contain stretches of DNA that help regulate nearby genes such as *ERG2*. This gene has a potential role in immune system function, and researchers are interested in studying the gene further to determine whether it is associated with Crohn disease risk.

other chromosomal conditions

Other changes in the number or structure of chromosome 10 can have a variety of effects. Intellectual disability, delayed growth and development, distinctive facial features, and heart defects are common features. Changes to chromosome 10 include an extra piece of the chromosome in each cell (partial trisomy), a missing segment of the chromosome in each cell (partial monosomy), and an abnormal structure called a ring chromosome 10. Ring chromosomes occur when a chromosome breaks in two places and the ends of the chromosome arms fuse together to form a circular structure. Rearrangements (translocations) of genetic material between chromosomes can also lead to extra or missing material from chromosome 10.

Is there a standard way to diagram chromosome 10?

Geneticists use diagrams called ideograms as a standard representation for chromosomes. Ideograms show a chromosome's relative size and its banding pattern. A banding pattern is the characteristic pattern of dark and light bands that appears when a chromosome is stained with a chemical solution and then viewed under a microscope. These bands are used to describe the location of genes on each chromosome.

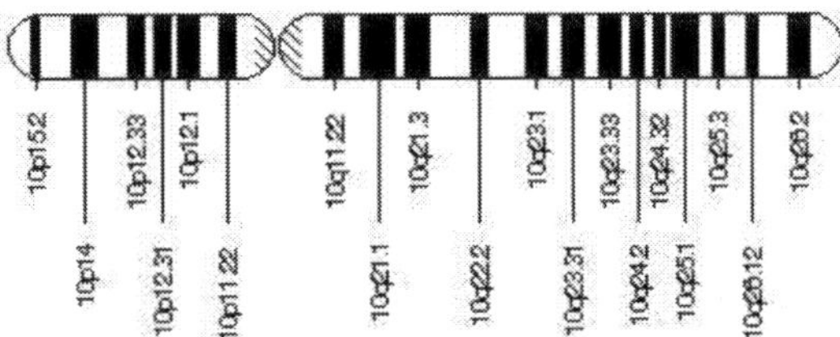

Conditions related to genes on chromosome 10

Genetics Home Reference includes these conditions related to genes on chromosome 10:

2-methylbutyryl-CoA dehydrogenase
deficiency age-related macular degeneration
Apert syndrome

ataxia neuropathy spectrum
autoimmune lymphoproliferative syndrome
autosomal dominant partial epilepsy with auditory features
Bannayan-Riley-Ruvalcaba syndrome
Beare-Stevenson cutis gyrata syndrome
cerebral autosomal recessive arteriopathy with subcortical infarcts and leukoencephalopathy
Charcot-Marie-Tooth disease
cholesteryl ester storage disease
CHST3-related skeletal dysplasia
Cockayne syndrome
congenital myasthenic syndrome
Cornelia de Lange syndrome
Cowden syndrome
Crouzon syndrome Diamond-
Blackfan anemia Dubin-
Johnson syndrome familial
dilated cardiomyopathy
familial hemophagocytic lymphohistiocytosis
familial thoracic aortic aneurysm and dissection
genitopatellar syndrome
gyrate atrophy of the choroid and retina
Hermansky-Pudlak syndrome
Hirschsprung disease
hypermethioninemia Imerslund-
Gräsbeck syndrome infantile-onset
spinocerebellar ataxia Jackson-Weiss
syndrome
junctional epidermolysis bullosa
juvenile polyposis syndrome lacrimo-
auriculo-dento-digital syndrome
mannose-binding lectin deficiency
metachromatic leukodystrophy multiple
endocrine neoplasia myofibrillar
myopathy
nonsyndromic deafness
nonsyndromic paraganglioma
Ochoa syndrome
Ohdo syndrome, Say-Barber-Biesecker-Young-Simpson variant
Pfeiffer syndrome
Pol III-related leukodystrophy

porphyria
progressive external ophthalmoplegia
Refsum disease
renal coloboma syndrome
tetrahydrobiopterin deficiency
Usher syndrome UV-sensitive
syndrome Wolman disease

Reviewed: *August 2007*

Published: *November 17, 2014*

Genes on chromosome 10

Genetics Home Reference includes these genes on chromosome 10:

ABCC2
ACADSB
ACTA2
ARMS2
BMPR1A
C10orf2
CDH23
CHAT
CHST3
COL17A1
CUBN
EGR2
ERCC6
FAS
FGFR2
HPS1
HPSE2
HTRA1
KAT6B
KLLN
LDB3
LGI1
LIPA
MAT1A
MBL2

OAT
PAX2
PCBD1
PCDH15
PHYH
POLR3A
PRF1
PSAP
PTEN
RET
RPS24
SMC3
UROS

State of Rhode Island
Chromosome Eleven

Rhode Island Alpert Medical School at Brown University (Providence) (Autism Neurexin 1)

Chromosome 11

Reviewed October 2012

What is chromosome 11?

Humans normally have 46 chromosomes in each cell, divided into 23 pairs. Two copies of chromosome 11, one copy inherited from each parent, form one of the pairs. Chromosome 11 spans about 135 million DNA building blocks (base pairs) and represents between 4 and 4.5 percent of the total DNA in cells.

Identifying genes on each chromosome is an active area of genetic research. Because researchers use different approaches to predict the number of genes on each chromosome, the estimated number of genes varies. Chromosome 11 likely contains 1,300 to 1,400 genes that provide instructions for making proteins. These proteins perform a variety of different roles in the body.

Genes on chromosome 11 are among the estimated 20,000 to 25,000 total genes in the human genome.

How are changes in chromosome 11 related to health conditions?

Many genetic conditions are related to changes in particular genes on chromosome 11. This list of disorders associated with genes on chromosome 11 provides links to additional information.

Changes in the structure or number of copies of a chromosome can also cause problems with health and development. The following chromosomal conditions are associated with such changes in chromosome 11.

Beckwith-Wiedemann syndrome

Beckwith-Wiedemann syndrome results from the abnormal regulation of genes on part of the short (p) arm of chromosome 11. The genes are located close together in a region designated 11p15.5 near one end of the chromosome.

People normally inherit one copy of chromosome 11 from each parent. For most genes on this chromosome, both copies of the gene are expressed, or "turned on," in cells. For some genes in the 11p15.5 region, however, only the copy inherited from a person's father (the paternal copy) is expressed. For other genes, only the copy inherited from a person's mother (the maternal copy) is expressed. These parent-specific differences in gene expression are caused by a phenomenon called genomic imprinting. Researchers have determined that changes in genomic imprinting disrupt the regulation of several genes located at 11p15.5, including *CDKN1C*, *H19*, *IGF2*, and *KCNQ1OT1* . Because these genes are involved in directing normal growth, problems with their regulation lead to overgrowth and the other characteristic features of Beckwith-Wiedemann syndrome.

Ten percent to 20 percent of cases of Beckwith-Wiedemann syndrome are caused by a genetic change known as paternal uniparental disomy (UPD). Paternal UPD causes people to have two active copies of paternally expressed imprinted genes rather than one active copy from the father and one inactive copy from the mother. People with paternal UPD are also missing genes that are active only on the maternal copy of the chromosome. In Beckwith-Wiedemann syndrome, paternal UPD usually occurs early in embryonic development and affects only some of the body's cells. This phenomenon is called mosaicism. Mosaic paternal UPD leads to an imbalance in active paternal and maternal genes on chromosome 11, which underlies the signs and symptoms of the disorder.

About 1 percent of all people with Beckwith-Wiedemann syndrome have a chromosomal abnormality such as a rearrangement (translocation) involving 11p15.5 or abnormal copying (duplication) of genetic material in this region. Like the other genetic changes responsible for Beckwith-Wiedemann syndrome, these changes disrupt the normal regulation of genes in this part of chromosome 11.

Emanuel syndrome

Emanuel syndrome is caused by the presence of extra genetic material from chromosome 11 and chromosome 22 in each cell. In addition to the usual 46 chromosomes, people with Emanuel syndrome have an extra (supernumerary) chromosome consisting of a piece of chromosome 22 attached to a piece of chromosome 11. The extra chromosome is known as a derivative 22 or der(22) chromosome.

People with Emanuel syndrome typically inherit the der(22) chromosome from an unaffected parent. The parent carries a chromosomal rearrangement between chromosomes 11 and 22 called a balanced translocation. No genetic material is gained or lost in a balanced translocation, so these chromosomal changes usually do not cause any health problems. As the translocation is passed to the next generation, it can become unbalanced. Individuals with Emanuel syndrome inherit an unbalanced translocation between chromosomes 11 and 22 in the form of a der(22) chromosome. These individuals have two normal copies of chromosome 11, two normal copies of chromosome 22, and extra genetic material from the der(22) chromosome.

As a result of the extra chromosome, people with Emanuel syndrome have three copies of some genes in each cell instead of the usual two copies. The excess genetic material disrupts the normal course of development, leading to intellectual disability and birth defects. Researchers are working to determine which genes are included on the der(22) chromosome and what role these genes play in development.

Ewing sarcoma

A translocation involving chromosome 11 can cause a type of cancerous tumor known as Ewing sarcoma. These tumors develop in bones or soft tissues, such as nerves and cartilage. This translocation, t(11;22), fuses part of the *EWSR1* gene from chromosome 22 with part of the *FLI1* gene from chromosome 11, creating the *EWSR1/FLI1* fusion gene. This mutation is acquired during a person's lifetime and is present only in tumor cells. This type of genetic change, called a somatic mutation, is not inherited.

The protein produced from the *EWSR1/FLI1* fusion gene, called EWS/FLI, has functions of the protein products of both genes. The FLI protein, produced from the *FLI1* gene, attaches (binds) to DNA and regulates an activity called transcription, which is the first step in the production of proteins from genes. The FLI protein controls the growth and development of some cell types by regulating the transcription of certain genes. The EWS protein, produced from the *EWSR1* gene, also regulates transcription. The EWS/FLI protein has the DNA-binding function of the FLI protein as well as the transcription regulation function of the EWS protein. It is thought that the EWS/FLI protein turns the transcription of a variety of genes on and off abnormally. This dysregulation of transcription leads to uncontrolled growth and division (proliferation) and abnormal maturation and survival of cells, causing tumor development.

Jacobsen syndrome

Jacobsen syndrome, which is also known as 11q terminal deletion disorder, is caused by a deletion of genetic material at the end (terminus) of the long (q) arm of chromosome 11. The size of the deletion varies among affected individuals, with most affected people missing from about 5 million to 16 million DNA building blocks (also written as 5 Mb to 16 Mb). In almost all affected people, the deletion includes the tip of chromosome 11. Larger deletions tend to cause more severe signs and symptoms than smaller deletions.

The features of Jacobsen syndrome are likely related to the loss of multiple genes on chromosome 11. Depending on its size, the deleted region can contain from about 170 to more than 340 genes. Many of these genes have not been well characterized. However, genes in this region appear to be critical for the normal development of many parts of the body, including the brain, facial features, and heart. Researchers are working to determine which genes contribute to the specific features of Jacobsen syndrome.

neuroblastoma

About 35 percent of people with neuroblastoma have a deletion of genetic material on the long (q) arm of chromosome 11 at a position designated 11q23. Neuroblastoma is a type of cancerous tumor composed of immature nerve cells (neuroblasts). The 11q23 deletion can occur in the body's cells after conception, which is called a somatic mutation, or it can be inherited from a parent. This deletion is associated with a more severe form of neuroblastoma. Researchers believe the deleted region could contain a gene that keeps cells from growing and dividing too quickly or in an uncontrolled way, called a tumor suppressor gene. When tumor suppressor genes are deleted, cancer can occur. However, no tumor suppressor genes have been identified in the deleted region of chromosome 11. It is unknown how deletion of this region contributes to the formation or progression of neuroblastoma.

Potocki-Shaffer syndrome

A condition called Potocki-Shaffer syndrome is caused by the deletion of a segment of the short (p) arm of chromosome 11 at a position described as 11p11.2. This condition is also known as proximal 11p deletion syndrome. The characteristic features of Potocki-Shaffer syndrome include enlarged openings in the two bones that make up much of the top and sides of the skull (enlarged parietal foramina), multiple noncancerous bone tumors called exostoses, intellectual disability, delayed development, a distinctive facial appearance, and problems with vision. Occasionally, people with this condition have defects in the heart, kidneys, and urinary tract. The features of Potocki-Shaffer syndrome result from the loss of several genes on the short arm of chromosome 11. In particular, the deletion of a gene called *ALX4* causes enlarged parietal foramina in people with this condition, while the loss of another gene, *EXT2*, underlies the multiple exostoses. Researchers are working to find genes on the short arm of chromosome 11 that are associated with the other features of Potocki-Shaffer syndrome.

Another condition called WAGR syndrome (described below) is caused by a deletion of genetic material from the short arm of chromosome 11 at a position described as 11p13. Occasionally, a deletion is large enough to include the 11p11.2 and 11p13 regions. Individuals with such a deletion have signs and symptoms of both Potocki-Shaffer syndrome and WAGR syndrome.

Russell-Silver syndrome

Like Beckwith-Wiedemann syndrome, Russell-Silver syndrome can result from changes in genes in the 11p15.5 region. Specifically, Russell-Silver syndrome has been associated with changes in genomic imprinting that affect the regulation of the *H19* and *IGF2* genes on chromosome 11. The changes are different from those seen in Beckwith-Wiedemann syndrome and have the opposite effect on growth. Although both disorders can be caused by abnormal regulation of these genes, the changes that cause Russell-Silver syndrome lead to slow growth and short stature instead of overgrowth.

Wilms tumor, aniridia, genitourinary anomalies, and mental retardation syndrome

Wilms tumor, aniridia, genitourinary anomalies, and mental retardation syndrome, more commonly known by the acronym WAGR syndrome, is caused by a deletion of genetic material on the short (p) arm of chromosome 11 at a position described as 11p13. The signs and symptoms of WAGR syndrome are related to the loss of multiple genes from this part of the chromosome. The size of the deletion varies among affected individuals. Researchers have identified genes on the short arm of chromosome 11 that are associated with particular features of WAGR syndrome. A loss of the *PAX6* gene disrupts normal eye development, leading to aniridia and other eye problems, and may also affect the development of the brain. Deletion of the *WT1* gene is responsible for the genitourinary abnormalities and the increased risk of Wilms tumor (a rare form of kidney cancer) in affected individuals. Researchers are working to identify additional genes deleted in people with WAGR syndrome and determine how their loss leads to the other features of the disorder.

other cancers

Changes in chromosome 11 have been identified in other types of cancer. These chromosomal changes are somatic, which means they are acquired during a person's lifetime and are present only in certain cells. In some cases, translocations of genetic material between chromosome 11 and other chromosomes have been associated with cancers of blood-forming cells (leukemias) and cancers of immune system cells (lymphomas).

other chromosomal conditions

Other changes in the number or structure of chromosome 11 can have a variety of effects, including intellectual disability, delayed development, slow growth, distinctive facial features, and weak muscle tone (hypotonia). Changes involving chromosome 11 include an extra piece of the chromosome in each cell (partial trisomy 11), a missing segment of the chromosome in each cell (partial monosomy 11), and a circular structure called a ring chromosome 11. Ring chromosomes occur when a chromosome breaks in two places and the ends of the chromosome arms fuse together to form a circular structure.

Is there a standard way to diagram chromosome 11?

Geneticists use diagrams called ideograms as a standard representation for chromosomes. Ideograms show a chromosome's relative size and its banding pattern. A banding pattern is the characteristic pattern of dark and light bands that appears when a chromosome is stained with a chemical solution and then viewed under a microscope. These bands are used to describe the location of genes on each chromosome.

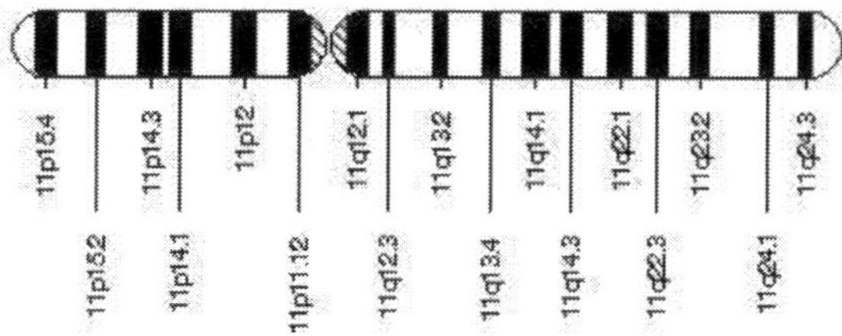

Conditions related to genes on chromosome 11

Genetics Home Reference includes these conditions related to genes on chromosome 11:

acatalasemia
age-related macular degeneration
Aicardi-Goutieres syndrome
amelogenesis imperfecta aniridia

ataxia-telangiectasia
Bardet-Biedl syndrome
Bartter syndrome
Beckwith-Wiedemann syndrome
Berardinelli-Seip congenital
lipodystrophy beta-ketothiolase deficiency
beta thalassemia
bladder cancer
breast cancer
carnitine palmitoyltransferase I deficiency
CATSPER1-related nonsyndromic male infertility
cerebral folate transport deficiency Charcot-
Marie-Tooth disease
cold-induced sweating
syndrome coloboma
congenital deafness with labyrinthine aplasia, microtia, and microdontia
congenital fibrosis of the extraocular
muscles congenital hyperinsulinism
congenital myasthenic syndrome
congenital neuronal ceroid
lipofuscinosis Costello syndrome
Cowden syndrome
cutis laxa
cytogenetically normal acute myeloid leukemia
Denys-Drash syndrome
distal hereditary motor neuropathy, type V
dopa-responsive dystonia
enlarged parietal
foramina Ewing sarcoma
familial atrial fibrillation familial
dilated cardiomyopathy
familial exudative vitreoretinopathy

familial HDL deficiency
familial hyperaldosteronism
familial hypertrophic cardiomyopathy
familial isolated hyperparathyroidism
familial isolated pituitary adenoma
familial Mediterranean fever
Frasier syndrome frontonasal
dysplasia gastrointestinal
stromal tumor Gillespie
syndrome
glycogen storage disease type I
glycogen storage disease type V
hereditary angioedema
hereditary multiple exostoses
hereditary paraganglioma-pheochromocytoma
horizontal gaze palsy with progressive scoliosis
intrauterine growth restriction, metaphyseal dysplasia, adrenal
    hypoplasia congenita, and genital anomalies
isobutyryl-CoA dehydrogenase deficiency
Jervell and Lange-Nielsen syndrome
juvenile Batten disease
juvenile primary osteoporosis
Kufs disease
lactate dehydrogenase deficiency late-
infantile neuronal ceroid lipofuscinosis
limb-girdle muscular dystrophy
Mabry syndrome
methemoglobinemia, beta-globin type
microphthalmia
Miyoshi myopathy
multiple endocrine neoplasia
multiple pterygium syndrome
neutral lipid storage disease with myopathy
Niemann-Pick disease
nonsyndromic deafness
nonsyndromic paraganglioma
oculocutaneous albinism
osteopetrosis osteoporosis-
pseudoglioma syndrome permanent
neonatal diabetes mellitus Peters
anomaly
porphyria

Potocki-Shaffer syndrome
prothrombin deficiency
prothrombin thrombophilia
pyruvate carboxylase deficiency
pyruvate dehydrogenase deficiency
retinitis pigmentosa Romano-Ward
syndrome Russell-Silver syndrome

short QT syndrome
sickle cell disease
Silver syndrome
Smith-Lemli-Opitz syndrome
spinal muscular *atrophy* with respiratory distress type 1
Stormorken syndrome
tetrahydrobiopterin deficiency
tubular aggregate myopathy
type 1 diabetes
tyrosine hydroxylase deficiency
Usher syndrome
vitelliform macular dystrophy
Wilms tumor, aniridia, genitourinary anomalies, and mental retardation syndrome

Reviewed: October 2012

Published: November 17, 2014

Genes on chromosome 11

Genetics Home Reference includes these genes on chromosome 11:

ABCC8
ACAD8
ACAT1
AIP
ALX4
ANO5
APOA1
ATM
BBS1
BDNF
BEST1

BSCL2
CAT
CATSPER1
CDKN1C
CLCF1
CPT1A
CTSD
CTSF
DHCR7
DLAT
EFEMP2
EXT2
F2
FGF3
FLI1
FOLR1
FZD4
H19
HBB
HMBS
HRAS
IGF2
IGHMBP2
INS
KCNJ1
KCNJ5
KCNJ11
KCNQ1
KCNQ1OT1
LDHA
LRP5
MEN1
MMP20
MTMR2
MYBPC3
MYO7A
PAX6
PC
PDHX
PGAP2
PHOX2A
PNPLA2

PTS
PYGM
RAPSN
RNASEH2C
ROBO3
SAA1
SBF2
SDHAF2
SDHD
SERPING1
SLC37A4
SMPD1
STIM1
TCIRG1
TECTA
TH
TPP1
TYR
USH1C
WT1

State of South Carolina
Chromosome Twelve

Medical University of South Carolina College of Medicine (Charleston) (Achondrogenesis Type 11)

University of South Carolina School of Medicine (Columbia Greenville) (Addison Disease)

Edward Via College of Osteopathic Medicine Carolinas Campus (Spartanburg) (Spondyloepimetaphyseal Dysplasia Strudwick Type)

Chromosome 12

Reviewed February 2013

What is chromosome 12?

Humans normally have 46 chromosomes in each cell, divided into 23 pairs. Two copies of chromosome 12, one copy inherited from each parent, form one of the pairs. Chromosome 12 spans almost 134 million DNA building blocks (base pairs) and represents between 4 and 4.5 percent of the total DNA in cells.

Identifying genes on each chromosome is an active area of genetic research. Because researchers use different approaches to predict the number of genes on each chromosome, the estimated number of genes varies. Chromosome 12 likely contains 1,100 to 1,200 genes that provide instructions for making proteins. These proteins perform a variety of different roles in the body.

Genes on chromosome 12 are among the estimated 20,000 to 25,000 total genes in the human genome.

How are changes in chromosome 12 related to health conditions?

Many genetic conditions are related to changes in particular genes on chromosome 12. This list of disorders associated with genes on chromosome 12 provides links to additional information.

Changes in the structure or number of copies of a chromosome can also cause problems with health and development. The following chromosomal conditions are associated with such changes in chromosome 12.

cancers

Changes in chromosome 12 have been identified in several types of cancer. These genetic changes are somatic, which means they are acquired during a person's lifetime and are present only in certain cells. For example, rearrangements (translocations) of genetic material between chromosome 12 and other chromosomes are often found in certain cancers of blood-forming cells (leukemias) and cancers of immune system cells (lymphomas). Additionally, somatic mutations may lead to an extra copy of chromosome 12 (trisomy 12) in cancer cells, specifically a type of leukemia called chronic lymphocytic leukemia.

Translocations involving chromosome 12 have also been found in solid tumors such as lipomas and liposarcomas, which are made up of fatty tissue. In these tumors, the most common chromosome 12 rearrangements involve the long (q) arm in a region designated q13-q15. Abnormalities of chromosome 12 have been identified in at least two other rare tumors, angiomatoid fibrous histiocytomas and clear cell sarcomas. Angiomatoid fibrous histiocytomas occur primarily in adolescents and young adults and are usually found in the arms and legs (extremities) . Clear cell sarcomas occur most often in young adults and tend to be associated with tendons and related structures called aponeuroses.

Researchers are working to determine which genes on chromosome 12 are disrupted by translocations, and they are studying how these chromosomal changes could contribute to the uncontrolled growth and division of tumor cells.

Pallister-Killian mosaic syndrome

Pallister-Killian mosaic syndrome is usually caused by the presence of an abnormal extra chromosome called an isochromosome 12p or i(12p). An isochromosome is a chromosome with two identical arms. Normal chromosomes have one long (q) arm and one short (p) arm, but isochromosomes have either two q arms or two p arms. Isochromosome 12p is a version of chromosome 12 made up of two p arms.

Cells normally have two copies of each chromosome, one inherited from each parent. In people with Pallister-Killian mosaic syndrome, cells have the two usual copies of chromosome 12, but some cells also have the isochromosome 12p. These cells have a total of four copies of all the genes on the p arm of chromosome 12. The extra genetic material from the isochromosome disrupts the normal course of development, causing the characteristic features of this disorder.

Although Pallister-Killian mosaic syndrome is usually caused by an isochromosome 12p, other, more complex chromosomal changes involving chromosome 12 are responsible for the disorder in rare cases.

PDGFRB-associated chronic eosinophilic leukemia

Translocations involving chromosome 12 are involved in a type of blood cell cancer called *PDGFRB*-associated chronic eosinophilic leukemia. This condition is characterized by an increased number of eosinophils, a type of white blood cell. The most common translocation that causes this condition fuses part of the *PDGFRB* gene from chromosome 5 with part of the *ETV6* gene from chromosome 12, written as t(5;12)(q31-33;p13). Translocations that fuse the *PDGFRB* gene with other genes can also cause *PDGFRB*-associated chronic eosinophilic leukemia, but these translocations are relatively uncommon. These translocations are acquired during a person's lifetime and are present only in cancer cells. This type of genetic change, called a somatic mutation, is not inherited.

The protein produced from the *ETV6-PDGFRB* fusion gene, called ETV6/PDGFRβ, functions differently than the proteins normally produced from the individual genes. The ETV6 protein normally turns off (represses) gene activity and the PDGFRβ protein plays a role in turning on (activating) signaling pathways. The ETV6/PDGFRβ protein is always turned on, activating signaling pathways and gene activity. When the *ETV6-PDGFRB* fusion gene mutation occurs in cells that develop into blood cells, the growth of eosinophils (and occasionally other white blood cells, such as neutrophils and mast cells) is poorly controlled, leading to *PDGFRB*-associated chronic eosinophilic leukemia. It is unclear why eosinophils are preferentially affected by this genetic change.

other chromosomal conditions

Other changes in the number or structure of chromosome 12 can have a variety of effects on health and development. These effects include intellectual disability, slow growth, distinctive facial features, weak muscle tone (hypotonia), skeletal abnormalities, and heart defects.

Several different changes involving chromosome 12 have been reported, including an extra piece of the chromosome in each cell (partial trisomy 12), a missing segment of the chromosome in each cell (partial monosomy 12), and a circular structure called a ring chromosome 12.

Ring chromosomes occur when a chromosome breaks in two places and the ends of the chromosome arms fuse together to form a circular structure.

Is there a standard way to diagram chromosome 12?

Geneticists use diagrams called ideograms as a standard representation for chromosomes. Ideograms show a chromosome's relative size and its banding pattern. A banding pattern is the characteristic pattern of dark and light bands that appears when a chromosome is stained with a chemical solution and then viewed under a microscope. These bands are used to describe the location of genes on each chromosome.

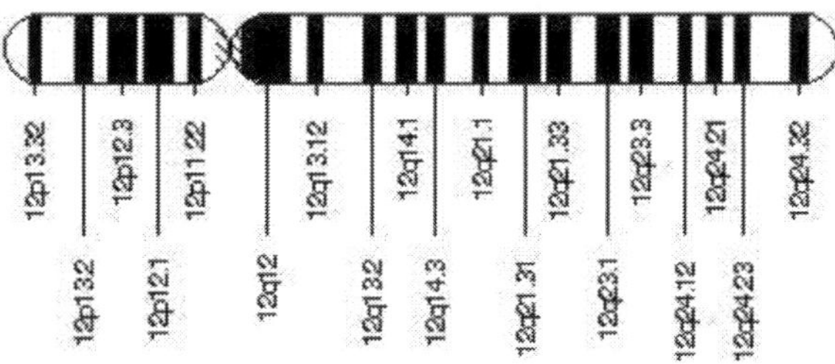

Conditions related to genes on chromosome 12

Genetics Home Reference includes these conditions related to genes on chromosome 12:

achondrogenesis
amyotrophic lateral sclerosis
arrhythmogenic right ventricular cardiomyopathy
autoimmune Addison disease
autoimmune lymphoproliferative syndrome
Bardet-Biedl syndrome Buschke-
Ollendorff syndrome
Cantú syndrome
cardiofaciocutaneous syndrome
Charcot-Marie-Tooth disease
coloboma
congenital fibrosis of the extraocular muscles
congenital stromal corneal dystrophy
core binding factor acute myeloid leukemia
cutis laxa
Czech dysplasia
Darier disease

dentatorubral-pallidoluysian atrophy
Diamond-Blackfan anemia
distal arthrogryposis type 1
distal hereditary motor neuropathy, type II
Dowling-Degos disease
epidermolysis bullosa simplex
epidermolytic hyperkeratosis
episodic ataxia
familial dilated cardiomyopathy
focal palmoplantar keratoderma
Fraser syndrome
frontonasal dysplasia glycogen
storage disease type 0
glycogen storage disease type VII
hereditary hemorrhagic telangiectasia
hereditary hypophosphatemic rickets
hereditary sensory and autonomic neuropathy type II
histidinemia
Holt-Oram syndrome
hyperphosphatemic familial tumoral
calcinosis hypochondrogenesis
IRAK-4 deficiency
isolated lissencephaly
sequence Joubert syndrome
Kabuki syndrome
Klippel-Feil syndrome
Kniest dysplasia
lactate dehydrogenase deficiency
Leber congenital amaurosis Legg-
Calvé-Perthes disease
leukoencephalopathy with vanishing white matter
lissencephaly with cerebellar hypoplasia
Meckel syndrome Meesmann
corneal dystrophy metatropic
dysplasia methylmalonic
acidemia mevalonate kinase
deficiency microphthalmia

monilethrix
mucolipidosis II alpha/beta
mucolipidosis III alpha/beta
mucopolysaccharidosis type III

multiple endocrine neoplasia
multiple lentigines syndrome
multiple sclerosis
myopathy with deficiency of iron-sulfur cluster assembly
enzyme nephrogenic diabetes insipidus
nonsyndromic deafness
Noonan syndrome
pachyonychia congenita
Parkinson disease
PDGFRB -associated chronic eosinophilic
leukemia persistent Müllerian duct syndrome
phenylketonuria
platyspondylic lethal skeletal dysplasia, Torrance type
Pol III-related leukodystrophy
PRICKLE1-related progressive myoclonus epilepsy with ataxia
pseudohypoaldosteronism type 1
pseudohypoaldosteronism type 2
Rotor syndrome
Senior-Løken syndrome
short-chain acyl-CoA dehydrogenase deficiency
spinocerebellar ataxia type 2
spondyloepimetaphyseal dysplasia, Strudwick type
spondyloepiphyseal dysplasia congenita
spondyloperipheral dysplasia
Stickler syndrome
Swyer syndrome
Timothy syndrome
triosephosphate isomerase deficiency
triple A syndrome
tumor necrosis factor receptor-associated periodic
syndrome tyrosinemia
vitamin D-dependent rickets
von Willebrand disease
Warsaw breakage syndrome
white sponge nevus

Reviewed: *February 2013*

Published: *November 17, 2014*

Genes on chromosome 12

Genetics Home Reference includes these genes on chromosome 12:

AAAS
ABCC9
ACADS
ACVRL1
ALX1
AMHR2
AQP2
ATN1
ATP2A2
ATP6V0A2
ATXN2
BBS10
CACNA1C
CDKN1B
CEP290
COL2A1
CYP27B1
DCN
DDX11
DHH
EIF2B1
ETV6
FGD4
FGF23
GDF3
GNPTAB
GNS
GRIP1
GYS2
HAL
HPD
HSPB8
IRAK4
ISCU
KCNA1
KIF21A
KMT2D
KRAS

KRT1
KRT3
KRT4
KRT5
KRT6A
KRT6B
KRT6C
KRT81
KRT83
KRT86
LDHB
LEMD3
LRRK2
MMAB
MVK
MYBPC1
MYO1A
PAH
PFKM
PKP2
POLR3B
PPP1R12A
PRICKLE1
PRPH
PTPN11
RPS26
SCNN1A
SLCO1B1
SLCO1B3
TBX5
TNFRSF1A
TPI1
TRPV4
TUBA1A
VDR
VWF
WNK1

Reviewed: *February 2013*

Published: *November 17, 2014*

State of South Dakota
Chromosome Thirteen

Sanford School of Medicine of the University of South Dakota (Vermillion) (Brain Small Vessel Disease)

What is chromosome 13?

Humans normally have 46 chromosomes in each cell, divided into 23 pairs. Two copies of chromosome 13, one copy inherited from each parent, form one of the pairs. Chromosome 13 is made up of about 115 million DNA building blocks (base pairs) and represents between 3.5 and 4 percent of the total DNA in cells.

Identifying genes on each chromosome is an active area of genetic research. Because researchers use different approaches to predict the number of genes on each chromosome, the estimated number of genes varies. Chromosome 13 likely contains 300 to 400 genes that provide instructions for making proteins. These proteins perform a variety of different roles in the body.

Genes on chromosome 13 are among the estimated 20,000 to 25,000 total genes in the human genome.

How are changes in chromosome 13 related to health conditions?

Many genetic conditions are related to changes in particular genes on chromosome 13. This list of disorders associated with genes on chromosome 13 provides links to additional information.

Changes in the structure or number of copies of a chromosome can also cause problems with health and development. The following chromosomal conditions are associated with such changes in chromosome 13.

8p11 myeloproliferative syndrome

A rearrangement (translocation) of genetic material involving chromosome 13 has been identified in most people with a rare blood cancer called 8p11 myeloproliferative syndrome. This condition is characterized by an increased number of white blood cells (myeloproliferative disorder) and the development of lymphoma, a blood-related cancer that causes tumor formation in the lymph nodes. The myeloproliferative disorder usually develops into another form of blood cancer called acute myeloid leukemia. 8p11 myeloproliferative syndrome most commonly results from a translocation between chromosome 13 and chromosome 8, written as t(8;13)(p11;q12). This genetic change fuses part of the *ZMYM2* gene on chromosome 13 with part of the *FGFR1* gene on chromosome 8. The translocation occurs only in cancer cells.

The protein produced from the normal *FGFR1* gene can turn on cellular signaling that helps the cell respond to its environment, for example by stimulating cell growth. The protein produced from the fused *ZMYM2-FGFR1* gene leads to constant FGFR1 signaling. The uncontrolled signaling promotes continuous cell growth and division, leading to cancer.

retinoblastoma

Retinoblastoma, a cancer of the light-sensing tissue at the back of the eye (the retina) that affects mostly children, is caused by abnormalities of a gene called *RB1*. This gene is located on a region of the long (q) arm of chromosome 13 designated 13q14. Although most retinoblastomas are caused by mutations within the *RB1* gene, a small percentage of retinoblastomas result from a deletion of the 13q14 region.

In addition to retinoblastoma, deletions of the 13q14 region may cause intellectual disability, slow growth, and characteristic facial features such as prominent eyebrows, a broad nasal bridge, a short nose, and ear abnormalities. A loss of several genes is likely responsible for these developmental problems, although researchers have not determined which other genes in the deleted region are involved.

trisomy 13

Trisomy 13 occurs when each cell in the body has three copies of chromosome 13 instead of the usual two copies. Trisomy 13 can also result from an extra copy of chromosome 13 in only some of the body's cells (mosaic trisomy 13).

In some cases, trisomy 13 occurs when part of chromosome 13 becomes attached (translocated) to another chromosome during the formation of reproductive cells (eggs and sperm) or very early in embryonic development. Affected individuals have two copies of chromosome 13, plus extra material from chromosome 13 attached to another chromosome. People with this genetic change are said to have translocation trisomy 13. The physical signs of translocation trisomy 13 may be different from those typically seen in trisomy 13 because only part of chromosome 13 is present in three copies.

Researchers believe that extra copies of some genes on chromosome 13 disrupt the course of normal development, causing the characteristic features of trisomy 13 and the increased risk of medical problems associated with this disorder.

other cancers

Changes in chromosome 13 have been associated with several types of cancer. These genetic changes are somatic, which means they are acquired during a person's lifetime and are present only in certain cells. The loss of genetic material from the middle of chromosome 13 is common in cancers of blood-forming cells (leukemias), cancers of immune system cells (lymphomas), and other related cancers.

other chromosomal conditions

Partial monosomy and partial trisomy of chromosome 13 occur when a portion of the long (q) arm of this chromosome is deleted or duplicated, respectively. The effect of missing or extra chromosome material varies with the size and location of the chromosome abnormality. Affected individuals may have developmental delay, intellectual disability, low birth weight, and other physical abnormalities.

Is there a standard way to diagram chromosome 13?

Geneticists use diagrams called ideograms as a standard representation for chromosomes. Ideograms show a chromosome's relative size and its banding pattern. A banding pattern is the characteristic pattern of dark and light bands that appears when a chromosome is stained with a chemical solution and then viewed under a microscope. These bands are used to describe the location of genes on each chromosome.

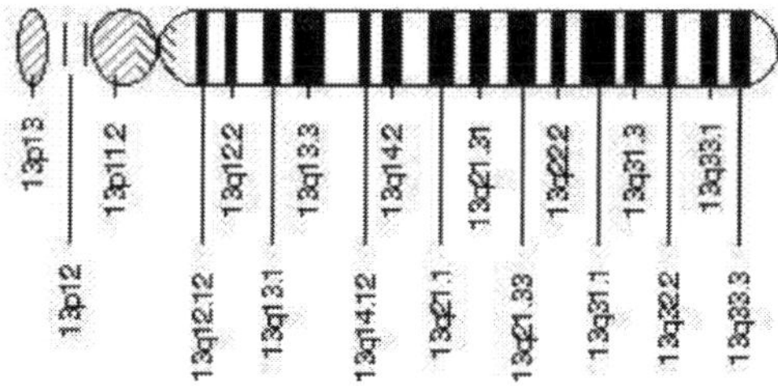

Conditions related to genes on chromosome 13

Genetics Home Reference includes these conditions related to genes on chromosome 13

8p11 myeloproliferative syndrome
Aicardi-Goutieres syndrome
autosomal recessive hypotrichosis
autosomal recessive spastic ataxia of Charlevoix-Saguenay
Bart-Pumphrey syndrome
bladder cancer
breast cancer
Clouston syndrome
COL4A1-related brain small-vessel disease
core binding factor acute myeloid leukemia
cytogenetically normal acute myeloid leukemia
familial porencephaly
Fanconi anemia
Fraser syndrome
hereditary angiopathy with nephropathy, aneurysms, and muscle cramps syndrome
hereditary cerebral amyloid
angiopathy Hirschsprung disease
hyperphosphatemic familial tumoral calcinosis
hystrix-like ichthyosis with deafness
infantile neuronal ceroid lipofuscinosis
juvenile Batten disease keratitis-ichthyosis-
deafness syndrome late-infantile neuronal
ceroid lipofuscinosis limb-girdle muscular
dystrophy nonsyndromic deafness

nonsyndromic holoprosencephaly
ornithine translocase deficiency
palmoplantar keratoderma with deafness
Peters plus syndrome
propionic acidemia
retinoblastoma
succinate-CoA ligase
deficiency Tourette syndrome
Treacher Collins syndrome
Troyer syndrome
Vohwinkel syndrome
Waardenburg syndrome

Wilson disease

Reviewed: *July 2013*

Published: *November 17, 2014*

Conditions related to genes on chromosome 13

Genetics Home Reference includes these conditions related to genes on chromosome 13:

8p11 myeloproliferative syndrome
Aicardi-Goutieres syndrome
autosomal recessive hypotrichosis
autosomal recessive spastic ataxia of Charlevoix-Saguenay
Bart-Pumphrey syndrome
bladder cancer
breast cancer
Clouston syndrome
COL4A1-related brain small-vessel disease
core binding factor acute myeloid leukemia
cytogenetically normal acute myeloid leukemia
familial porencephaly
Fanconi anemia
Fraser syndrome
hereditary angiopathy with nephropathy, aneurysms, and muscle cramps syndrome
hereditary cerebral amyloid
angiopathy Hirschsprung disease
hyperphosphatemic familial tumoral calcinosis
hystrix-like ichthyosis with deafness
infantile neuronal ceroid lipofuscinosis
juvenile Batten disease keratitis-
ichthyosis-deafness syndrome late-
infantile neuronal ceroid lipofuscinosis
limb-girdle muscular dystrophy
nonsyndromic deafness
nonsyndromic holoprosencephaly
ornithine translocase deficiency
palmoplantar keratoderma with deafness
Peters plus syndrome
propionic acidemia
retinoblastoma
succinate-CoA ligase deficiency

Tourette syndrome
Treacher Collins syndrome
Troyer syndrome
Vohwinkel syndrome
Waardenburg syndrome
Wilson disease

Reviewed: *July 2013*

Published: *November 17, 2014*

Genes on chromosome 13

Genetics Home Reference includes these genes on chromosome 13:

ATP7B
B3GALTL
BRCA2
CLN5
COL4A1
EDNRB
FLT3
FREM2
GJB2
GJB6
ITM2B
KL
LPAR6
PCCA
POLR1D
RB1
RNASEH2B
SACS
SGCG
SLC25A15
SLITRK1
SPG20
SUCLA2
ZIC2
ZMYM2

Reviewed: *July 2013*

Published: *November 17, 2014*

State of Tennesse
Chromosome 14

East Tennesse State University James H Quillen College of Medicine (Johnson City) (Multiple Myeloma)

Meharry Medical College School of Medicine (Nashville) (Limb Girdle Muscular Dystrophy)

Lincoln Memorial University DeBusk College of Osteopathic Medicine (Harrogate) (Familial Cardiomyopathy)

University of Tennesse College of Medicine (Memphis) (Cranioectodermal Dysplasia)

Vanderbilt University School of Medicine (Nashville) (Age Related Macular Degeneration)

Chromosome Fourteen

Reviewed December 2013

What is chromosome 14?

Humans normally have 46 chromosomes in each cell, divided into 23 pairs. Two copies of chromosome 14, one copy inherited from each parent, form one of the pairs. Chromosome 14 spans more than 107 million DNA building blocks (base pairs) and represents about 3.5 percent of the total DNA in cells.

Identifying genes on each chromosome is an active area of genetic research. Because researchers use different approaches to predict the number of genes on each chromosome, the estimated number of genes varies. Chromosome 14 likely contains 800 to 900 genes that provide instructions for making proteins. These proteins perform a variety of different roles in the body.

Genes on chromosome 14 are among the estimated 20,000 to 25,000 total genes in the human genome.

How are changes in chromosome 14 related to health conditions?

Many genetic conditions are related to changes in particular genes on chromosome 14. This list of disorders associated with genes on chromosome 14 provides links to additional information.

Changes in the structure or number of copies of a chromosome can also cause problems with health and development. The following chromosomal conditions are associated with such changes in chromosome 14.

cancers

Rearrangements (translocations) of genetic material between chromosome 14 and other chromosomes have been associated with several types of cancer. These chromosome abnormalities are somatic, which means they are acquired during a person's lifetime and are present only in certain cells. Studies show that these translocations disrupt genes that are critical for keeping cell growth and division under control. Unregulated cell division can lead to the development of cancer.

Translocations involving chromosome 14 have been found in cancers of blood-forming cells (leukemias), cancers of immune system cells (lymphomas), and several related diseases. For example, Burkitt lymphoma, a cancer of white blood cells that occurs most often in children and young adults, is related to a translocation between chromosomes 8 and 14. Another type of lymphoma, called follicular lymphoma, is often associated with a translocation between chromosomes 14 and 18. In a cancer of white blood cells called multiple myeloma, the presence of a translocation between chromosomes 4 and 14 is associated with a more aggressive form of the disease.

FOXG1 syndrome

A deletion of genetic material from part of the long (q) arm of chromosome 14 can cause *FOXG1* syndrome, which is a rare disorder characterized by impaired development and structural brain abnormalities. The region of chromosome 14 that is deleted includes the *FOXG1* gene as well as several neighboring genes. Depending on which genes are involved, affected individuals may have additional signs and symptoms, including distinctive facial features and a missing connection between the left and right halves of the brain (a structure called the corpus callosum).

The protein normally produced from the *FOXG1* gene plays an important role in brain development before birth, particularly in a region of the embryonic brain known as the telencephalon. The telencephalon ultimately develops into several critical structures, including the the largest part of the brain (the cerebrum), which controls most voluntary activity, language, sensory perception, learning, and memory. A loss of the *FOXG1* gene disrupts normal brain development starting before birth, which appears to underlie the structural brain abnormalities and severe developmental problems characteristic of *FOXG1* syndrome. It is unclear how the loss of additional genes contributes to the signs and symptoms of the condition.

ring chromosome 14 syndrome

Ring chromosome 14 syndrome is caused by a chromosomal abnormality known as a ring chromosome 14 or r(14). A ring chromosome is a circular structure that occurs when a chromosome breaks in two places and its broken ends fuse together. People with ring chromosome 14 syndrome have one copy of this abnormal chromosome in some or all of their cells.

Researchers believe that several critical genes near the end of the long (q) arm of chromosome 14 are lost when the ring chromosome forms. The loss of these genes is likely responsible for several of the major features of ring chromosome 14 syndrome, including intellectual disability and delayed development. Researchers are still working to determine which missing genes contribute to the signs and symptoms of this disorder.

Epilepsy is a common feature of ring chromosome syndromes, including ring chromosome 14. There may be something about the ring structure itself that causes epilepsy. Seizures may occur because certain genes on the ring chromosome 14 are less active than those on the normal chromosome 14. Alternately, seizures might result from instability of the ring chromosome in some cells.

other chromosomal conditions

A rare condition known as terminal deletion 14 syndrome causes signs and symptoms similar to those of ring chromosome 14 syndrome. Terminal deletion 14 syndrome is caused by the loss of several genes at the end (terminus) of the long (q) arm of chromosome 14. In addition, some people with terminal deletion 14 syndrome have a loss or gain of genetic material from another chromosome. People with this condition may have weak muscle tone (hypotonia), a small head (microcephaly), frequent respiratory infections, developmental delay, and learning difficulties.

Other changes in the number or structure of chromosome 14 can have a variety of effectsFincluding delayed growth and development, distinctive facial features, and other health problems. Several different changes involving chromosome 14 have been reported. These include an extra copy of a segment of chromosome 14 in every cell (partial trisomy 14), an extra copy of the entire chromosome in only some of the body's cells (mosaic trisomy 14), and deletions or duplications of part of chromosome 14. Full trisomy 14, an extra copy of the entire chromosome 14 in all of the body's cells, is not compatible with life.

Health problems can also result from a chromosome abnormality called uniparental disomy (UPD). UPD occurs when people inherit both copies of a chromosome from one parent instead of one copy from each parent. The long arm of chromosome 14 contains some genes that are active only when inherited from the mother, and other genes that are active only when inherited from the father. Therefore, people who have two paternal copies or two maternal copies of chromosome 14 are missing some functional genes and have an extra copy of others.

When both copies of chromosome 14 are inherited from the mother, the phenomenon is known as maternal UPD 14. Maternal UPD 14 is associated with premature birth, slow growth before and after birth, short stature, developmental delay, small hands and feet, and early onset of puberty. When both copies of the chromosome are inherited from the father, the phenomenon is known as paternal UPD 14. Paternal UPD 14 is associated with an excess of amniotic fluid (which surrounds the baby before birth); an opening in the wall of the abdomen; distinctive facial features; a small, bell-shaped chest with short ribs; and developmental delay. Both maternal UPD 14 and paternal UPD 14 appear to be rare.

Is there a standard way to diagram chromosome 14?

Geneticists use diagrams called ideograms as a standard representation for chromosomes. Ideograms show a chromosome's relative size and its banding pattern. A banding pattern is the characteristic pattern of dark and light bands that appears when a chromosome is stained with a chemical solution and then viewed under a microscope. These bands are used to describe the location of genes on each chromosome.

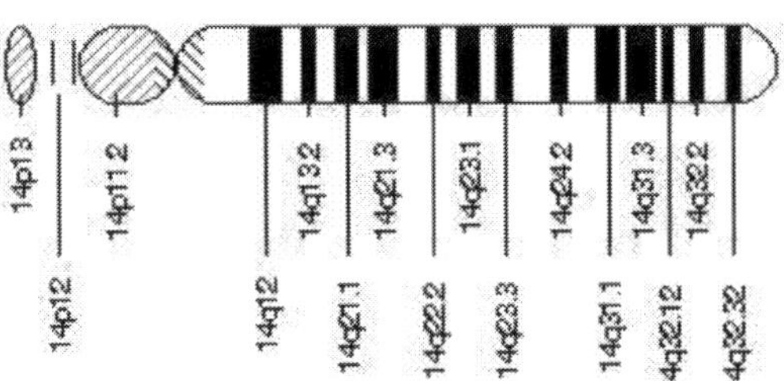

What is chromosome 14?

Humans normally have 46 chromosomes in each cell, divided into 23 pairs. Two copies of chromosome 14, one copy inherited from each parent, form one of the pairs. Chromosome 14 spans more than 107 million DNA building blocks (base pairs) and represents about 3.5 percent of the total DNA in cells.

Identifying genes on each chromosome is an active area of genetic research. Because researchers use different approaches to predict the number of genes on each chromosome, the estimated number of genes varies. Chromosome 14 likely contains 800 to 900 genes that provide instructions for making proteins. These proteins perform a variety of different roles in the body.

Genes on chromosome 14 are among the estimated 20,000 to 25,000 total genes in the human genome.

How are changes in chromosome 14 related to health conditions?

Many genetic conditions are related to changes in particular genes on chromosome 14. This list of disorders associated with genes on chromosome 14 provides links to additional information.

Changes in the structure or number of copies of a chromosome can also cause problems with health and development. The following chromosomal conditions are associated with such changes in chromosome 14.

cancers

Rearrangements (translocations) of genetic material between chromosome 14 and other chromosomes have been associated with several types of cancer. These chromosome abnormalities are somatic, which means they are acquired during a person's lifetime and are present only in certain cells. Studies show that these translocations disrupt genes that are critical for keeping cell growth and division under control. Unregulated cell division can lead to the development of cancer.

Translocations involving chromosome 14 have been found in cancers of blood-forming cells (leukemias), cancers of immune system cells (lymphomas), and several related diseases. For example, Burkitt lymphoma, a cancer of white blood cells that occurs most often in children and young adults, is related to a translocation between chromosomes 8 and 14. Another type of lymphoma, called follicular lymphoma, is often associated with a translocation between chromosomes 14 and 18. In a cancer of white blood cells called multiple myeloma, the presence of a translocation between chromosomes 4 and 14 is associated with a more aggressive form of the disease.

FOXG1 syndrome

A deletion of genetic material from part of the long (q) arm of chromosome 14 can cause *FOXG1* syndrome, which is a rare disorder characterized by impaired development and structural brain abnormalities. The region of chromosome 14 that is deleted includes the *FOXG1* gene as well as several neighboring genes. Depending on which genes are involved, affected individuals may have additional signs and symptoms, including distinctive facial features and a missing connection between the left and right halves of the brain (a structure called the corpus callosum).

The protein normally produced from the *FOXG1* gene plays an important role in brain development before birth, particularly in a region of the embryonic brain known as the telencephalon. The telencephalon ultimately develops into several critical structures, including the the largest part of the brain (the cerebrum), which controls most voluntary activity, language, sensory perception, learning, and memory. A loss of the *FOXG1* gene disrupts normal brain development starting before birth, which appears to underlie the structural brain abnormalities and severe developmental problems characteristic of *FOXG1* syndrome. It is unclear how the loss of additional genes contributes to the signs and symptoms of the condition.

ring chromosome 14 syndrome

Ring chromosome 14 syndrome is caused by a chromosomal abnormality known as a ring chromosome 14 or r(14). A ring chromosome is a circular structure that occurs when a chromosome breaks in two places and its broken ends fuse together. People with ring chromosome 14 syndrome have one copy of this abnormal chromosome in some or all of their cells.

Researchers believe that several critical genes near the end of the long (q) arm of chromosome 14 are lost when the ring chromosome forms. The loss of these genes is likely responsible for several of the major features of ring chromosome 14 syndrome, including intellectual disability and delayed development. Researchers are still working to determine which missing genes contribute to the signs and symptoms of this disorder.

Epilepsy is a common feature of ring chromosome syndromes, including ring chromosome 14. There may be something about the ring structure itself that causes epilepsy. Seizures may occur because certain genes on the ring chromosome 14 are less active than those on the normal chromosome 14. Alternately, seizures might result from instability of the ring chromosome in some cells.

other chromosomal conditions

A rare condition known as terminal deletion 14 syndrome causes signs and symptoms similar to those of ring chromosome 14 syndrome. Terminal deletion 14 syndrome is caused by the loss of several genes at the end (terminus) of the long (q) arm of chromosome 14. In addition, some people with terminal deletion 14 syndrome have a loss or gain of genetic material from another chrom

Conditions related to genes on chromosome 14

Genetics Home Reference includes these conditions related to genes on chromosome 14:

2-hydroxyglutaric aciduria age-
related macular degeneration
alpha-1 antitrypsin deficiency
Alzheimer disease
amyotrophic lateral sclerosis
anhidrotic ectodermal dysplasia with immune
deficiency branchiootorenal syndrome
Charcot-Marie-Tooth
disease coloboma
combined pituitary hormone deficiency
congenital hypothyroidism
corticosteroid-binding globulin
deficiency cranioectodermal dysplasia
cutis laxa DICER1
syndrome
dopa-responsive dystonia
dyskeratosis congenita
familial dilated cardiomyopathy
familial hypertrophic
cardiomyopathy familial restrictive
cardiomyopathy FOXG1 syndrome

glycogen storage disease type VI
Graves disease
hereditary hyperekplexia
hidradenitis suppurativa
Imerslund-Gräsbeck syndrome
Krabbe disease
Laing distal myopathy
lamellar ichthyosis
leukoencephalopathy with vanishing white matter
limb-girdle muscular dystrophy
lysinuric protein
intolerance microphthalmia
molybdenum cofactor deficiency
myosin storage myopathy Niemann-
Pick disease nonsyndromic deafness
oculopharyngeal muscular dystrophy
ophthalmo-acromelic syndrome
pontocerebellar hypoplasia

Proteus syndrome
purine nucleoside phosphorylase deficiency
septo-optic dysplasia
sick sinus syndrome
spastic paraplegia type 3A
spastic paraplegia type 15
spinal muscular atrophy
spinocerebellar ataxia type 3
tetrahydrobiopterin deficiency
Walker-Warburg syndrome
Winchester syndrome

Reviewed: *December 2013*

Published: *November 17, 2014*

Genes on chromosome 14

Genetics Home Reference includes these genes on chromosome 14:

AKT1
AMN
ANG
ATL1
ATXN3

COCH
DICER1
DYNC1H1
EIF2B2
FBLN5
FOXG1
GALC
GCH1
GPHN
IFT43
L2HGDH
MMP14
MYH6
MYH7
NFKBIA
NPC2
OTX2
PABPN1
PNP
POMT2
PSEN1
PYGL
SERPINA1
SERPINA6
SIX1
SLC7A7
SMOC1
TGM1
TINF2
TSHR
VRK1
ZFYVE26

State of Texas
Chromosome 15

Baylor College of Medicine (Houston) (Oculocutaneous Albanism) (Marfans Syndrome)

Texas A and M Health Science Center College of Medicine (College Station) (Acute Promyelocytic Leukemia)

University of North Texas Health Science Center Texas College of Osteopathic Medicine (Fort Worth) (Primary Myelofibrosis)

Texas Tech University Health Sciences Center Paul L Foster School of Medicine (El Paso) (Ataxia Neuropathy Spectrum)

University of Texas Medical School at Houston (Houston) (Sensorineural Deafness and Male Infertility)

Texas Tech University Health Sciences Center School of Medicine (Lubbock) (Spastic ParaplegiaType 11)

University of Texas Medical Branch School of Medicine (Galveston) (AR Congenital Stationary Night Blindness)

University of Texas Medical School at San Antonio (San Antonio) (Spondylothoracic Dysostosis)

University of Texas Southwestern Medical School at Dallas (Dallas) (Progressive External Ophthalmoplegia)

Chromosome Fifteen

What is chromosome 15?

Humans normally have 46 chromosomes in each cell, divided into 23 pairs. Two copies of chromosome 15, one copy inherited from each parent, form one of the pairs. Chromosome 15 spans more than 102 million DNA building blocks (base pairs) and represents more than 3 percent of the total DNA in cells.

Identifying genes on each chromosome is an active area of genetic research. Because researchers use different approaches to predict the number of genes on each chromosome, the estimated number of genes varies. Chromosome 15 likely contains 600 to 700 genes that provide instructions for making proteins. These proteins perform a variety of different roles in the body.

Genes on chromosome 15 are among the estimated 20,000 to 25,000 total genes in the human genome.

How are changes in chromosome 15 related to health conditions?

Many genetic conditions are related to changes in particular genes on chromosome 15. This list of disorders associated with genes on chromosome 15 provides links to additional information.

Changes in the structure or number of copies of a chromosome can also cause problems with health and development. The following chromosomal conditions are associated with such changes in chromosome 15.

15q13.3 microdeletion

15q13.3 microdeletion is a chromosomal change in which a small piece of chromosome 15 is deleted in each cell. The deletion occurs on the long (q) arm of the chromosome at a position designated q13.3. Most people with a 15q13.3 microdeletion are missing a sequence of about 2 million DNA building blocks (base pairs), also written as 2 megabases (Mb). The exact size of the deleted region varies, but it typically contains at least six genes. It is unclear how a loss of these genes increases the risk of intellectual disability, seizures, behavioral problems, and psychiatric disorders in some individuals with a 15q13.3 microdeletion.

Other people with a 15q13.3 microdeletion have no obvious signs or symptoms related to the chromosomal change. In these individuals, the microdeletion is often detected when they undergo genetic testing because they have an affected relative. It is unknown why 15q13.3 microdeletion causes cognitive and behavioral problems in some individuals but few or no health problems in others. Researchers believe that additional genetic or environmental factors may be involved.

15q24 microdeletion

15q24 microdeletion is a chromosomal change in which a small piece of chromosome 15 is deleted in each cell. Specifically, affected individuals are missing between 1.7 Mb and 6.1 Mb of DNA at position q24 on chromosome 15. The exact size of the deletion varies, but all individuals are missing the same 1.2 Mb region. This region contains several genes that are thought to be important for normal development. It is unclear how a loss of these genes leads to intellectual disability, distinctive facial features, and other abnormalities often seen in people with a 15q24 microdeletion.

acute promyelocytic leukemia

A type of blood cancer known as acute promyelocytic leukemia is caused by a rearrangement (translocation) of genetic material between chromosomes 15 and 17. This translocation, written as t(15;17), fuses part of the *PML* gene from chromosome 15 with part of the *RARA* gene from chromosome 17. This mutation is acquired during a person's lifetime and is present only in certain cells. This type of genetic change, called a somatic mutation, is not inherited. The t(15;17) translocation is called a balanced reciprocal translocation because the pieces of chromosome are exchanged with each other (reciprocal) and no genetic material is gained or lost (balanced). The protein produced from this fused gene is known as PML-RARα.

The PML-RARα protein functions differently than the protein products from the normal *PML* and *RARA* genes. The *PML* gene on chromosome 15 provides instructions for a protein that acts as a tumor suppressor, which means it prevents cells from growing and dividing too rapidly or in an uncontrolled way. The PML protein blocks cell growth and division (proliferation) and induces self-destruction (apoptosis) in combination with other proteins. The *RARA* gene on chromosome 17 provides instructions for making a transcription factor called the retinoic acid receptor alpha (RARα). A transcription factor is a protein that attaches (binds) to specific regions of DNA and helps control the activity of particular genes. Normally, the RARα protein controls the activity (transcription) of genes important for the maturation (differentiation) of immature white blood cells beyond a particular stage called the promyelocyte. The PML-RARα protein interferes with the normal function of both the PML and the RARα proteins. As a result, blood cells are stuck at the promyelocyte stage, and they proliferate abnormally. Excess promyelocytes accumulate in the bone marrow and normal white blood cells cannot form, leading to acute promyelocytic leukemia.

Angelman syndrome

Angelman syndrome results from a loss of gene activity (expression) in a specific part of chromosome 15 in each cell. This region is located on the long (q) arm of the chromosome and is designated 15q11-q13. This region contains a gene called *UBE3A* that, when mutated or absent, likely causes the characteristic neurologic features of Angelman syndrome.

People normally inherit one copy of the *UBE3A* gene from each parent, and both copies of this gene are turned on (active) in many of the body's tissues. In certain areas of the brain, however, only the copy inherited from a person's mother (the maternal copy) is active. This parent-specific gene activation results from a phenomenon called genomic imprinting. If the maternal copy is lost because of a chromosomal change or a gene mutation, a person will have no working copies of the *UBE3A* gene in some parts of the brain.

In most cases (about 70 percent), Angelman syndrome results from a deletion in the maternal copy of chromosome 15. This chromosomal change deletes the region of chromosome 15 that includes the *UBE3A* gene. Because the copy of the *UBE3A* gene inherited from a person's father (the paternal copy) is normally inactive in certain parts of the brain, a deletion in the maternal chromosome 15 leaves no active copies of the *UBE3A* gene in these brain regions.

In 3 percent to 7 percent of cases of Angelman syndrome, the condition results when a person inherits two copies of chromosome 15 from his or her father instead of one copy from each parent. This phenomenon is called paternal uniparental disomy (UPD). People with paternal UPD for chromosome 15 have two copies of the *UBE3A* gene, but they are both inherited from the father and are therefore inactive in the brain.

About 10 percent of cases of Angelman syndrome are caused by a mutation in the *UBE3A* gene, and another 3 percent results from a defect in the DNA region that controls the activation of the *UBE3A* gene and other genes on the maternal copy of chromosome 15. In a small percentage of cases, Angelman syndrome is caused by a chromosomal rearrangement (translocation) or by a mutation in a gene other than *UBE3A*. These genetic changes abnormally inactivate the *UBE3A* gene.

isodicentric chromosome 15 syndrome

Isodicentric chromosome 15 syndrome results from the presence of an abnormal extra chromosome, called an isodicentric chromosome 15, in each cell. An isodicentric chromosome contains mirror-image segments of genetic material and has two constriction points (centromeres), rather than one centromere as in normal chromosomes. In isodicentric chromosome 15 syndrome, the isodicentric chromosome is made up of two extra copies of a segment of genetic material from chromosome 15, attached end-to-end. Typically this copied genetic material includes the 15q11-q13 region.

Cells normally have two copies of each chromosome, one inherited from each parent. In people with isodicentric chromosome 15 syndrome, cells have the usual two copies of chromosome 15 plus the two extra copies of the segment of genetic material in the isodicentric chromosome. The extra genetic material disrupts the normal course of development, causing the characteristic features of this disorder. These features include weak muscle tone (hypotonia), intellectual disability, recurrent seizures (epilepsy), characteristics of autism or related developmental disorders affecting communication and social interaction, and other behavioral problems. Some individuals with isodicentric chromosome 15 whose copied genetic material does not include the 15q11-q13 region do not show signs or symptoms of the condition.

Prader-Willi syndrome

Prader-Willi syndrome is caused by a loss of active genes in a region of chromosome 15. This region is located on the long (q) arm of the chromosome and is designated 15q11-q13. It is the same part of chromosome 15 that is usually affected in people with Angelman syndrome, although different genes are associated with the two disorders. People can have either Prader-Willi syndrome or Angelman syndrome, but they typically cannot have both.

People normally inherit one copy of chromosome 15 from each parent. Some genes on this chromosome are turned on (active) only on the copy inherited from a person's father (the paternal copy). This parent-specific gene activation results from a phenomenon called genomic imprinting.

In about 70 percent of cases, Prader-Willi syndrome occurs when the 15q11-q13 region of the paternal chromosome 15 is deleted in each cell. A person with this chromosomal change will be missing certain critical genes in this region because the genes on the paternal copy have been deleted, and the genes on the maternal copy are turned off (inactive). Researchers are working to identify which missing genes are associated with the characteristic features of Prader-Willi syndrome.

In about 25 percent of cases, people with Prader-Willi syndrome inherit two copies of chromosome 15 from their mother instead of one copy from each parent. This phenomenon is called maternal uniparental disomy. A person with two maternal copies of chromosome 15 will have no active copies of certain genes in the 15q11-q13 region.

In a small percentage of cases, Prader-Willi syndrome is caused by a chromosomal rearrangement called a translocation. Rarely, the condition results from a mutation or other defect that abnormally inactivates genes on the paternal copy of chromosome 15.

sensorineural deafness and male infertility

Sensorineural deafness and male infertility is caused by a deletion of genetic material on the long (q) arm of chromosome 15. The symptoms of sensorineural deafness and male infertility are related to the loss of multiple genes in this region. The size of the deletion varies among affected individuals. Researchers have determined that the loss of a particular gene on chromosome 15, *STRC*, is responsible for hearing loss in affected individuals. The loss of another gene, *CATSPER2*, in the same region of chromosome 15 is responsible for sperm abnormalities, which lead to an inability to father children (infertility) in affected males. Researchers are working to determine how the loss of additional genes in the deleted region affects people with sensorineural deafness and male infertility.

other chromosomal conditions

Other changes in the number or structure of chromosome 15 can cause intellectual disability, delayed growth and development, hypotonia, and characteristic facial features. These changes include an extra copy of part of chromosome 15 in each cell (partial trisomy 15), a missing segment of the chromosome in each cell (partial monosomy 15), and a circular structure called ring chromosome 15. A ring chromosome occurs when a chromosome breaks in two places and the ends of the chromosome arms fuse together to form a circular structure.

Is there a standard way to diagram chromosome 15?

Geneticists use diagrams called ideograms as a standard representation for chromosomes. Ideograms show a chromosome's relative size and its banding pattern. A banding pattern is the characteristic pattern of dark and light bands that appears when a chromosome is stained with a chemical solution and then viewed under a microscope. These bands are used to describe the location of genes on each chromosome.

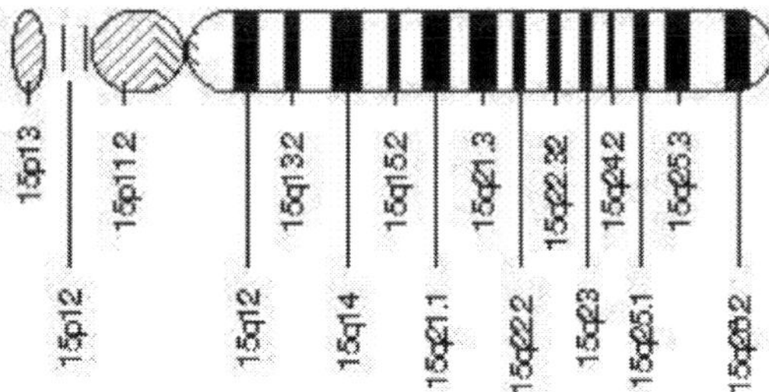

Conditions related to genes on chromosome 15

Genetics Home Reference includes these conditions related to genes on chromosome 15:

2-hydroxyglutaric aciduria
acral peeling skin syndrome
acute promyelocytic leukemia
Alpers-Huttenlocher syndrome
amyotrophic lateral sclerosis
Andermann syndrome
Angelman syndrome
arginine:glycine amidinotransferase deficiency
aromatase deficiency
aromatase excess syndrome
ataxia neuropathy spectrum
autosomal recessive congenital stationary night blindness
Bartter syndrome
Bloom syndrome
breast cancer
cardiofaciocutaneous syndrome
childhood myocerebrohepatopathy
spectrum congenital dyserythropoietic
anemia congenital hypothyroidism
cytogenetically normal acute myeloid leukemia
Diamond-Blackfan anemia
familial osteochondritis dissecans
familial thoracic aortic aneurysm and
dissection geleophysic dysplasia
glutaric acidemia type II
Griscelli syndrome
isolated hyperchlorhidrosis
isovaleric acidemia
Kufs disease
late-infantile neuronal ceroid lipofuscinosis
Legius syndrome
limb-girdle muscular dystrophy
Loeys-Dietz syndrome
Marfan syndrome
myoclonic epilepsy myopathy sensory ataxia
nonsyndromic deafness
oculocutaneous albinism
Prader-Willi syndrome

primary myelofibrosis
progressive external ophthalmoplegia
sensorineural deafness and male
infertility Shprintzen-Goldberg syndrome
sick sinus syndrome spastic
paraplegia type 11
spondylocostal dysostosis
spondylothoracic
dysostosis Tay-Sachs
disease tyrosinemia
Weill-Marchesani syndrome

Reviewed: *September 2012*

Published: *November 17, 2014*

Genes on chromosome 15

Genetics Home Reference includes these genes on chromosome 15:

ACAN
BLM
CA12
CAPN3
CATSPER2
CDAN1
CLN6
CYP19A1
DUOX2
ETFA
FAH
FBN1
GATM
HCN4
HEXA
IDH2
IVD
MAP2K1
MESP2
MYO5A
OCA2
PML
POLG
RAB27A

RAD51
RPS17
SLC12A1
SLC12A6
SMAD3
SPG11
SPRED1
STRC
TGM5
TRPM1
UBE3A

State of Utah
Chromosome 16

University of Utah School of Medicine (Salt Lake City) (Hereditary Diffuse Gastric Cancer)

Chromosome 16

What is chromosome 16?

Humans normally have 46 chromosomes in each cell, divided into 23 pairs. Two copies of chromosome 16, one copy inherited from each parent, form one of the pairs. Chromosome 16 spans more than 90 million DNA building blocks (base pairs) and represents almost 3 percent of the total DNA in cells.

Identifying genes on each chromosome is an active area of genetic research. Because researchers use different approaches to predict the number of genes on each chromosome, the estimated number of genes varies. Chromosome 16 likely contains 800 to 900 genes that provide instructions for making proteins. These proteins perform a variety of different roles in the body.

Genes on chromosome 16 are among the estimated 20,000 to 25,000 total genes in the human genome.

How are changes in chromosome 16 related to health conditions?

Many genetic conditions are related to changes in particular genes on chromosome 16. This list of disorders associated with genes on chromosome 16 provides links to additional information.

Changes in the structure or number of copies of a chromosome can also cause problems with health and development. The following chromosomal conditions are associated with such changes in chromosome 16.

16p11.2 deletion syndrome

16p11.2 deletion syndrome is caused by a deletion of about 600,000 DNA building blocks (base pairs), also written as 600 kilobases (kb), at position 11.2 on the short (p) arm of chromosome 16. This deletion affects one of the two copies of chromosome 16 in each cell. The 600 kb region contains more than 25 genes, and in many cases little is known about their function. Researchers are working to determine how the missing genes contribute to the features of 16p11.2 deletion syndrome, which include delayed development, intellectual disability, and developmental disorders that affect communication and social interaction (autism spectrum disorders). Obesity is another common feature of 16p11.2 deletion syndrome, and affected individuals also have an increased risk of seizures. Most people with the deletion have some of these symptoms, but others do not. Although some people have this deletion without serious consequences, they can still pass it to their children, who may be more severely affected.

16p11.2 duplication

A 16p11.2 duplication is an extra copy of the same 600 kb segment of chromosome 16 that is missing in 16p11.2 deletion syndrome (described above). A 16p11.2 duplication may result in similar symptoms as the deletion in some affected individuals, including features of autism spectrum disorders; however, being underweight is common in people with the duplication, while obesity often occurs with the deletion.

The 16p11.2 duplication appears to have a milder effect than the deletion, with a higher proportion of individuals with this chromosomal change showing no apparent problems. These individuals can still pass along the duplication to their children, who may have symptoms related to the chromosomal change. Researchers are working to determine how the extra genetic material contributes to the features that occur in some people with a 16p11.2 duplication, and why duplication or deletion of the same chromosomal region can have some similar effects.

alveolar capillary dysplasia with misalignment of pulmonary veins

Alveolar capillary dysplasia with misalignment of pulmonary veins (ACD/MPV) is a disorder that affects the development of blood vessels in the lungs. It can be caused by a deletion of genetic material on chromosome 16 in a region known as 16q24.1. This region includes several genes, including the *FOXF1* gene. The protein produced from the *FOXF1* gene is a transcription factor, which means that it attaches (binds) to specific regions of DNA and helps control the activity of many other genes. The FOXF1 protein helps regulate the development of the lungs and the gastrointestinal tract. Genetic changes that result in a nonfunctional FOX1 protein interfere with the development of pulmonary blood vessels and cause ACD/MPV. Affected infants may also have gastrointestinal abnormalities.

Researchers suggest that deletions resulting in the loss of other genes in this region of chromosome 16 probably cause the additional abnormalities seen in some infants with this disorder. Like *FOXF1*, these genes also provide instructions for making transcription factors that regulate development of various body systems before birth.

cancers

Changes in the structure of chromosome 16 are associated with several types of cancer. These genetic changes are somatic, which means they are acquired during a person's lifetime and are present only in certain cells. In some cases, chromosomal rearrangements called translocations disrupt the region of chromosome 16 that contains the *CREBBP* gene. The protein produced from this gene normally plays a role in regulating cell growth and division, which helps prevent the development of cancers.

Researchers have found a translocation between chromosome 8 and chromosome 16 that disrupts the *CREBBP* gene in some people with a cancer of blood-forming cells called acute myeloid leukemia (AML). Another translocation involving the *CREBBP* gene, which rearranges pieces of chromosomes 11 and 16, has been found in some people who have undergone cancer treatment. This chromosomal change is associated with the later development of AML and two other cancers of blood-forming tissues (chronic myeloid leukemia and myelodysplastic syndrome). These are sometimes described as treatment-related cancers because the translocation between chromosomes 11 and 16 occurs following chemotherapy for other forms of cancer.

core binding factor acute myeloid leukemia

Another type of blood cancer known as core binding factor acute myeloid leukemia (CBF-AML) is associated with rearrangements of genetic material on chromosome 16. The most common of these rearrangements is an inversion of a region of chromosome 16 (written as inv(16)). An inversion involves breakage of the chromosome in two places; the resulting piece of DNA is reversed and reinserted into the chromosome. Less commonly, a translocation occurs between the two copies of chromosome 16 (written as t(16;16)). Both types of genetic rearrangement result in the fusion of two genes found on chromosome 16, *CBFB* and *MYH11*. These genetic changes are associated with 5 to 8 percent of cases of AML in adults. These mutations are acquired during a person's lifetime and are present only in certain cells. This type of genetic change, called a somatic mutation, is not inherited.

The protein produced from the normal *CBFB* gene interacts with another protein called RUNX1 to form a complex called core binding factor (CBF). This complex attaches to specific areas of DNA and turns on genes that are involved in the development of blood cells. The protein produced from the fusion gene, CBFβ-MYH11, can still bind to RUNX1. However, the function of CBF is impaired. The presence of CBFβ-MYH11 may block binding of CBF to DNA, impairing its ability to control gene activity. Alternatively, the MYH11 portion of the fusion protein may interact with other proteins that prevent the complex from controlling gene activity. The change in gene activity blocks the maturation (differentiation) of blood cells, which leads to the production of abnormal, immature white blood cells called myeloid blasts and to a shortage of normal, mature blood cell types. However, one or more additional genetic changes are typically needed for the myeloid blasts to develop into cancerous leukemia cells.

Rubinstein-Taybi syndrome

Some cases of severe Rubinstein-Taybi syndrome (also known as chromosome 16p13.3 deletion syndrome) have resulted from a deletion of genetic material from the short (p) arm of chromosome 16. When this deletion is present in all of the body's cells, it can cause serious complications such as a failure to gain weight and grow at the expected rate (failure to thrive) and an increased risk of life-threatening infections. Affected individuals also have many of the typical features of Rubinstein-Taybi syndrome, including intellectual disability, distinctive facial features, and broad thumbs and first toes. Infants born with the severe form of this disorder usually survive only into early childhood.

Several genes are missing as a result of the deletion in the short arm of chromosome 16. The deleted region includes the *CREBBP* gene, which is often mutated or missing in people with the typical features of Rubinstein-Taybi syndrome. Researchers believe that the loss of additional genes in this region probably accounts for the serious complications associated with severe Rubinstein-Taybi syndrome.

other chromosomal conditions

Trisomy 16 occurs when cells have three copies of chromosome 16 instead of the usual two copies. Full trisomy 16, which occurs when all of the body's cells contain an extra copy of chromosome 16, is not compatible with life. A similar but less severe condition called mosaic trisomy 16 occurs when only some of the body's cells have an extra copy of chromosome 16. The signs and symptoms of mosaic trisomy 16 vary widely and can include slow growth before birth (intrauterine growth retardation), delayed development, and heart defects.

Other changes in the number or structure of chromosome 16 can have a variety of effects. Intellectual disability, delayed growth and development, distinctive facial features, weak muscle tone (hypotonia), heart defects, and other medical problems are common. Frequent changes to chromosome 16 include an extra segment of the short (p) or long (q) arm of the chromosome in each cell (partial trisomy 16p or 16q) and a missing segment of the long arm of the chromosome in each cell (partial monosomy 16q).

Is there a standard way to diagram chromosome 16?

Geneticists use diagrams called ideograms as a standard representation for chromosomes. Ideograms show a chromosome's relative size and its banding pattern. A banding pattern is the characteristic pattern of dark and light bands that appears when a chromosome is stained with a chemical solution and then viewed under a microscope. These bands are used to describe the location of genes on each chromosome.

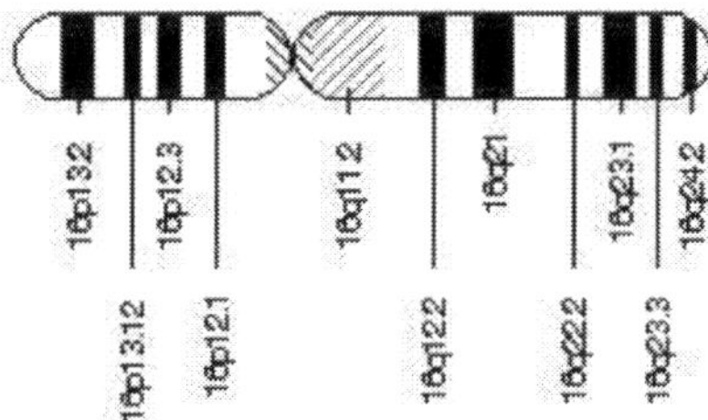

Conditions related to genes on chromosome 16

Genetics Home Reference includes these conditions related to genes on chromosome 16:

adenine phosphoribosyltransferase deficiency
alpha thalassemia
alveolar capillary dysplasia with misalignment of pulmonary
veins amyotrophic lateral sclerosis
Blau syndrome
breast cancer
Brody myopathy
Brooke-Spiegler syndrome
Charcot -Marie-Tooth disease
chronic granulomatous disease
combined malonic and methylmalonic aciduria

complete LCAT deficiency
congenital disorder of glycosylation type Ia
core binding factor acute myeloid leukemia
Crohn disease
Ewing sarcoma familial
cylindromatosis
familial hemiplegic migraine
familial Mediterranean fever
familial paroxysmal kinesigenic dyskinesia
familial thoracic aortic aneurysm and
dissection Fanconi anemia
fatty acid hydroxylase-associated neurodegeneration
fish-eye disease
Floating-Harbor syndrome
giant axonal neuropathy
Gitelman syndrome
glycogen storage disease type IX
hereditary diffuse gastric cancer
Huntington disease-like syndrome
infantile neuronal ceroid lipofuscinosis
juvenile Batten disease
Liddle syndrome
lymphangioleiomyomatosis
lymphedema-distichiasis syndrome
Mainzer-Saldino syndrome malonyl-
CoA decarboxylase deficiency Meier-
Gorlin syndrome
Miller syndrome mucolipidosis
III gamma
mucopolysaccharidosis type IV
multicentric osteolysis, nodulosis, and arthropathy
multiple familial trichoepithelioma
North American Indian childhood
cirrhosis oculocutaneous albinism
osteopetrosis
polycystic kidney disease
polymicrogyria
pseudohypoaldosteronism type 1
pseudoxanthoma elasticum
Rubinstein-Taybi syndrome
spastic paraplegia type 7
surfactant dysfunction

TK2-related mitochondrial DNA depletion syndrome, myopathic form
Townes-Brocks Syndrome
tuberous sclerosis complex
tyrosinemia
uromodulin-associated kidney disease

Reviewed: *October 2014*

Published: *November 17, 2014*

Genes on chromosome 16

Genetics Home Reference includes these genes on chromosome 16:

ABCA3
ABCC6
ACSF3
APRT
ATP2A1
CBFB
CDH1
CDT1
CIRH1A
CLCN7
CLN3
CREBBP
CYBA
CYLD
DHODH
FA2H
FANCA
FOXC2
FOXF1
FUS
GALNS
GAN
GNPTG
GPR56
HBA1
HBA2
IFT140
JPH3
LCAT

LITAF
MC1R
MEFV
MLYCD
MMP2
MYH11
NOD2
ORC6
PALB2
PHKB
PHKG2
PKD1
PMM2
PRRT2
SALL1
SCNN1B
SCNN1G
SLC12A3
SPG7
SRCAP
TAT
TK2
TSC2
UMOD

State of Vermont
Chromosome Seventeen

University of Vermont College of Medicine (Burlington) (Dermatofibrosarcoma Protuberans)

Chromosome 17

What is chromosome 17?

Humans normally have 46 chromosomes in each cell, divided into 23 pairs. Two copies of chromosome 17, one copy inherited from each parent, form one of the pairs. Chromosome 17 spans about 81 million DNA building blocks (base pairs) and represents between 2.5 and 3 percent of the total DNA in cells.

Identifying genes on each chromosome is an active area of genetic research. Because researchers use different approaches to predict the number of genes on each chromosome, the estimated number of genes varies. Chromosome 17 likely contains 1,200 to 1,300 genes that provide instructions for making proteins. These proteins perform a variety of different roles in the body.

Genes on chromosome 17 are among the estimated 20,000 to 25,000 total genes in the human genome.

How are changes in chromosome 17 related to health conditions?

Many genetic conditions are related to changes in particular genes on chromosome 17. This list of disorders associated with genes on chromosome 17 provides links to additional information.

Changes in the structure or number of copies of a chromosome can also cause problems with health and development. The following chromosomal conditions are associated with such changes in chromosome 17.

acute promyelocytic leukemia

A type of blood cancer known as acute promyelocytic leukemia is caused by a rearrangement (translocation) of genetic material between chromosomes 15 and 17. This translocation, written as t(15;17), fuses part of the *PML* gene from chromosome 15 with part of the *RARA* gene from chromosome 17. This mutation is acquired during a person's lifetime and is present only in certain cells. This type of genetic change, called a somatic mutation, is not inherited. The t(15;17) translocation is called a balanced reciprocal translocation because the pieces of chromosome are exchanged with each other (reciprocal) and no genetic material is gained or lost (balanced). The protein produced from this fused gene is known as PML-RARα.

The PML-RARα protein functions differently than the protein products from the normal *PML* and *RARA* genes. The *RARA* gene on chromosome 17 provides instructions for making a transcription factor called the retinoic acid receptor alpha (RARα). A transcription factor is a protein that attaches (binds) to specific regions of DNA and helps control the activity (transcription) of particular genes. Normally, the RARα protein controls the activity of genes important for the maturation (differentiation) of immature white blood cells beyond a particular stage called the promyelocyte. The *PML* gene on chromosome 15 provides instructions for a protein that acts as a tumor suppressor, which means it prevents cells from growing and dividing too rapidly or in an uncontrolled way. The PML protein blocks cell growth and division (proliferation) and induces self-destruction (apoptosis) in combination with other proteins. The PML-RARα protein interferes with the normal function of both the PML and the RARα proteins. As a result, blood cells are stuck at the promyelocyte stage, and they proliferate abnormally. Excess promyelocytes accumulate in the bone marrow and normal white blood cells cannot form, leading to acute promyelocytic leukemia.

dermatofibrosarcoma protuberans

Translocation of genetic material between chromosomes 17 and 22, written as t(17;22), causes a rare type of skin cancer known as dermatofibrosarcoma protuberans. This translocation fuses part of the *COL1A1* gene from chromosome 17 with part of the *PDGFB* gene from chromosome 22. The translocation is found on one or more extra chromosomes that can be either linear or circular. When circular, the extra chromosomes are known as supernumerary ring chromosomes. This mutation is acquired during a person's lifetime and is present only in certain cells. This type of genetic change, called a somatic mutation, is not inherited.

The fused *COL1A1-PDGFB* gene provides instructions for making a combined (fusion) protein that researchers believe ultimately functions like the active PDGFB protein. In the translocation, the *PDGFB* gene loses the part of its DNA that limits its activity, and production of the COL1A1-PDGFB fusion protein is controlled by *COL1A1* gene sequences. As a result, the gene fusion leads to the production of a larger amount of active PDGFB protein than normal. Active PDGFB protein signals for cell growth and division (proliferation) and maturation (differentiation) . Excess PDGFB protein abnormally stimulates cells to proliferate and differentiate, leading to the tumor formation seen in dermatofibrosarcoma protuberans.

Koolen-de Vries syndrome

Deletion of a small amount of genetic material (a microdeletion) on chromosome 17 can cause Koolen-de Vries syndrome. This disorder is characterized by developmental delay, intellectual disability, a cheerful and sociable disposition, and a variety of physical abnormalities.

Most people with Koolen-de Vries syndrome are missing a sequence of about 500,000 base pairs, also written as 500 kilobases (kb), at position q21.31 on chromosome 17. The exact size of the deletion varies among affected individuals, but it contains at least six genes including *KANSL1*. This deletion affects one of the two copies of chromosome 17 in each cell.

Because mutations in the *KANSL1* gene cause the same signs and symptoms as the deletion, researchers have concluded that the loss of this gene accounts for the features of Koolen-de Vries syndrome. The protein produced from the *KANSL1* gene is involved in controlling the activity of other genes and plays an important role in the development and function of many parts of the body. Although the loss of this gene impairs normal development and function, its relationship to the specific features of Koolen-de Vries syndrome is unclear.

While Koolen-de Vries syndrome is usually not inherited, most individuals with the condition caused by a deletion have had at least one parent with a common variant of the 17q21.31 region of chromosome 17 called the H2 lineage. This variant is found in 20 percent of people of European and Middle Eastern descent, although it is rare in other populations. In the H2 lineage, a 900 kb segment of DNA, which includes the region deleted in most cases of Koolen-de Vries syndrome, has undergone an inversion. An inversion involves two breaks in a chromosome; the resulting piece of DNA is reversed and reinserted into the chromosome.

People with the H2 lineage have no health problems related to the inversion. However, genetic material can be lost or duplicated when the inversion is passed to the next generation. Researchers believe that a parental inversion is probably necessary for a child to have the 17q21.31 microdeletion most often associated with Koolen-de Vries syndrome, but other, unknown factors are also thought to play a role. So while the inversion is very common, only an extremely small percentage of parents with the inversion have a child affected by Koolen-de Vries syndrome.

Miller-Dieker syndrome

Miller-Dieker syndrome is caused by a deletion of genetic material near the end of the short (p) arm of chromosome 17. The signs and symptoms of Miller-Dieker syndrome are related to the loss of multiple genes in this region. The size of the deletion varies among affected individuals. The loss of a particular gene on chromosome 17, called *PAFAH1B1*, is responsible for the syndrome's characteristic sign of lissencephaly, a problem with brain development in which the surface of the brain is abnormally smooth. The loss of another gene, called *YWHAE*, in the same region of chromosome 17 increases the severity of lissencephaly in people with Miller -Dieker syndrome. Additional genes in the deleted region contribute to the varied features of Miller-Dieker syndrome.

Smith-Magenis syndrome

Most people with Smith-Magenis syndrome have a deletion of genetic material from a specific part of chromosome 17 called the Smith-Magenis syndrome critical region. This region is located on the short (p) arm of chromosome 17 at position 11.2 (written as 17p11.2). Although this region contains multiple genes, researchers believe that the loss of one particular gene, *RAI1*, in each cell is responsible for most of the physical, mental, and behavioral features of Smith-Magenis syndrome. The loss of other genes in the deleted region may help explain why the signs and symptoms of this condition vary among affected individuals.

other cancers

Changes in chromosome 17 have been identified in several additional types of human cancer. These genetic changes are somatic, which means they are acquired during a person's lifetime and are present only in certain cells. A particular chromosomal abnormality called an isochromosome 17q occurs frequently in some cancers. This abnormal version of chromosome 17 has two long (q) arms instead of one long arm and one short (p) arm. As a result, the chromosome has an extra copy of some genes and is missing copies of other genes.

An isochromosome 17q is commonly found in a cancer of blood-forming tissue called chronic myeloid leukemia (CML). It also has been identified in certain solid tumors, including a type of brain tumor called a medulloblastoma and tumors of the brain and spinal cord known as primitive neuroectodermal tumors. Although an isochromosome 17q probably plays a role in both the development and progression of these cancers, the specific genetic changes related to cancer growth are unknown.

other chromosomal conditions

Other changes in the number or structure of chromosome 17 can have a variety of effects, including intellectual disability, delayed development, characteristic facial features, weak muscle tone (hypotonia), and short stature. These changes include an extra piece of chromosome 17 in each cell (partial trisomy 17), a missing segment of the chromosome in each cell (partial monosomy 17), and a circular structure called a ring chromosome 17. Ring chromosomes occur when a chromosome breaks in two places and the ends of the chromosome arms fuse together to form a circular structure.

Is there a standard way to diagram chromosome 17?

Geneticists use diagrams called ideograms as a standard representation for chromosomes. Ideograms show a chromosome's relative size and its banding pattern. A banding pattern is the characteristic pattern of dark and light bands that appears when a chromosome is stained with a chemical solution and then viewed under a microscope. These bands are used to describe the location of genes on each chromosome.

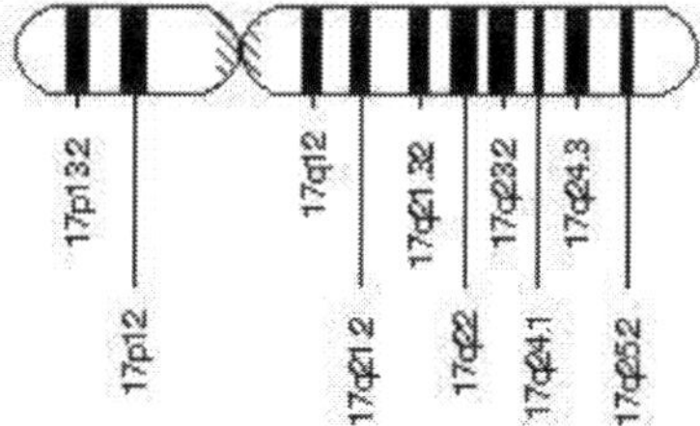

Conditions related to genes on chromosome 17

Genetics Home Reference includes these conditions related to genes on chromosome 17:

acute promyelocytic
leukemia Alexander disease
Amish lethal microcephaly

Andersen-Tawil syndrome
autoimmune Addison disease
Baraitser-Winter syndrome
Birt-Hogg-Dubé syndrome
bladder cancer
bradyopsia
breast cancer
Caffey disease
campomelic dysplasia
Canavan disease
Carney complex
Charcot-Marie-Tooth
disease Coats plus syndrome
Coffin-Siris syndrome
common variable immune
deficiency congenital myasthenic
syndrome cystinosis
dermatofibrosarcoma protuberans
dyskeratosis congenita Ehlers-
Danlos syndrome epidermolysis
bullosa simplex
epidermolysis bullosa with pyloric atresia
epidermolytic hyperkeratosis
familial atrial fibrillation
familial hemophagocytic lymphohistiocytosis
familial pityriasis rubra pilaris
Fanconi anemia
focal palmoplantar keratoderma
Freeman-Sheldon syndrome
frontotemporal dementia with parkinsonism-
17 galactosemia
glycogen storage disease type I GRN-
related frontotemporal dementia
hereditary folate malabsorption
hereditary neuralgic amyotrophy
hereditary neuropathy with liability to pressure palsies
hereditary spherocytosis
hyperkalemic periodic paralysis
hypokalemic periodic paralysis
ichthyosis with confetti
isolated growth hormone deficiency
isolated lissencephaly sequence

isolated Pierre Robin sequence
Job syndrome
Koolen-de Vries syndrome
Kuskokwim syndrome
Leber congenital amaurosis
Li-Fraumeni syndrome
limb-girdle muscular dystrophy
mandibulofacial dysostosis with
microcephaly Meesmann corneal dystrophy
Meier-Gorlin syndrome Miller-Dieker
syndrome mucopolysaccharidosis
type III N-acetylglutamate synthase
deficiency
Naegeli-Franceschetti-Jadassohnsyndrome/dermatopathia pigmentosa
reticularis
neuroblastoma
neurofibromatosis type 1
nonbullous congenital ichthyosiform erythroderma
nonsyndromic deafness
osteogenesis imperfecta
pachyonychia congenita
paramyotonia congenita
peroxisomal acyl-CoA oxidase deficiency
Pompe disease
pontocerebellar hypoplasia
potassium-aggravated myotonia
primary spontaneous pneumothorax
progressive supranuclear palsy
pseudohypoaldosteronism type 2
psoriatic arthritis
pyridoxal 5'-phosphate-dependent epilepsy
renal tubular dysgenesis
short QT syndrome
Sjögren-Larsson syndrome
SLC4A1-associated distal renal tubular acidosis
Smith-Magenis syndrome
SOST-related sclerosing bone dysplasia
spondylocostal dysostosis
steatocystoma multiplex
tarsal-carpal coalition syndrome
T-cell immunodeficiency, congenital alopecia, and nail dystrophy
tetra-amelia syndrome

Usher syndrome
very long-chain acyl-CoA dehydrogenase deficiency
vitiligo
white sponge nevus

Reviewed: *March 2013*

Published: *November 17, 2014*

Genes on chromosome 17

Genetics Home Reference includes these genes on chromosome 17:

ACADVL
ACE
ACOX1
ACTG1
ALDH3A2
ALOX12B
ALOXE3
ASPA
BRCA1
BRIP1
CARD14
CDC6
CHRNE
COL1A1
CTC1
CTNS
EFTUD2
ERBB2
FKBP10
FLCN
FOXN1
G6PC
GAA
GALK1
GFAP
GH1 GRN
GUCY2D
HES7
ITGB4
KANSL1

KCNJ2
KRT10
KRT12
KRT13
KRT14
KRT16
KRT17
MAPT
MYH3
MYO15A
NAGLU
NAGS
NF1
NLRP1
NOG
PAFAH1B1
PMP22
PNPO
PRKAR1A
RAI1
RARA
RGS9
SCN4A
SEPT9
SGCA
SGSH
SLC4A1
SLC25A19
SLC46A1
SMARCE1
SOST
SOX9
STAT3
TNFRSF13B
TP53
TSEN54
UNC13D
USH1G
WNK4
WNT3
YWHAE

State of Virginia
Chromosome 18

Eastern Virginia Medical School (Norfolk) (Fascioscapulohumeral Muscular Dystrophy)

Edward Via College of Osteopathic Medicine (Blacksburg) (Hereditary Hemorrhagic Telangiectasia)

Liberty University College of Osteopathic Medicine (Lynchburg) (Progessive Familial Intrahepatic Cholestasis)

University of Virginia School of Medicine (Charlottesville) (Congenital Cataracts)

Medical College of Virginia Health Sciences Division (Richmond) (Amyloidosis)

Virginia Tech Carilion School of Medicine (Roanoke) (Junctional Epidermololysis Bullosa)

Chromosome 18

What is chromosome 18?

Humans normally have 46 chromosomes in each cell, divided into 23 pairs. Two copies of chromosome 18, one copy inherited from each parent, form one of the pairs. Chromosome 18 spans about 78 million DNA building blocks (base pairs) and represents approximately 2.5 percent of the total DNA in cells.

Identifying genes on each chromosome is an active area of genetic research. Because researchers use different approaches to predict the number of genes on each chromosome, the estimated number of genes varies. Chromosome 18 likely contains 200 to 300 genes that provide instructions for making proteins. These proteins perform a variety of different roles in the body.

Genes on chromosome 18 are among the estimated 20,000 to 25,000 total genes in the human genome.

How are changes in chromosome 18 related to health conditions?

Many genetic conditions are related to changes in particular genes on chromosome 18. This list of disorders associated with genes on chromosome 18 provides links to additional information.

Changes in the structure or number of copies of a chromosome can also cause problems with health and development. The following chromosomal conditions are associated with such changes in chromosome 18.

18q deletion syndrome

18q deletion syndrome is caused by a deletion of genetic material from the long (q) arm of chromosome 18. This chromosomal change is written as 18q-. The signs and symptoms of 18q deletion syndrome are probably related to the loss of multiple genes in this region. Researchers are working to determine how the loss of these genes leads to the signs and symptoms of the disorder, which is characterized by neurological abnormalities including intellectual disability or learning problems and a wide variety of other features. Some affected individuals have a loss of tissue called white matter in the brain and spinal cord (leukodystrophy).

18q deletion syndrome is often categorized into two types: individuals with deletions near the end of the long arm of chromosome 18 are said to have distal 18q deletion syndrome, and those with deletions in the part of the long arm near the center of chromosome 18 are said to have proximal 18q deletion syndrome. The signs and symptoms of these two types of the condition are overlapping, with certain features being more common in one form of the disorder than in the other. For example, hearing loss and heart abnormalities are more common in people with distal 18q deletion syndrome, while seizures occur more often in people with proximal 18q deletion syndrome.

tetrasomy 18p

Tetrasomy 18p results from the presence of an abnormal extra chromosome, called an isochromosome 18p, in each cell. An isochromosome is a chromosome with two identical arms. Normal chromosomes have one long (q) arm and one short (p) arm, but isochromosomes have either two q arms or two p arms. Isochromosome 18p is a version of chromosome 18 made up of two p arms.

Cells normally have two copies of each chromosome, one inherited from each parent. In people with tetrasomy 18p, cells have the usual two copies of chromosome 18 plus an isochromosome 18p. As a result, each cell has four copies of the short arm of chromosome 18. (The word "tetrasomy" is derived from "tetra," the Greek word for "four.") The extra genetic material from the isochromosome disrupts the normal course of development, causing intellectual disability, delayed development, and the other characteristic features of this disorder.

trisomy 18

Trisomy 18 occurs when each cell in the body has three copies of chromosome 18 instead of the usual two copies, causing severe intellectual disability and multiple birth defects that are usually fatal by early childhood. (The word "trisomy" comes from "tri," the Greek word for "three.") In some cases, the extra copy of chromosome 18 is present in only some of the body's cells. This condition is known as mosaic trisomy 18.

Rarely, trisomy 18 is caused by an extra copy of only a piece of chromosome 18. This condition is known as partial trisomy 18. Partial trisomy 18 occurs when part of the long (q) arm of chromosome 18 becomes attached (translocated) to another chromosome during the formation of reproductive cells (eggs and sperm) or very early in embryonic development. Affected individuals have two copies of chromosome 18, plus the extra material from chromosome 18 attached to another chromosome. If only part of the q arm is present in three copies, the physical signs of partial trisomy 18 may be less severe than those typically seen in trisomy 18. If the entire q arm is present in three copies, individuals may be as severely affected as if they had three full copies of chromosome 18.

Researchers believe that extra copies of some genes on chromosome 18 disrupt the course of normal development, causing the characteristic features of trisomy 18 and the health problems associated with this disorder.

other chromosomal conditions

Other disorders associated with chromosome 18 occur when pieces of the short (p) arm of this chromosome are missing or when extra genetic material from chromosome 18 is present. Researchers are uncertain how missing or extra pieces of chromosome 18 lead to the specific features of these disorders.

Partial monosomy of chromosome 18p (18p-) occurs when a piece of the short arm of this chromosome is deleted. Individuals with this condition often have short stature, a round face, large ears, a shortened space between the nose and mouth (philtrum), droopy eyelids (ptosis), and mild to moderate intellectual disability. About 10 to 15 percent of people with this condition have serious abnormalities of the brain and spinal cord (central nervous system). The lifespan of individuals with partial monosomy of chromosome 18p is typically not reduced, except when severe brain abnormalities are present.

Some people have a chromosome 18 with a circular structure, which is called a ring chromosome 18. This type of chromosome is formed when breaks occur at both ends of the chromosome and the broken ends join together to form a ring. Individuals with this chromosome abnormality often have intellectual disability, an unusually small head (microcephaly), widely spaced eyes (hypertelorism), low-set ears, and speech problems. The signs and symptoms associated with ring chromosome 18 depend on how much genetic material is lost from each arm of the chromosome.

Is there a standard way to diagram chromosome 18?

Geneticists use diagrams called ideograms as a standard representation for chromosomes. Ideograms show a chromosome's relative size and its banding pattern. A banding pattern is the characteristic pattern of dark and light bands that appears when a chromosome is stained with a chemical solution and then viewed under a microscope. These bands are used to describe the location of genes on each chromosome.

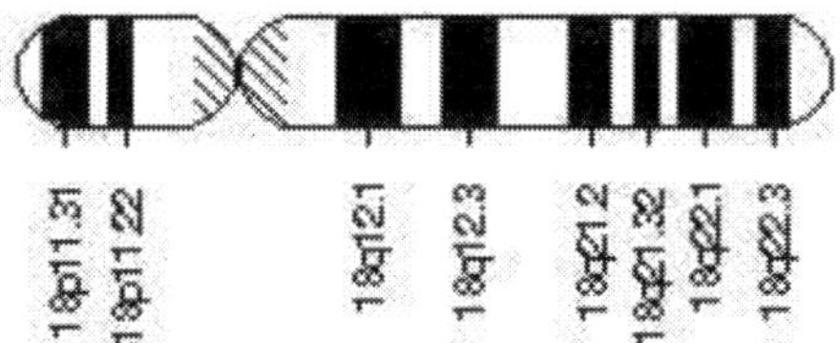

Chromosomes - chromosome 18

Conditions related to genes on chromosome 18

Genetics Home Reference includes these conditions related to genes on chromosome 18:

autosomal recessive hypotrichosis benign
recurrent intrahepatic cholestasis
congenital cataracts, facial dysmorphism, and
neuropathy facioscapulohumeral muscular dystrophy
Fuchs endothelial dystrophy
Hennekam syndrome
hereditary hemorrhagic telangiectasia

junctional epidermolysis bullosa
juvenile polyposis syndrome laryngo
-onycho-cutaneous syndrome
Majeed syndrome
microvillus inclusion disease
monilethrix
Myhre syndrome
Niemann-Pick disease
nonsyndromic holoprosencephaly
osteopetrosis
Paget disease of bone
Pitt-Hopkins syndrome
porphyria
progressive familial intrahepatic cholestasis
Schinzel-Giedion syndrome
transthyretin amyloidosis

Reviewed: *June 2014*

Published: *November 17, 2014*

Genes on chromosome 18

Genetics Home Reference includes these genes on chromosome 18:

ATP8B1
CCBE1
CTDP1
DSG4
FECH
LAMA3
LPIN2
MYO5B
NPC1
SETBP1
SMAD4
SMCHD1
TCF4
TGIF1
TNFRSF11A
TTR

State of Washington
Chromosome Nineteen

Pacific Northwest University of Health Sciences (Yakima) (Familial Hemiplegic Migraine)

University of Washinton School of Medicine (Seattle) (Breast Cancer) (Atypical Hemolytic Uremic Syndrome)

Chromosome 19

What is chromosome 19?

Humans normally have 46 chromosomes in each cell, divided into 23 pairs. Two copies of chromosome 19, one copy inherited from each parent, form one of the pairs. Chromosome 19 spans about 59 million base pairs (the building blocks of DNA) and represents almost 2 percent of the total DNA in cells.

Identifying genes on each chromosome is an active area of genetic research. Because researchers use different approaches to predict the number of genes on each chromosome, the estimated number of genes varies. Chromosome 19 likely contains about 1,500 genes that provide instructions for making proteins. These proteins perform a variety of different roles in the body.

Genes on chromosome 19 are among the estimated 20,000 to 25,000 total genes in the human genome.

How are changes in chromosome 19 related to health conditions?

Many genetic conditions are related to changes in particular genes on chromosome 19. This list of disorders associated with genes on chromosome 19 provides links to additional information.

Is there a standard way to diagram chromosome 19?

Geneticists use diagrams called ideograms as a standard representation for chromosomes. Ideograms show a chromosome's relative size and its banding pattern. A banding pattern is the characteristic pattern of dark and light bands that appears when a chromosome is stained with a chemical solution and then viewed under a microscope. These bands are used to describe the location of genes on each chromosome.

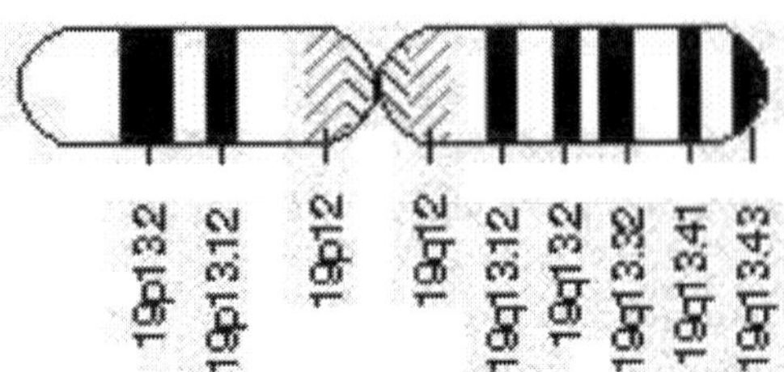

Chromosome 19

Conditions related to genes on chromosome 19

Genetics Home Reference includes these conditions related to genes on chromosome 19:

Adams-Oliver syndrome age-
related macular degeneration
Aicardi-Goutieres syndrome
alpha-mannosidosis
alternating hemiplegia of childhood
Alzheimer disease
atypical hemolytic-uremic syndrome
bradyopsia
branchiootorenal syndrome
breast cancer Camurati-
Engelmann disease
cardiofaciocutaneous syndrome
Carpenter syndrome
central core disease
centronuclear myopathy
cerebral autosomal dominant arteriopathy with subcortical infarcts and leukoencephalopathy
Charcot-Marie-Tooth disease
Coffin-Siris syndrome cold-
induced sweating syndrome
congenital fiber-type disproportion
congenital hypothyroidism Costeff
syndrome

cyclic neutropenia
cystinuria
cytogenetically normal acute myeloid leukemia
dense deposit disease
Diamond-Blackfan anemia
episodic ataxia
essential thrombocythemia
ethylmalonic encephalopathy
familial acute myeloid leukemia with mutated CEBPA
familial cold autoinflammatory syndrome
familial dilated cardiomyopathy
familial erythrocytosis
familial hemiplegic migraine familial
hypertrophic cardiomyopathy familial
restrictive cardiomyopathy
glucose phosphate isomerase deficiency
glutaric acidemia type I
glutaric acidemia type II
glycogen storage disease type 0
guanidinoacetate methyltransferase deficiency
hemochromatosis
hereditary sensory and autonomic neuropathy type IE
hidradenitis suppurativa
hypercholesterolemia
hyperferritinemia-cataract syndrome
Kawasaki disease
limb-girdle muscular dystrophy
malignant hyperthermia
maple syrup urine disease
mitochondrial membrane protein-associated neurodegeneration
mucolipidosis type IV
multiminicore disease
multiple epiphyseal dysplasia
myotonic dystrophy
neuroferritinopathy
persistent Müllerian duct syndrome
Peutz-Jeghers syndrome
polycystic lipomembranous osteodysplasia with sclerosing
leukoencephalopathy
pontocerebellar hypoplasia
primary myelofibrosis
prolidase deficiency

pseudoachondroplasia rapid-onset
dystonia parkinsonism recurrent
hydatidiform mole severe
congenital neutropenia
spinocerebellar ataxia type 6
spondylocostal dysostosis
spondyloenchondrodysplasia with immune dysregulation
sporadic hemiplegic migraine
trichothiodystrophy Walker-
Warburg syndrome Weill-
Marchesani syndrome
xeroderma pigmentosum

Published: *November 17, 2014*

Genes on chromosome 19

Genetics Home Reference includes these genes on chromosome 19:

ACP5
ADAMTS10
AMH APOE
ATP1A3
BCKDHA
C3 C19orf12
CACNA1A
CALR
CEBPA
COMP
CRLF1
DLL3
DMPK
DNM2
DNMT1
DOCK6
ELANE
EPOR
ERCC2
ETFB

ETHE1
FKRP
FTL
GAMT
GCDH
GPI
GYS1
HAMP
ITPKC
LDLR
MAN2B1
MAP2K2
MCOLN1
MEGF8
NLRP7
NLRP12
NOTCH3
OPA3
PEPD
PRX
PSENEN
RGS9BP
RNASEH2A
RPS19
RYR1
SIX5
SLC5A5
SLC7A9
SMARCA4
STK11
TGFB1
TNNI3
TSEN34
TYROBP

State of West Virginia
Chromosome Twenty

West Virginia University School of Medicine (Morgantown) (Benign Familial Neonatal Seizures)

West Virginia School of Osteopathic Medicine (Lewisburg) (Neurohypophyseal Diabetes Insipitus)

Joan C Edwards School of Medicine at Marshall University (Huntington) (Multiple Epiphyseal Dysplasia)

Chromosome 20

What is chromosome 20?

Humans normally have 46 chromosomes in each cell, divided into 23 pairs. Two copies of chromosome 20, one copy inherited from each parent, form one of the pairs. Chromosome 20 spans about 63 million DNA building blocks (base pairs) and represents approximately 2 percent of the total DNA in cells.

Identifying genes on each chromosome is an active area of genetic research. Because researchers use different approaches to predict the number of genes on each chromosome, the estimated number of genes varies. Chromosome 20 likely contains 500 to 600 genes that provide instructions for making proteins. These proteins perform a variety of different roles in the body.

Genes on chromosome 20 are among the estimated 20,000 to 25,000 total genes in the human genome.

How are changes in chromosome 20 related to health conditions?

Many genetic conditions are related to changes in particular genes on chromosome 20. This list of disorders associated with genes on chromosome 20 provides links to additional information.

Changes in the structure or number of copies of a chromosome can also cause problems with health and development. The following chromosomal conditions are associated with such changes in chromosome 20.

Alagille syndrome

Approximately 7 percent of individuals with Alagille syndrome have small deletions of genetic material on chromosome 20, in a region known as 20p12. This region includes the *JAG1* gene, which is involved in signaling between neighboring cells during embryonic development. This signaling influences how the cells are used to build body structures in the developing embryo. Loss of the *JAG1* gene probably disrupts the signaling pathway. As a result, errors may occur during development, especially affecting the heart, bile ducts in the liver, the spinal column, and certain facial features.

cancers

Changes in chromosome 20 have been identified in several types of cancer. These chromosome abnormalities are somatic, which means they are acquired during a person's lifetime and are present only in certain cells. Deletions involving the long (q) arm of chromosome 20 appear to be common in blood-related cancers such as leukemia and lymphoma. Deletions of this chromosomal region have also been identified in other disorders of the blood and bone marrow, including polycythemia vera (which causes an overproduction of red blood cells) and myelodysplastic syndrome (which leads to a shortage of healthy blood cells).

Researchers are working to determine which genes on chromosome 20 are disrupted in these conditions. Studies suggest that some genes on the long arm of the chromosome may play critical roles in controlling the growth and division of cells.

ring chromosome 20 syndrome

Ring chromosome 20 syndrome is caused by a chromosomal abnormality known as a ring chromosome 20 or r(20). A ring chromosome is a circular structure that occurs when a chromosome breaks in two places and its broken ends fuse together. People with ring chromosome 20 syndrome have one copy of this abnormal chromosome in some or all of their cells.

It is not well understood how the ring chromosome causes the signs and symptoms of this syndrome. In some affected individuals, genes near the ends of chromosome 20 are deleted when the ring chromosome forms. Researchers suspect that the loss of these genes may be responsible for epilepsy and other health problems. However, other affected individuals do not have gene deletions associated with the ring chromosome. In these people, the ring chromosome may change the activity of certain genes on chromosome 20, or it may be unable to copy (replicate) itself normally during cell division. Researchers are still working to determine the precise relationship between the ring chromosome 20 and the characteristic features of the syndrome.

other chromosomal conditions

Deletions or duplications of genetic material from chromosome 20 can have a variety of effects, including intellectual disability, delayed development, distinctive facial features, skeletal abnormalities, and heart defects. Several different changes in the structure of chromosome 20 have been reported. These include an extra segment of the short (p) or long (q) arm of the chromosome in each cell (partial trisomy 20p or 20q) or a missing segment of the short or long arm of the chromosome in each cell (partial monosomy 20p or 20q).

Is there a standard way to diagram chromosome 20?

Geneticists use diagrams called ideograms as a standard representation for chromosomes. Ideograms show a chromosome's relative size and its banding pattern. A banding pattern is the characteristic pattern of dark and light bands that appears when a chromosome is stained with a chemical solution and then viewed under a microscope. These bands are used to describe the location of genes on each chromosome.

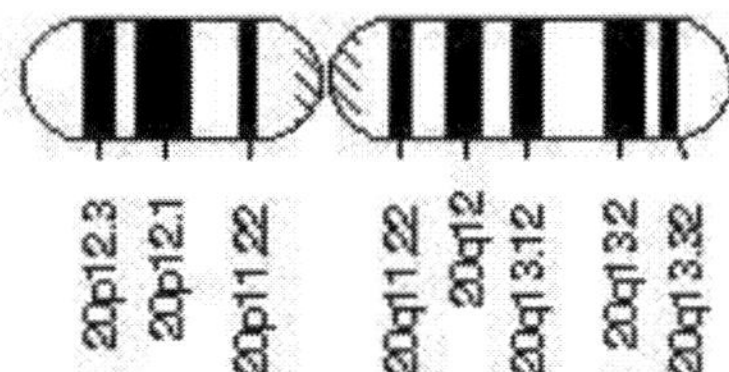

Chromosomes > chromosome 20 >

Conditions related to genes on chromosome 20

Genetics Home Reference includes these conditions related to genes on chromosome 20:

adenosine deaminase deficiency
Aicardi-Goutieres syndrome
Alagille syndrome amyotrophic
lateral sclerosis arterial
tortuosity syndrome auriculo-
condylar syndrome
autosomal dominant nocturnal frontal lobe epilepsy

Bardet-Biedl syndrome
benign familial neonatal seizures
Brown-Vialetto-Van Laere
syndrome congenital
dyserythropoietic anemia Duane-
radial ray syndrome galactosialidosis
glutathione synthetase deficiency
hereditary cerebral amyloid
angiopathy Hirschsprung disease
Huntington disease-like syndrome
hypermethioninemia
Kallmann syndrome
Kufs disease
McCune-Albright syndrome
McKusick-Kaufman syndrome
multiple epiphyseal dysplasia
neurohypophyseal diabetes insipidus
pantothenate kinase-associated
neurodegeneration periventricular heterotopia
prion disease
progressive osseous
heteroplasia spinal muscular
atrophy Waardenburg syndrome
Wilson disease

Genes on chromosome 20

Genetics Home Reference includes these genes on chromosome 20:

ADA
AHCY
ARFGEF2
AVP
CHRNA4
COL9A3
CST3
CTSA
DNAJC5
EDN3
GNAS
GSS
JAG1
KCNQ2

MKKS
PANK2
PLCB4
PRNP
PROKR2
SALL4
SAMHD1
SEC23B
SLC2A10
SLC52A3
VAPB

State of Wisconsin
Chromosome Twenty One

Medical College of Wisconsin (Milwaukee) (Down Syndrome)

University of Wisconsin School of Medicine and Pubic Health (Madison) (Acute Lymphoblastic Leukemia)

Chromosome 21

Reviewed November 2013

What is chromosome 21?

Humans normally have 46 chromosomes in each cell, divided into 23 pairs. Two copies of chromosome 21, one copy inherited from each parent, form one of the pairs. Chromosome 21 is the smallest human chromosome, spanning about 48 million base pairs (the building blocks of DNA) and representing 1.5 to 2 percent of the total DNA in cells.

In 2000, researchers working on the Human Genome Project announced that they had determined the sequence of base pairs that make up this chromosome. Chromosome 21 was the second human chromosome to be fully sequenced.

Identifying genes on each chromosome is an active area of genetic research. Because researchers use different approaches to predict the number of genes on each chromosome, the estimated number of genes varies. Chromosome 21 likely contains 200 to 300 genes that provide instructions for making proteins. These proteins perform a variety of different roles in the body.

Genes on chromosome 21 are among the estimated 20,000 to 25,000 total genes in the human genome.

How are changes in chromosome 21 related to health conditions?

Many genetic conditions are related to changes in particular genes on chromosome 21. This list of disorders associated with genes on chromosome 21 provides links to additional information.

Changes in the structure or number of copies of a chromosome can also cause problems with health and development. The following chromosomal conditions are associated with such changes in chromosome 21.

core binding factor acute myeloid leukemia

A genetic rearrangement (translocation) involving chromosome 21 is associated with a type of blood cancer known as core binding factor acute myeloid leukemia (CBF-AML). This rearrangement occurs in approximately 7 percent of acute myeloid leukemia cases in adults. The translocation, written as t(8;21), fuses part of the *RUNX1* gene from chromosome 21 with part of the *RUNX1T1* gene (also known as *ETO*) from chromosome 8. This mutation is acquired during a person's lifetime and is present only in certain cells. This type of genetic change, called a somatic mutation, is not inherited.

The fusion protein produced from the t(8;21) translocation, called RUNX1-ETO, retains some functions of the two individual proteins. The normal RUNX1 protein, produced from the *RUNX1* gene, is part of a protein complex called core binding factor (CBF) that attaches (binds) to DNA and turns on genes involved in blood cell development. The normal ETO protein, produced from the *RUNX1T1* gene, turns off gene activity. The RUNX1-ETO fusion protein forms CBF and attaches to DNA, but instead of turning on genes that stimulate the development of blood cells, it turns those genes off. This change in gene activity blocks the maturation (differentiation) of blood cells and leads to the production of abnormal, immature white blood cells called myeloid blasts. While t(8;21) is important for leukemia development, one or more additional genetic changes are typically needed for the myeloid blasts to develop into cancerous leukemia cells.

Down syndrome

Down syndrome is a chromosomal condition that is associated with intellectual disability, a characteristic facial appearance, and weak muscle tone (hypotonia) in infancy. This condition is most often caused by trisomy 21. Trisomy 21 means that each cell in the body has three copies of chromosome 21 instead of the usual two copies.

Less commonly, Down syndrome occurs when part of chromosome 21 becomes attached (translocated) to another chromosome during the formation of reproductive cells (eggs and sperm) or very early in fetal development. Affected people have two copies of chromosome 21 plus extra material from chromosome 21 attached to another chromosome, resulting in three copies of genetic material from chromosome 21.

Affected individuals with this genetic change are said to have translocation Down syndrome.

In a very small percentage of cases, Down syndrome results from an extra copy of chromosome 21 in only some of the body's cells. In these people, the condition is called mosaic Down syndrome.

Researchers believe that having extra copies of genes on chromosome 21 disrupts the course of normal development, causing the characteristic features of Down syndrome and the increased risk of health problems associated with this condition.

other cancers

Translocations of genetic material between chromosome 21 and other chromosomes have been associated with several types of cancer. For example, acute lymphoblastic leukemia (a type of blood cancer most often diagnosed in childhood) has been associated with a translocation between chromosomes 12 and 21.

other chromosomal conditions

Other changes in the number or structure of chromosome 21 have a variety of effects on health and development. Chromosome 21 abnormalities can cause intellectual disability, delayed development, and characteristic facial features. In some cases, the signs and symptoms are similar to those of Down syndrome.

Changes involving chromosome 21 can include a missing segment of the chromosome in each cell (partial monosomy 21) and a circular structure called ring chromosome 21. A ring chromosome occurs when a chromosome breaks in two places and the ends of the chromosome arms fuse together to form a circular structure.

Is there a standard way to diagram chromosome 21?

Geneticists use diagrams called ideograms as a standard representation for chromosomes. Ideograms show a chromosome's relative size and its banding pattern. A banding pattern is the characteristic pattern of dark and light bands that appears when a chromosome is stained with a chemical solution and then viewed under a microscope. These bands are used to describe the location of genes on each chromosome.

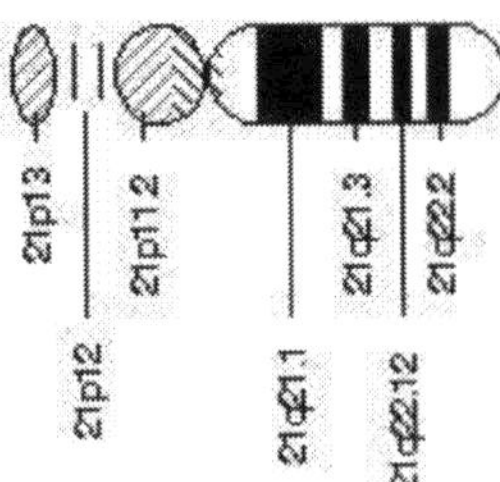

Conditions related to genes on chromosome 21

Genetics Home Reference includes these conditions related to genes on chromosome 21:

- Alzheimer disease
- amyotrophic lateral sclerosis
- autoimmune polyglandular syndrome, type 1
- Bethlem myopathy
- core binding factor acute myeloid leukemia
- cytogenetically normal acute myeloid leukemia
- familial atrial fibrillation
- glutamate formiminotransferase deficiency
- hereditary cerebral amyloid angiopathy
- holocarboxylase synthetase deficiency
- homocystinuria
- Jervell and Lange-Nielsen syndrome
- Knobloch syndrome
- leukocyte adhesion deficiency type 1
- microcephalic osteodysplastic primordial dwarfism type II
- nonsyndromic deafness
- rheumatoid arthritis
- Romano-Ward syndrome
- Ullrich congenital muscular dystrophy
- Unverricht-Lundborg disease

Reviewed: November 2013

Published: November 17, 2014

Genes on chromosome 21

Genetics Home Reference includes these genes on chromosome 21:

- AIRE
- APP
- CBS

CLDN14
COL6A1
COL6A2
COL18A1
CSTB
FTCD
HLCS
ITGB2
KCNE1
KCNE2
PCNT
RUNX1
SOD1
TMPRSS3

Chromosome Twenty Two
Puerto Rico and Wyoming

University of Puerto Rico School of Medicine (San Juan) (Methemoglobinemia)

Ponce School of Medicine (Ponce) (Schizophrenia) (COMT Gene Mental Illness)

San Juan Bautista School of Medicine (Caguas) (Breast Cancer) (Glucose /Galactose Malabsorption)

Universidad Central del Caribe School of Medicine (Bayamon) (Amyotropic Lateral Sclerosis)

Chromosome 22

What is chromosome 22?

Humans normally have 46 chromosomes (23 pairs) in each cell. Two copies of chromosome 22, one copy inherited from each parent, form one of the pairs. Chromosome 22 is the second smallest human chromosome, spanning more than 51 million DNA building blocks (base pairs) and representing between 1.5 and 2 percent of the total DNA in cells.

In 1999, researchers working on the Human Genome Project announced they had determined the sequence of base pairs that make up this chromosome. Chromosome 22 was the first human chromosome to be fully sequenced.

Identifying genes on each chromosome is an active area of genetic research. Because researchers use different approaches to predict the number of genes on each chromosome, the estimated number of genes varies. Chromosome 22 likely contains 500 to 600 genes that provide instructions for making proteins. These proteins perform a variety of different roles in the body.

Genes on chromosome 22 are among the estimated 20,000 to 25,000 total genes in the human genome.

How are changes in chromosome 22 related to health conditions?

Many genetic conditions are related to changes in particular genes on chromosome 22. This list of disorders associated with genes on chromosome 22 provides links to additional information.

Changes in the structure or number of copies of a chromosome can also cause problems with health and development. The following chromosomal conditions are associated with such changes in chromosome 22.

22q11.2 deletion syndrome

Most people with 22q11.2 deletion syndrome are missing about 3 million base pairs on one copy of chromosome 22 in each cell. The deletion occurs near the middle of the chromosome at a location designated as q11.2. This region contains 30 to 40 genes, but many of these genes have not been well characterized. A small percentage of affected individuals have shorter deletions in the same region.

The loss of a particular gene, *TBX1*, is thought to be responsible for many of the characteristic features of 22q11.2 deletion syndrome such as heart defects, an opening in the roof of the mouth (a cleft palate), distinctive facial features, and low calcium levels. Some studies suggest that a deletion of this gene may contribute to behavioral problems as well. The loss of another gene, *COMT*, in the same region of chromosome 22 may also help explain the increased risk of behavioral problems and mental illness. Additional genes in the deleted region likely contribute to the varied signs and symptoms of 22q11.2 deletion syndrome.

22q11.2 duplication

22q11.2 duplication is caused by an extra copy of some genetic material at position q11.2 on chromosome 22. In most cases, this extra genetic material consists of a sequence of about 3 million DNA building blocks (base pairs), also written as 3 megabases (Mb). This sequence is the same one that is missing in 22q11.2 deletion syndrome. A small percentage of affected individuals have a shorter duplication in the same region. The duplication affects one of the two copies of chromosome 22 in each cell. Researchers are working to determine the genes that may contribute to the developmental delay and other problems that affect some people with this duplication.

22q13.3 deletion syndrome

22q13.3 deletion syndrome, which is also commonly known as Phelan-McDermid syndrome, is caused by a deletion near the end of the long (q) arm of chromosome 22. A ring chromosome 22 can also cause 22q13.3

deletion syndrome. A ring chromosome is a circular structure that occurs when a chromosome breaks in two places, the tips of the chromosome are lost, and the broken ends fuse together. People with ring chromosome 22 have one copy of this abnormal chromosome in some or all of their cells. Researchers believe that several critical genes near the end of the long (q) arm of chromosome 22 are lost when the ring chromosome 22 forms. If the break point on the long arm is at chromosome position 22q13.3, people with ring chromosome 22 will experience similar signs and symptoms as those with a simple deletion.

The signs and symptoms of 22q13.3 deletion syndrome are probably related to the loss of multiple genes at the end of chromosome 22. The size of the deletion varies among affected individuals. The loss of a particular gene, *SHANK3*, is thought to be responsible for many of the characteristic features of 22q13.3 deletion syndrome, such as developmental delay, intellectual disability, and absent or severely delayed speech. Additional genes in the deleted region likely contribute to the signs and symptoms of 22q13.3 deletion syndrome.

dermatofibrosarcoma protuberans

A rearrangement (translocation) of genetic material between chromosomes 17 and 22, written as t(17;22), causes a rare type of skin cancer known as dermatofibrosarcoma protuberans. This translocation fuses part of the *PDGFB* gene from chromosome 22 with part of the *COL1A1* gene from chromosome 17. The translocation is found on one or more extra chromosomes that can be either linear or circular. When circular, the extra chromosomes are known as supernumerary ring chromosomes. This mutation is acquired during a person's lifetime and is present only in certain cells. This type of genetic change, called a somatic mutation, is not inherited.

The fused *COL1A1-PDGFB* gene provides instructions for making a combined (fusion) protein that researchers believe ultimately functions like the active PDGFB protein. In the translocation, the *PDGFB* gene loses the part of its DNA that limits its activity, and production of the COL1A1-PDGFB fusion protein is controlled by *COL1A1* gene sequences. As a result, the gene fusion leads to the production of a larger amount of active PDGFB protein than normal. Active PDGFB protein signals for cell growth and division (proliferation) and maturation (differentiation) . Excess PDGFB protein abnormally stimulates cells to proliferate and differentiate, leading to the tumor formation seen in dermatofibrosarcoma protuberans.

Emanuel syndrome

Emanuel syndrome is caused by the presence of extra genetic material from chromosome 11 and chromosome 22 in each cell. In addition to the usual 46 chromosomes, people with Emanuel syndrome have an extra (supernumerary) chromosome consisting of a piece of chromosome 11 attached to a piece of chromosome 22. The extra chromosome is known as a derivative 22 or der(22) chromosome.

People with Emanuel syndrome typically inherit the der(22) chromosome from an unaffected parent. The parent carries a chromosomal rearrangement between chromosomes 11 and 22 called a balanced translocation. No genetic material is gained or lost in a balanced translocation, so these chromosomal changes usually do not cause any health problems. As this translocation is passed to the next generation, it can become unbalanced. Individuals with Emanuel syndrome inherit an unbalanced translocation between chromosomes 11 and 22 in the form of a der(22) chromosome. These individuals have two normal copies of chromosome 11, two normal copies of chromosome 22, and extra genetic material from the der(22) chromosome.

As a result of the extra chromosome, people with Emanuel syndrome have three copies of some genes in each cell instead of the usual two copies. The excess genetic material disrupts the normal course of development, leading to intellectual disability and birth defects. Researchers are working to determine which genes are included on the der(22) chromosome and what role these genes play in development.

Ewing sarcoma

Translocations involving chromosome 22 are also involved in a type of cancerous tumor known as Ewing sarcoma. These tumors develop in the bones or soft tissues, such as cartilage and nerves. The most common translocation, t(11;22), fuses part of the *EWSR1* gene from chromosome 22 with part of the *FLI1* gene from chromosome 11, creating the *EWSR1/FLI1* fusion gene. Translocations that fuse the *EWSR1* gene with other genes that are related to the *FLI1* gene can also cause Ewing sarcomas, although these alternative translocations are relatively uncommon. The mutations that cause these cancers are acquired during a person's lifetime and are present only in tumor cells. This type of genetic change, called a somatic mutation, is not inherited.

The protein produced from the *EWSR1/FLI1* fusion gene, called EWS/FLI, has functions of the protein products of both genes. The FLI protein, produced from the FLI1 gene, attaches (binds) to DNA and regulates an activity called transcription, which is the first step in the production of proteins from genes. The FLI protein controls the growth and development of some cell types by regulating the transcription of certain genes. The EWS protein, produced from the EWSR1 gene, also regulates transcription. The EWS/FLI protein has the DNA-binding function of the FLI protein as well as the transcription regulation function of the EWS protein. It is thought that the EWS/FLI protein turns the transcription of a variety of genes on and off abnormally. This dysregulation of transcription leads to uncontrolled growth and division (proliferation) and abnormal maturation and survival of cells, causing tumor development.

Opitz G/BBB syndrome

The autosomal dominant form of Opitz G/BBB syndrome is caused by a deletion in one copy of chromosome 22 in each cell. This condition is considered part of 22q11.2 deletion syndrome because affected people usually have a deletion in the same region of chromosome 22. These cases occur in people with no history of the disorder in their family. It is not yet known which deleted gene(s) cause the signs and symptoms of Opitz G/BBB syndrome.

other cancers

Several types of blood cancer known as leukemias are associated with a translocation of genetic material between chromosomes 9 and 22. This chromosomal abnormality, which is commonly called the Philadelphia chromosome, is found only in cancer cells. The translocation that results in the Philadelphia chromosome is somatic, which means it is acquired during a person's lifetime. This translocation fuses part of a specific gene from chromosome 22 (the *BCR* gene) with part of another gene from chromosome 9 (the *ABL1* gene). The protein produced from this fused gene abnormally signals tumor cells to continue dividing and prevents them from adequately repairing DNA damage.

The Philadelphia chromosome has been identified in most cases of a slowly progressing form of blood cancer called chronic myeloid leukemia (CML). It also has been found in some cases of more rapidly progressing blood cancers known as acute leukemias. The presence of the Philadelphia chromosome can help predict how a cancer will progress and provides a target for molecular therapies.

other chromosomal conditions

Other changes in the number or structure of chromosome 22 can have a variety of effects. Intellectual disability, delayed development, delayed or absent speech, distinctive facial features, and behavioral problems are common features. Frequent changes to chromosome 22 include an extra piece of the chromosome in each cell (partial trisomy), a missing segment of the chromosome in each cell (partial monosomy), and a ring chromosome 22. Rearrangements (translocations) of genetic material between chromosomes can also lead to extra or missing material from chromosome 22. The most common of these translocations involves chromosomes 11 and 22.

Cat-eye syndrome is a rare disorder most often caused by a chromosomal change called an inverted duplicated 22. In people with this condition, each cell has at least one small extra chromosome made up of genetic material from chromosome 22 that has been abnormally copied (duplicated). The extra genetic material causes the characteristic signs and symptoms of cat-eye syndrome, including an eye abnormality called an iris coloboma (a gap or split in the colored part of the eye), small skin tags or pits in front of the ear, unusually formed ears, heart defects, kidney problems, malformations of the anus, and, in some cases, delayed development.

Is there a standard way to diagram chromosome 22?

Geneticists use diagrams called ideograms as a standard representation for chromosomes. Ideograms show a chromosome's relative size and its banding pattern. A banding pattern is the characteristic pattern of dark and light bands that appears when a chromosome is stained with a chemical solution and then viewed under a microscope. These bands are used to describe the location of genes on each chromosome.

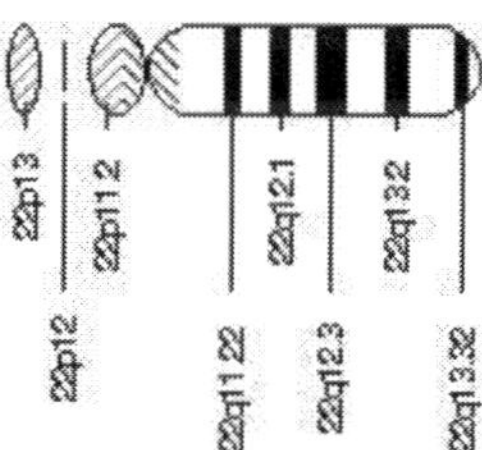

Conditions related to genes on chromosome 22

Genetics Home Reference includes these conditions related to genes on chromosome 22:

2-hydroxyglutaric aciduria
22q11.2 deletion syndrome

22q13.3 deletion syndrome
amyotrophic lateral sclerosis
beta-ureidopropionase deficiency
breast cancer
chronic granulomatous disease
Coffin-Siris syndrome
dermatofibrosarcoma protuberans
Ewing sarcoma glucose-
galactose malabsorption
hyperprolinemia
infantile neuroaxonal dystrophy iron-
refractory iron deficiency anemia Li-
Fraumeni syndrome
megalencephalic leukoencephalopathy with subcortical
cysts metachromatic leukodystrophy
mitochondrial neurogastrointestinal encephalopathy
disease MYH9-related disorder
neurofibromatosis type 2
nonsyndromic deafness
Rubinstein-Taybi
syndrome Schindler disease
transcobalamin deficiency
Waardenburg syndrome
Walker-Warburg syndrome

Reviewed: *May 2012*

Published: *November 17, 2014*

Genes on chromosome 22

Genetics Home Reference includes these genes on chromosome 22:

ARSA
CHEK2
COMT
EP300
EWSR1
LARGE
MLC1
MYH9
NAGA
NCF4
NEFH
NF2
PDGFB
PLA2G6
PRODH
SHANK3
SLC5A1
SLC25A1
SMARCB1
SOX10
TBX1
TCN2
TMPRSS6
TRIOBP
TYMP
UPB1

The State of Alaska and Montana are Assigned Chromosome X in addition to the newly opened Touro College of Osteopathic Medicine in Middletown New York

Chromosome X
Alaska and Montana

Newly Opened Touro College of Osteopathic Medicine (Middletown New York) (Turner Syndrome) (Paroxysmal Noctural Hemoglobinuria)

X chromosome

What is the X chromosome?

The X chromosome is one of the two sex chromosomes in humans (the other is the Y chromosome). The sex chromosomes form one of the 23 pairs of human chromosomes in each cell. The X chromosome spans about 155 million DNA building blocks (base pairs) and represents approximately 5 percent of the total DNA in cells.

Each person normally has one pair of sex chromosomes in each cell. Females have two X chromosomes, while males have one X and one Y chromosome. Early in embryonic development in females, one of the two X chromosomes is randomly and permanently inactivated in cells other than egg cells. This phenomenon is called X-inactivation or Lyonization. X-inactivation ensures that females, like males, have one functional copy of the X chromosome in each body cell. Because X-inactivation is random, in normal females the X chromosome inherited from the mother is active in some cells, and the X chromosome inherited from the father is active in other cells.

Some genes on the X chromosome escape X -inactivation. Many of these genes are located at the ends of each arm of the X chromosome in areas known as the pseudoautosomal regions. Although many genes are unique to the X chromosome, genes in the pseudoautosomal regions are present on both sex chromosomes. As a result, men and women each have two functional copies of these genes. Many genes in the pseudoautosomal regions are essential for normal development.

Identifying genes on each chromosome is an active area of genetic research. Because researchers use different approaches to predict the number of genes on each chromosome, the estimated number of genes varies. The X chromosome likely contains 800 to 900 genes that provide instructions for making proteins. These proteins perform a variety of different roles in the body.

Genes on the X chromosome are among the estimated 20,000 to 25,000 total genes in the human genome.

How are changes in the X chromosome related to health conditions?

Many genetic conditions are related to changes in particular genes on the X chromosome. This list of disorders associated with genes on the X chromosome provides links to additional information.

Changes in the structure or number of copies of a chromosome can also cause problems with health and development. The following chromosomal conditions are associated with such changes in the X chromosome.

46,XX testicular disorder of sex development

In most individuals with 46,XX testicular disorder of sex development, the condition results from an abnormal exchange of genetic material between chromosomes (trnslocation). This exchange occurs as a random event during the formation of sperm cells in the affected person's father. The translocation affects the gene responsible for development of a fetus into a male (the *SRY* gene). The *SRY* gene, which is normally found on the Y chromosome, is misplaced in this disorder, almost always onto an X chromosome. A fetus with an X chromosome that carries the *SRY* gene will develop as a male despite not having a Y chromosome.

48,XXYY syndrome

48,XXYY syndrome is caused by the presence of an extra X chromosome and an extra Y chromosome in a male's cells. Extra genetic material from the X chromosome interferes with male sexual development, preventing the testes from functioning normally and reducing the levels of testosterone in adolescent and adult males. Extra copies of genes from the pseudoautosomal regions of the extra X and Y chromosome contribute to the signs and symptoms of 48,XXYY syndrome; however, the specific genes have not been identified.

intestinal pseudo-obstruction

Intestinal pseudo-obstruction, a condition characterized by impairment of the muscle contractions that move food through the digestive tract (peristalsis), can be caused by genetic changes within the X chromosome.

Some individuals with intestinal pseudo-obstruction have mutations, duplications, or deletions of genetic material in the X chromosome that affect the *FLNA* gene. Researchers believe that these genetic changes may impair the function of the filamin A protein, causing abnormalities in the cytoskeleton of nerve cells (neurons) in the gastrointestinal tract. These abnormalities result in impaired peristalsis, which causes abdominal pain and the other gastrointestinal symptoms of intestinal pseudo-obstruction.

Deletions or duplications of genetic material that affect the *FLNA* gene can also include adjacent genes on the X chromosome. Changes in adjacent genes may account for some of the other signs and symptoms, such as neurological abnormalities and unusual facial features, that occur in some affected individuals.

Klinefelter syndrome

Klinefelter syndrome is caused by the presence of one or more extra copies of the X chromosome in a male's cells. Extra genetic material from the X chromosome interferes with male sexual development, preventing the testes from functioning normally and reducing the levels of testosterone (a hormone that directs male sexual development). A shortage of testosterone can lead to delayed or incomplete puberty, genital abnormalities, breast enlargement (gynecomastia), reduced facial and body hair, and an inability to have biological children (infertility). Children with Klinefelter syndrome may also have learning disabilities, delayed speech and language development, and a shy and unassuming personality.

Typically, people with Klinefelter syndrome have one extra copy of the X chromosome in each cell, for a total of two X chromosomes and one Y chromosome (47,XXY). Less commonly, affected individuals may have two or three extra X chromosomes (48,XXXY or 49,XXXXY) . As the number of extra sex chromosomes increases, so does the risk of learning problems, intellectual disability, birth defects, and other health issues.

Some people with features of Klinefelter syndrome have the extra X chromosome in only some of their cells; in these individuals, the condition is described as mosaic Klinefelter syndrome (46,XY/47,XXY). Individuals with mosaic Klinefelter syndrome may have milder signs and symptoms, depending on how many cells have an additional X chromosome.

microphthalmia with linear skin defects syndrome

A deletion of genetic material in a region of the X chromosome called Xp22 causes microphthalmia with linear skin defects syndrome. This region includes a gene called *HCCS*, which carries instructions for producing an enzyme called holocytochrome c-type synthase. This enzyme helps produce a molecule called cytochrome c. Cytochrome c is involved in a process called oxidative phosphorylation, by which mitochondria generate adenosine triphosphate (ATP), the cell's main energy source. It also plays a role in the self-destruction of cells (apoptosis).

A deletion of genetic material that includes the *HCCS* gene prevents the production of the holocytochrome c-type synthase enzyme. In females (who have two X chromosomes), some cells produce a normal amount of the enzyme and other cells produce none. The resulting overall reduction in the amount of this enzyme leads to the signs and symptoms of microphthalmia with linear skin defects syndrome.

In males (who have only one X chromosome), a deletion that includes the *HCCS* gene results in a total loss of the holocytochrome c-type synthase enzyme. A lack of this enzyme appears to be lethal very early in development, so almost no males are born with microphthalmia with linear skin defects syndrome. A few affected individuals with male appearance but who have two X chromosomes have been identified.

A reduced amount of the holocytochrome c-type synthase enzyme can damage cells by impairing their ability to generate energy. In addition, without the holocytochrome c-type synthase enzyme, the damaged cells may not be able to undergo apoptosis. These cells may instead die in a process called necrosis that causes inflammation and damages neighboring cells. During early development this spreading cell damage may lead to the eye and skin abnormalities characteristic of microphthalmia with linear skin defects syndrome.

triple X syndrome

Triple X syndrome (also called 47,XXX or trisomy X) results from an extra copy of the X chromosome in each of a female's cells. Females with triple X syndrome have three X chromosomes, for a total of 47 chromosomes per cell. An extra copy of the X chromosome is associated with tall stature, learning problems, and other features in some girls and women.

Some females with triple X syndrome have an extra X chromosome in only some of their cells. This phenomenon is called 46,XX/47,XXX mosaicism.

Females with more than one extra copy of the X chromosome (48,XXXX or 49,XXXXX) have been identified, but these chromosomal changes are rare. As the number of extra sex chromosomes increases, so does the risk of learning problems, intellectual disability, birth defects, and other health issues.

Turner syndrome

Turner syndrome results when one normal X chromosome is present in a female's cells and the other sex chromosome is missing or structurally altered. The missing genetic material affects development before and after birth, leading to short stature, ovarian malfunction, and the other features of Turner syndrome.

About half of individuals with Turner syndrome have monosomy X (45,X), which means each cell in an individual's body has only one copy of the X chromosome instead of the usual two sex chromosomes. Turner syndrome can also occur if one of the sex chromosomes is partially missing or rearranged rather than completely absent.

Some women with Turner syndrome have a chromosomal change in only some of their cells, which is known as mosaicism. Some cells have the usual two sex chromosomes (either two X chromosomes or one X chromosome and one Y chromosome), and other cells have only one copy of the X chromosome. Women with Turner syndrome caused by X chromosome mosaicism (45,X/46,XX or 45,X/46,XY) are said to have mosaic Turner syndrome.

Researchers have not determined which genes on the X chromosome are responsible for most of the features of Turner syndrome. They have, however, identified one gene called *SHOX* that is important for bone development and growth. The *SHOX* gene is located in the pseudoautosomal regions of the sex chromosomes. Missing one copy of this gene likely causes short stature and skeletal abnormalities in women with Turner syndrome.

other chromosomal conditions

Chromosomal conditions involving the sex chromosomes often affect sex determination (whether a person has the sexual characteristics of a male or a female), sexual development, and the ability to have children (fertility). The signs and symptoms of these conditions vary widely and range from mild to severe. They can be caused by missing or extra copies of the sex chromosomes or by structural changes in the chromosomes.

Is there a standard way to diagram the X chromosome?

Geneticists use diagrams called ideograms as a standard representation for chromosomes. Ideograms show a chromosome's relative size and its banding pattern. A banding pattern is the characteristic pattern of dark and light bands that appears when a chromosome is stained with a chemical solution and then viewed under a microscope. These bands are used to describe the location of genes on each chromosome.

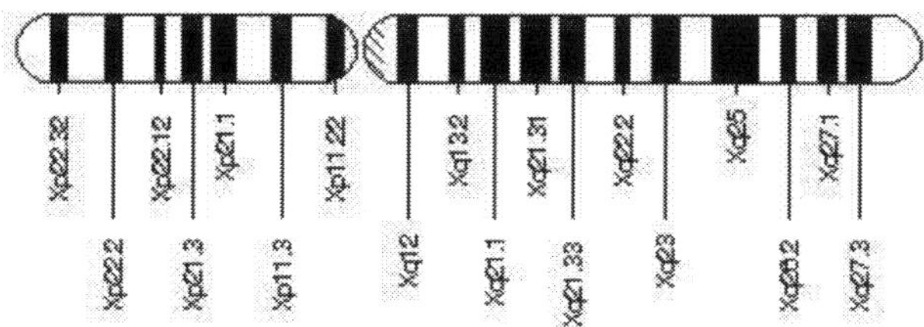

Conditions related to genes on the X chromosome

Genetics Home Reference includes these conditions related to genes on the X chromosome:

17β-hydroxysteroid dehydrogenase type 10
deficiency Aarskog-Scott syndrome
Allan-Herndon-Dudley syndrome
alpha thalassemia X-linked intellectual disability
syndrome Alport syndrome
amelogenesis imperfecta
androgenetic alopecia
androgen insensitivity syndrome
anhidrotic ectodermal dysplasia with immune
deficiency Arts syndrome
Barth syndrome
breast cancer
CASK-related intellectual
disability Charcot-Marie -Tooth
disease choroideremia
Christianson syndrome chronic
granulomatous disease Coffin-
Lowry syndrome color vision
deficiency
congenital hemidysplasia with ichthyosiform erythroderma and
limb defects
Cornelia de Lange
syndrome cutis laxa
Danon disease
deafness-dystonia-optic neuronopathy syndrome

Dent disease
DMD-associated dilated cardiomyopathy
Duchenne and Becker muscular dystrophy
dyserythropoietic anemia and thrombocytopenia
dyskeratosis congenita
Emery-Dreifuss muscular
dystrophy Fabry disease
familial dilated cardiomyopathy
familial exudative vitreoretinopathy
FG syndrome
focal dermal hypoplasia
fragile X-associated primary ovarian insufficiency
fragile X-associated tremor/ataxia syndrome
fragile XE syndrome
fragile X syndrome
frontometaphyseal dysplasia
glucose-6 -phosphate dehydrogenase
deficiency glycogen storage disease type IX
hemophilia
hereditary hypophosphatemic rickets
hypohidrotic ectodermal dysplasia
immune dysregulation, polyendocrinopathy, enteropathy, X-linked syndrome
incontinentia pigmenti
inherited thyroxine-binding globulin deficiency
intestinal pseudo-obstruction
isolated growth hormone deficiency
isolated lissencephaly sequence
Joubert syndrome
Kabuki syndrome
Kallmann syndrome
L1 syndrome
Langer mesomelic
dysplasia Leigh syndrome
Lenz microphthalmia syndrome
Léri-Weill dyschondrosteosis
Lesch-Nyhan syndrome
Lowe syndrome
Lujan syndrome
McLeod neuroacanthocytosis syndrome
MECP2 duplication syndrome
MECP2-related severe neonatal encephalopathy

Melnick-Needles syndrome
Menkes syndrome
microphthalmia
microphthalmia with linear skin defects syndrome
mucopolysaccharidosis type II
nephrogenic diabetes
insipidus nonsyndromic
deafness Norrie disease
ocular albinism
oculofaciocardiodental syndrome
Ohdo syndrome, Maat-Kievit-Brunner
type Opitz G/BBB syndrome
oral-facial-digital syndrome ornithine
transcarbamylase deficiency
osteopetrosis
otopalatodigital syndrome type 1
otopalatodigital syndrome type 2
paroxysmal nocturnal hemoglobinuria
Partington syndrome Pelizaeus-
Merzbacher disease periventricular
heterotopia phosphoglycerate kinase
deficiency
phosphoribosylpyrophosphate synthetase superactivity
porphyria
PPM-X syndrome
primary ciliary dyskinesia
pyruvate dehydrogenase deficiency
Renpenning syndrome
retinitis pigmentosa
Rett syndrome
severe congenital neutropenia
Simpson-Golabi-Behmel syndrome
Snyder -Robinson syndrome
spastic paraplegia type 2
spinal and bulbar muscular atrophy
spinal muscular atrophy
Swyer syndrome
Turner syndrome
type 1 diabetes
Wiskott-Aldrich syndrome
X-linked adrenal hypoplasia congenita
X-linked adrenoleukodystrophy

X-linked agammaglobulinemia X-
linked chondrodysplasia punctata 1 X-
linked chondrodysplasia punctata 2
X-linked congenital stationary night blindness
X-linked creatine deficiency
X-linked dystonia-parkinsonism
X-linked hyper IgM syndrome
X-linked immunodeficiency with magnesium defect, Epstein-Barr virus infection, and neoplasia
X-linked infantile nystagmus X-
linked infantile spasm syndrome
X-linked juvenile retinoschisis
X-linked lissencephaly with abnormal genitalia
X-linked lymphoproliferative disease X-linked
myotubular myopathy
X-linked severe combined immunodeficiency
X-linked sideroblastic anemia
X-linked sideroblastic anemia and ataxia X-
linked spondyloepiphyseal dysplasia tarda
X-linked thrombocytopenia

Reviewed: *January 2012*

Published: *November 17, 2014*

Genes on the X chromosome

Genetics Home Reference includes these genes on the X chromosome:

ABCB7
ABCD1
AFF2
ALAS2
AMELX
AR
ARSE
ARX
ATP7A
ATRX
AVPR2
BCOR
BTK
CACNA1F
CASK

CD40LG
CDKL5
CHM
CLCN5
COL4A5
CYBB
DCX
DKC1
DMD
EBP
EDA
EMD
F8
F9
FGD1
FLNA
FMR1
FOXP3
FRMD7
G6PD
GATA1
GJB1
GLA
GPC3
GPR143
HCCS
HPRT1
HSD17B10
IDS
IKBKG
IL2RG
KAL1
KDM6A
L1CAM
LAMP2
MAGT1
MECP2
MED12
MID1
MTM1
NDP
NR0B1

NSDHL
NYX
OCRL
OFD1
OPN1LW
OPN1MW
OTC
PDHA1
PGK1
PHEX
PHKA1
PHKA2
PIGA
PLP1
PORCN
POU3F4
PQBP1
PRPS1
RP2
RPGR
RPS6KA3
RS1
SERPINA7
SH2D1A
SHOX
SLC6A8
SLC9A6
SLC16A2
SMC1A
SMS
TAF1
TAZ
TIMM8A
TRAPPC2
UBA1
WAS
XIAP
XK

The State of Hawaii and Oklahoma are Assigned the Y Chromosome

State of Oklahoma

University of Oklahoma (Oklahoma City) (Langer Mesomelic Dysplasia)

Y chromosome

What is the Y chromosome?

The Y chromosome is one of the two sex chromosomes in humans (the other is the X chromosome). The sex chromosomes form one of the 23 pairs of human chromosomes in each cell. The Y chromosome spans more than 59 million building blocks of DNA (base pairs) and represents almost 2 percent of the total DNA in cells.

Each person normally has one pair of sex chromosomes in each cell. The Y chromosome is present in males, who have one X and one Y chromosome, while females have two X chromosomes.

Identifying genes on each chromosome is an active area of genetic research. Because researchers use different approaches to predict the number of genes on each chromosome, the estimated number of genes varies. The Y chromosome likely contains 50 to 60 genes that provide instructions for making proteins. Because only males have the Y chromosome, the genes on this chromosome tend to be involved in male sex determination and development. Sex is determined by the *SRY* gene, which is responsible for the development of a fetus into a male. Other genes on the Y chromosome are important for male fertility.

Many genes are unique to the Y chromosome, but genes in areas known as pseudoautosomal regions are present on both sex chromosomes. As a result, men and women each have two functional copies of these genes. Many genes in the pseudoautosomal regions are essential for normal development.

Genes on the Y chromosome are among the estimated 20,000 to 25,000 total genes in the human genome.

How are changes in the Y chromosome related to health conditions?

Many genetic conditions are related to changes in particular genes on the Y chromosome. This list of disorders associated with genes on the Y chromosome provides links to additional information.

Changes in the structure or number of copies of a chromosome can also cause problems with health and development. The following chromosomal conditions are associated with such changes in the Y chromosome.

46,XX testicular disorder of sex development

In most individuals with 46,XX testicular disorder of sex development, the condition results from an abnormal exchange of genetic material between chromosomes (translocation). This exchange occurs as a random event during the formation of sperm cells in the affected person's father. The translocation affects the gene responsible for development of a fetus into a male (the *SRY* gene). The *SRY* gene, which is normally found on the Y chromosome, is misplaced in this disorder, almost always onto an X chromosome. A fetus with an X chromosome that carries the *SRY* gene will develop as a male despite not having a Y chromosome.

47,XYY syndrome

Males with 47,XYY syndrome have one X chromosome and two Y chromosomes in each cell, for a total of 47 chromosomes. It is unclear why an extra copy of the Y chromosome is associated with tall stature, learning problems, and other features in some boys and men.

Some males with 47,XYY syndrome have an extra Y chromosome in only some of their cells. This phenomenon is called 46,XY/47,XYY mosaicism.

48,XXYY syndrome

48,XXYY syndrome is caused by the presence of an extra X chromosome and an extra Y chromosome in a male's cells. Extra genetic material from the X chromosome interferes with male sexual development, preventing the testes from functioning normally and reducing the levels of testosterone (a hormone that directs male sexual development) in adolescent and adult males. Extra copies of genes from the pseudoautosomal region of the extra X and Y chromosome contribute to the signs and symptoms of 48,XXYY syndrome; however, the specific genes have not been identified.

Y chromosome infertility

Y chromosome infertility is usually caused by deletions of genetic material in regions of the Y chromosome called azoospermia factor (AZF) A, B, or C. Genes in these regions are believed to provide instructions for making proteins involved in sperm cell development, although the specific functions of these proteins are unknown.

Deletions in the AZF regions may affect several genes. The missing genetic material likely prevents production of a number of proteins needed for normal sperm cell development, resulting in an inability to father children.

other chromosomal conditions

Chromosomal conditions involving the sex chromosomes often affect sex determination (whether a person has the sexual characteristics of a male or a female), sexual development, and the ability to have children (fertility). The signs and symptoms of these conditions vary widely and range from mild to severe. They can be caused by missing or extra copies of the sex chromosomes or by structural changes in these chromosomes.

Rarely, males may have more than one extra copy of the Y chromosome in every cell (polysomy Y). For example, the presence of two extra Y chromosomes is written as 48,XYYY. The extra genetic material in these cases can lead to skeletal abnormalities, decreased IQ, and delayed development, but the features of these conditions are variable.

Is there a standard way to diagram the Y chromosome?

Geneticists use diagrams called ideograms as a standard representation for chromosomes. Ideograms show a chromosome's relative size and its banding pattern. A banding pattern is the characteristic pattern of dark and light bands that appears when a chromosome is stained with a chemical solution and then viewed under a microscope. These bands are used to describe the location of genes on each chromosome.

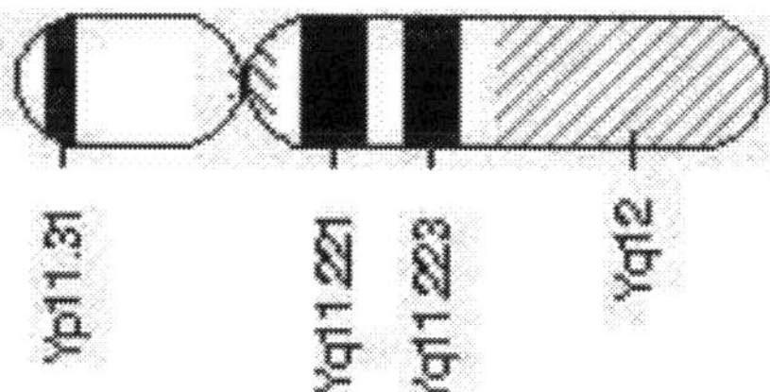

Conditions related to genes on the Y chromosome

Genetics Home Reference includes these conditions related to genes on the Y chromosome:

46,XX testicular disorder of sex
development Langer mesomelic dysplasia
Léri-Weill dyschondrosteosis
Swyer syndrome

Y chromosome infertility

Reviewed: *January 2010*

Published: *November 17, 2014*

Chromosomes > Y chromosome >

Genes on the Y chromosome

Genetics Home Reference includes these genes on the Y chromosome:

SHOX
SRY
USP9Y

Reviewed: *January 2010*

Published: *November 17, 2014*

Kinase Inhibitors

Tyrosine kinase inhibitors (TKIs) are a class of chemotherapy medications that inhibit, or block, the enzyme tyrosine kinase. TKIs were created out of modern genetics- the understanding of DNA, the cell cycle, and molecular signaling pathways- and thus represent a change from general to molecular methods of cancer treatment. This allows for targeted treatment of specific cancers, which lessens the risk of damage to healthy cells and increases treatment success.

To understand how TKIs function, one must look to the cellular level. Receptor Tyrosine Kinases (RTKs) are a family of tyrosine protein kinases. RTKs span the cell membrane with an intracellular (internal) and extracellular (external) portion. The intracellular portion removes a phosphate group, a process called phosphorylation, from the coenzyme messenger ATP. The extracellular portion has sites to which signal-sending proteins and hormones can bind. Many of these signaling binders are growth factors.

Growth factors are involved in the initialization and regulation of cell cycles. The type of growth factor determines its effects on the cell. There are three primary growth factors that relate to tyrosine kinase. The receptors of these growth factors are members of the RTK family. Epidermal growth factors (EGF) help regulate cell growth and differentiation. Platelet-derived growth factor (PDGF) regulates cell growth and development. Vascular endothelial growth factors (VEGFR) are involved in the creation of blood vessels.

The growth factors, and the kinases, act as though they are attached to an "on/off" switch. The removal of a phosphate group changes the shape and actions of the protein. This essentially "turns on" the cellular action (or actions). When the cellular action(s) is completed, the phosphate group is removed and that protein is "turned off". This "on/off" process can become disrupted, often thanks to a mutated kinase, and actions can become unregulated. An unregulated RTK bound to EGF, for example, could lead to uncontrolled growth and division in the cell. The rapid cell growth could then lead to cancer. Mutations of RTKs often lead to oncogenes, which are genes that help turn a healthy cell into a cancerous cell.

Tyrosine kinase inhibitors treat cancer by correcting this deregulation. How this is done varies depending on the medication. Imatinib (brand name: Gleevac), for example, blocks a kinase gene from binding to ATP, preventing the phosphorylation that would benefit the cancerous cell and promote cell division. The medication gefinitib (Iressa) inhibits EGFRs, preventing that signal from being stuck "on" and creating uncontrolled proliferation.

Eight TKI medications, including imatinib and gefinitib, have been approved by the Food and Drug Administration for use in humans. One TKI, toceranib (Palladia), was recently approved for the treatment of cancer in dogs. The human medications may inhibit one or more tyrosine kinases. Erlotinib (Tarceva), like gefitinib, inhibits EGFR. Lapatinib (Tykerb) is a dual inhibitor of EGFR and a subclass called Human EGFR type 2. EGFR isn't the only growth factor targeted. Sunitinib (Sutent) is multi-targeted, inhibiting PDGFR and VEGF.

Other tyrosine kinase inhibitors are more specialized. Sorafenib (Nexavar) targets a complex pathway that would lead to a kinase signaling cascade. Nilotinib (Tasinga) inhibits the fusion protein bcr-abl and is typically prescribed when a patient has shown resistance to imatinib.

More TKIs are currently in development, though the process is slow and more drugs end up being abandoned during clinical phases than get approved. Three TKIs are currently showing promise in clinical trials. Bosutinib targets abl and src kinases. Neratinib, like lapatinib, inhibits EGFR and Human EGFR type 2. Vatalanib inhibits both VEGFR and PDGFR.

Scientific research is being focused on TKIs because of their uniqueness compared to previous chemotherapy methods. All chemotherapy drugs seek to stop cells division and growth. They also attempt to kill cancerous cells without destroying the healthy cells. An inherent weakness in cancerous cells is that a failure of cell repair mechanisms is what turned the cells cancerous. The cell is therefore unable to repair damaged or changed DNA effectively.

Radiation therapy treatments interfered with the DNA to create a problem the cancer cell cannot repair, leading to cell death. This interference can be done by modifying the building blocks, the bases, by adding to them to prevent bonding. Or it may be done through mechanical processes that separate the strands of DNA. These methods are rather non-specific, both in terms of the cancers they treat and the cells the act upon. This means that healthy cells may also pay the price of treatment.

Destruction of healthy cells is one of the main problems with traditional chemo treatments. TKIs, however, are targeted in that the different types can be cancer-specific, acting upon pathways that have gone awry in the specific cancer. Furthermore, it is often a case that the kinase itself is slightly different from the normal version, meaning that the inhibitor can work specifically on cancerous cells. The only other current treatment that works in a similar way is using monoclonal antibodies, which can be used to target cancerous cells in preference over healthy ones.

The specificity of TKIs means there will be fewer side effects and less time in the hospital needed for the patient. The genetic basis for these cancers that can be treated with TKIs can now be screened effectively and early treatment is by far the most effective.

Proteins from the National Center for Biotechnology Information (NCBI)

Databases

BioProject (formerly Genome Project)

A collection of genomics, functional genomics, and genetics studies and links to their resulting datasets. This resource describes project scope, material, and objectives and provides a mechanism to retrieve datasets that are often difficult to find due to inconsistent annotation, multiple independent submissions, and the varied nature of diverse data types which are often stored in different databases.

BioSystems

Database that groups biomedical literature, small molecules, and sequence data in terms of biological relationships.

Conserved Domain Database (CDD)

A collection of sequence alignments and profiles representing protein domains conserved in molecular evolution. It also includes alignments of the domains to known 3-dimensional protein structures in the MMDB database.

HIV-1, Human Protein Interaction Database

A database of known interactions of HIV-1 proteins with proteins from human hosts. It provides annotated bibliographies of published reports of protein interactions, with links to the corresponding PubMed records and sequence data.

Protein Clusters

A collection of related protein sequences (clusters), consisting of Reference Sequence proteins encoded by complete prokaryotic and organelle plasmids and genomes. The database provides easy access to annotation information, publications, domains, structures, external links, and analysis tools.

Protein Database

A database that includes protein sequence records from a variety of sources, including GenPept, RefSeq, Swiss-Prot, PIR, PRF, and PDB.

Reference Sequence (RefSeq)

A collection of curated, non-redundant genomic DNA, transcript (RNA), and protein sequences produced by NCBI. RefSeqs provide a stable reference for genome annotation, gene identification and characterization, mutation and polymorphism analysis, expression studies, and comparative analyses. The RefSeq collection is accessed through the Nucleotide and Protein databases.

Downloads

BLAST (Stand-alone)

BLAST executables for local use are provided for Solaris, LINUX, Windows, and MacOSX systems. See the README file in the ftp directory for more information. Pre-formatted databases for BLAST nucleotide, protein, and translated searches also are available for downloading under the db subdirectory.

FTP: BLAST Databases

Sequence databases for use with the stand-alone BLAST programs. The files in this directory are pre-formatted databases that are ready to use with BLAST.

FTP: CDD

This site provides full data records for CDD, along with individual Position Specific Scoring Matrices (PSSMs), mFASTA sequences and annotation data for each conserved domain. See the README file for full details.

FTP: FASTA BLAST Databases

Sequence databases in FASTA format for use with the stand-alone BLAST programs. These databases must be formatted using formatdb before they can be used with BLAST.

FTP: GenPept

The protein sequences corresponding to the translations of coding sequences (CDS) in GenBank are collected for each GenBank release..Please see the README file in the directory for more information.

FTP: Protein Clusters

This site contains data from the Protein Clusters database arranged by release date. See the README files for more information.

FTP: RefSeq

This site contains all nucleotide and protein sequence records in the Reference Sequence (RefSeq) collection. The ""release"" directory contains the most current release of the complete collection, while data for selected organisms (such as human, mouse and rat) are available in separate directories. Data are available in FASTA and flat file formats. See the README file for details.

Submissions

BioProject Submission

An online form that provides an interface for researchers, consortia and organizations to register their BioProjects. This serves as the starting point for the submission of genomic and genetic data for the study. The data does not need to be submitted at the time of BioProject registration.

Tools

BLAST Link (BLink)

A link option on protein records that displays the results of a pre-computed BLAST search of that protein against all other protein sequences at NCBI.

Basic Local Alignment Search Tool (BLAST)

Finds regions of local similarity between biological sequences. The program compares nucleotide or protein sequences to sequence databases and calculates the statistical significance of matches. BLAST can be used to infer functional and evolutionary relationships between sequences as well as to help identify members of gene families.

Batch Entrez

Allows you to retrieve records from many Entrez databases by uploading a file of GI or accession numbers from the Nucleotide or Protein databases, or a file of unique identifiers from other Entrez databases. Search results can be saved in various formats directly to a local file on your computer.

COBALT

COBALT is a protein multiple sequence alignment tool that finds a collection of pairwise constraints derived from conserved domain database, protein motif database, and sequence similarity, using RPS-BLAST, BLASTP, and PHI-BLAST.

Cn3D

A stand-alone application for viewing 3-dimensional structures from NCBI's Entrez retrieval service. Cn3D runs on Windows, Macintosh, and UNIX and can be configured to receive data from most popular web browsers. Cn3D simultaneously displays structure, sequence, and alignment, and has powerful annotation and alignment editing features.

Concise Microbial Protein BLAST

A specialized BLAST service in which the queried database consists of all proteins from complete microbial (prokaryotic) genomes. NCBI has precalculated clusters of similar proteins at the genus-level and one representative is chosen from each cluster in order to reduce the dataset, thereby reducing search time and providing a broader taxonomic view.

Conserved Domain Search Service (CD Search)

Identifies the conserved domains present in a protein sequence. CD-Search uses RPS-BLAST (Reverse Position-Specific BLAST) to compare a query sequence against position-specific score matrices that have been prepared from conserved domain alignments present in the Conserved Domain Database (CDD).

E-Utilities

Tools that provide access to data within NCBI's Entrez system outside of the regular web query interface. They provide a method of automating Entrez tasks within software applications. Each utility performs a specialized retrieval task, and can be used simply by writing a specially formatted URL.

ProSplign

A utility for computing alignment of proteins to genomic nucleotide sequence. It is based on a variation of the Needleman Wunsch global alignment algorithm and specifically accounts for introns and splice signals. Due to this algorithm, ProSplign is accurate in determining splice sites and tolerant to sequencing errors.

Sequence Viewer

Provides a configurable graphical display of a nucleotide or protein sequence and features that have been annotated on that sequence. In addition to use on NCBI sequence database pages, this viewer is available as an embeddable webpage component. Detailed documentation including an API Reference guide is available for developers wishing to embed the viewer in their own pages.

How To

Save text searches and set up automated searches with E-mailed results
Submit data to NCBI
Retrieve all sequences for an organism or taxon
Find the function of a gene or gene product
Find a curated version of a sequence record (NCBI Reference Sequence)
Align two or more 3D structures to a given structure
Find published information on a gene or sequence
Find transcript sequences for a gene
Link from an object on a map to another resource
View the 3D structure of a protein
View a mutation site in a 3D structure
Download a large, custom set of records from NCBI

Quick Links

BioProject (formerly Genome Project)
BioSystems
Conserved Domain Database (CDD)
Protein Clusters
Protein Database
Reference Sequence (RefSeq)
BLAST (Stand-alone)
BLAST Link (BLink)
Basic Local Alignment Search Tool (BLAST)
Cn3D
Conserved Domain Search Service (CD Search)
E-Utilities

ProSplign

NCBI Home
Resource List (A-Z)

All Resources
Chemicals & Bioassays
Data & Software
DNA & RNA
Domains & Structures
Genes & Expression
Genetics & Medicine
Genomes & Maps
Homology
Literature
Proteins
Sequence Analysis
Taxonomy
Training & Tutorialses
Variation

NHGRI)

National Human Genome Research Institute genome.gov/director Eric Green, MD, PhD

Current Topics in Genomic Analysis 2014 series Eric Green, MD, PhD invites us to Subscribe to The Genomic Landscape genome.gov/director

This is a monthly update from the NHGRI (National Human Genome Research Institute) director Eric Green, MD, PhD who states we should follow his link genome.gov/director

Cancer Immunotherapy

Monoclonal Antibodies

Medication Pharmaceutical Company	Target
Alemtuzumab (Campath) Millennium Pharmaceuticals	CD52
Bevacizumab (Avastin) Genentech	VEGF
Brentuximab (Adcetris) Seattle Genetics	CD30
Cetuximab (Erbitux) Bristol/Myers/Squib	EGFR
Gemtuzumab (Mylotarg) Wyeth	CD33
Ibritumomab (Zevalin) Biogen/IDEC	CD29
Ipilimumab (Vervoy) Medarex/BMS	CTLA4
Ofatumumab (Arzerra) Genmab	CD20
Panitumumab (Vectibix) Amgen	EGFR
Rituximab (Rituxan) IDEC	CD20
Tositumomab (Bexxar) Corixa/Glaxo/Smith/Kline	CD20
Trastuzumab (Herceptin) Roche	ErbB2

Vision of Queen Atossa's Time Travel Journey to 2017

If Queen Atossa found her way to my office in 2017, after I introduced myself as Clarisse Clemons Ferrara, MD joyfully stating that if I am Breathing I am Bragging, after showing the copy of my National Board Certificate, New York State License, and MD Degree, (MeharryMedical College Class of 1978), yes, 39 years ago, depending on the time, with the Glee of Thank You Lordie instead of Cum Laude, I might mention my journey from the first women accepted into Orthopaedic Surgery at Howard University Hospital (ortho training remains incomplete,Regina Hillsman, MD first to finish). Right up to Grandfathering into Emergency Medicine, as well as Addiction Medicine in 2014, presently studying for the Emergency Medicine Board Exams this year Nov, 2017. (I completed the 2015 LLSA exam with 96% as a requirement to be scheduled for the ABEM exam)

Eventually framed pictures reveal my marriage to Joseph Ferrara, MD graduate of Cornell University Medical School 1975, yes 42 years ago who passed his recertification exam given by the American Board of Surgery, as well as the fact that I am the mother of our twins together Angelo Joseph and Angela Jolyn and step mother to Ann, Joey(Lauren) the proud parents of Joseph Salvatore, Jr one of 5% of babies born spontaneously on his due date 6/24/16) and Maja(Frank) who were blessed with the birth of my first step grandson Antonio Joseph on 3/30/15.

Then, I state the fact that when the blood was drawn, several individual tumor markers, Carcinogenic Embryonic Antigen (CEA) (multiple cancers), Alpha Feto Protein (AFP), (liver cancer), CA 27-29/ CA15-3 (mucinous cancers), CA 125 ovarian cancer (women) and PSA prostate specific antigen (men) and CA19-9 (pancreatic cancer). Now, I mention to stay tuned as we shot footage, including participants that I filled out the paperwork for foundation one to sequence the 300 genomic sequence variant signatures known to be present in malignancies.

The fact that 50% of men and 33% of women develop cancer in their lifetime and Environmental Toxins, Dietary Carcinogens, Infectious Disease Irritants ie (HPV/Hepatitis C), as well as the 10 to15% of Hereditary Germ Line Mutations are identifiable with Complete Genomic Sequencing, should necessitate the cooperation of the Environmental Protection Agency (EPA), Food and Drug Administration (FDA), the Department of Public Health (DOH), and the American

College of Medical Genetics to publish reliable comparison studies that will lead to wise policy decisions.

When I attended the Current Topics in Genomic Analysis via NHGRI (National Human Genome Research Institute) Lecture Series in 2014, it was Stated that Personalized Medicine Via Complete Genomic Sequencing followed by Gene Based Targeted Therapy needs to go from the Bench to the Bedside and from the Helix to Health !!!!! I want to add that The More You Know The More You Know there is More to Know and of course, Early Detection is the Key to Longevity!!!! My life's work is to have regional Gene Theme Scene franchise opportunities to create jobs and save lives with Multiple Blended Edu-tainment Television Series that include the Complete Genomic Sequencing and an introduction to Gene Based Targeted Therapies that are either on the market like Gleevec and Kevetrin or are in Clinical Trials, as this will drive the cost down with increased volume.

I will begin a monthly outreach tour to invite people to join the Gene Theme Scene Franchise by making appearances in High Schools, Ice and Roller Skating Rinks, and Ballroom and Line-Dance Studios as we will create jobs and save lives with training to become either videographers or a career with-in the genetic field, especially when we convince people to undergo the training needed to become a research oriented molecular geneticist.

National Cancer Institute (NCI) Brain Cancer Overview

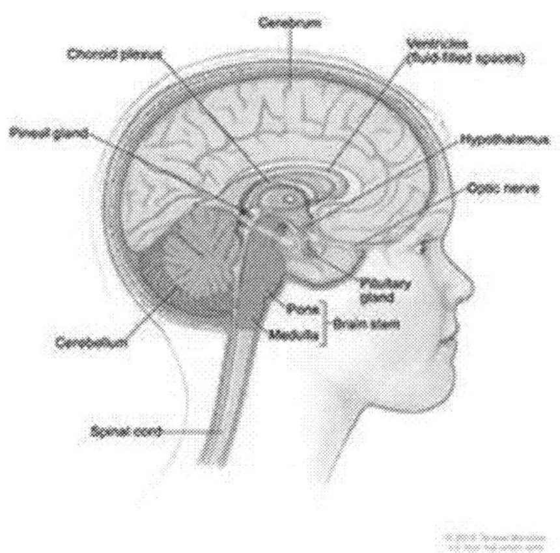

Anatomy of the inside of the brain.

The brain and spinal cord make up the central nervous system (CNS). Brain and spinal cord tumors are growths of abnormal cells in tissues of the brain or spinal cord. Tumors that start in the brain are called primary brain tumors. A tumor that starts in another part of the body and spreads to the brain is called a metastatic brain tumor.

Brain and spinal cord tumors may be either benign (not cancer) or malignant (cancer).

Both benign and malignant tumors cause signs and symptoms and need treatment. Benign brain and spinal cord tumors grow and press on nearby areas of the brain but rarely spread into other parts of the brain. Malignant brain and spinal cord tumors are likely to grow quickly and spread into other parts of the brain.

There are many types of brain and spinal cord tumors. They form in different cell types and different areas of the brain and spinal cord. The signs and symptoms of brain and spinal cord tumors depend on where the tumor forms, its size, how fast it is growing, and the age of the patient.

Brain and spinal cord tumors can occur in both adults and children. The types of tumors that form and the way they are treated are different in children and adults. In adults, anaplastic astrocytomas and glioblastomas make up about one-third of brain tumors. In children, astrocytomas are the most common type of brain tumor.

The prognosis (chance of recovery) depends on many factors, including age, tumor size, tumor type, and where the tumor is in the CNS.

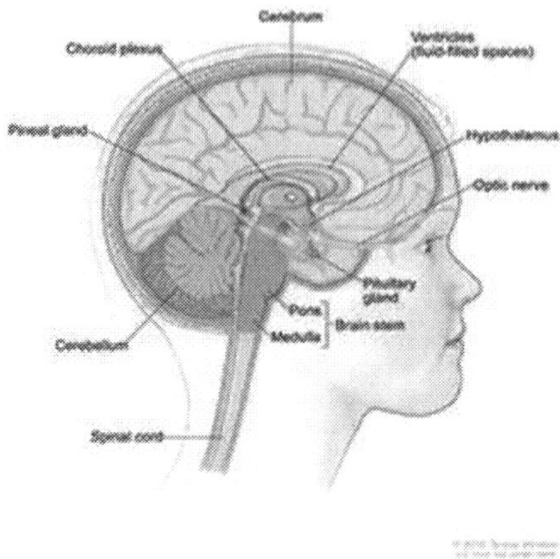

Anatomy of the inside of the brain

Gene Theme Scene Synopsis of Gene Based Targeted Therapies

Glioblastoma patients with the Vascular Endothelial Growth Factor (VEGF) Target

Bevacizumab (Avastin) Pharmaceutical Company Genentech

NCI Head and Neck Cancer Overview

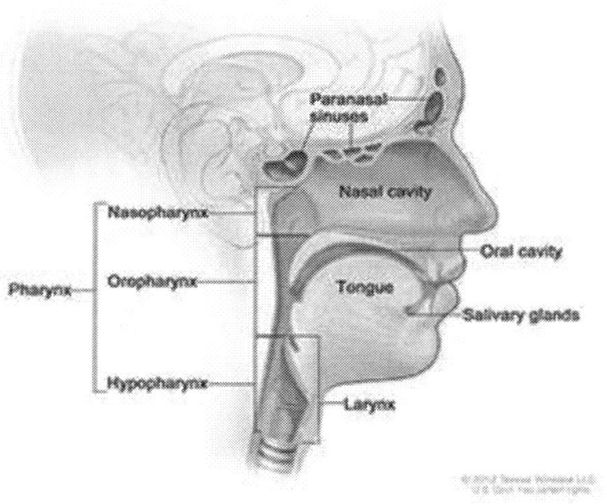

Areas where head and neck cancer may occur

Head and neck cancers are cancers that start in the tissues and organs of the head and neck. They include cancers of the larynx (voice box), throat, lips, mouth, nose, and salivary glands.

Most types of head and neck cancer begin in squamous cells that line the moist surfaces inside the head and neck (for example, the mouth, nose, and throat).

Tobacco use, heavy alcohol use, and infection with the human papillomavirus (HPV) increase the risk of many types of head and neck cancer.

To get the right information about treatment and prognosis, you need to know exactly what type of head and neck cancer you have and what stage it is.

The Head and Neck Cancers fact sheet has additional basic information.

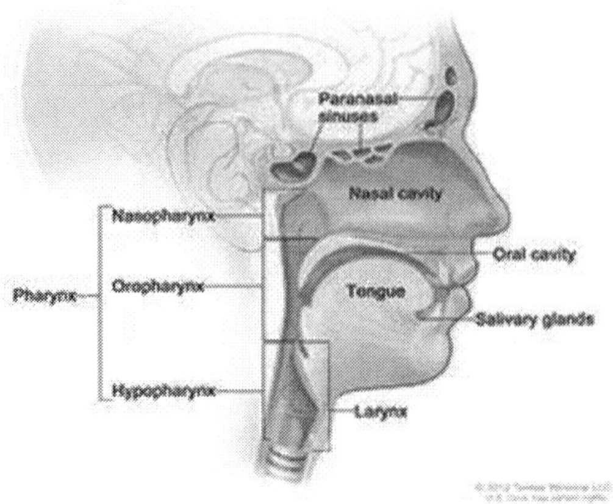

Areas where head and neck cancer may occur

Gene Theme Scene Synopsis of Gene Based Targeted Therapies

Metastatic Thyroid Cancer Patients with the C MET or VEGFR target
Cabazantinib (Cometriq) Pharmaceutical Company Exelixis
Advanced Medullary Thyroid Cancer Patients with the VEGFR, EGFR, RET, BRK target
Vandetanib (Caprelsa) Pharmaceutical Company AstraZeneca
Advanced Squamous Cell Carcinoma of the Head and Neck with the EGFR target
Cetuximab (Erbitux) Pharmaceutical Company Bristol-Myers Squibb

NCI Lung Cancer Overview

The lungs are a pair of cone-shaped breathing organs inside the chest. The lungs bring oxygen into the body when breathing in and send carbon dioxide out of the body when breathing out. Each lung has sections called lobes. Two tubes called bronchi lead from the trachea (windpipe) to the lungs.

The two main types of lung cancer are non-small cell lung cancer and small cell lung cancer. The types are based on the way the cells look under a microscope. Non-small cell lung cancer is much more common than small cell lung cancer.

Tobacco smoking is the most common cause of lung cancer. Lung cancer is the leading cause of death from cancer in the U.S. and the number of deaths from lung cancer in women is increasing.

For most patients with lung cancer, current treatments do not cure the cancer.

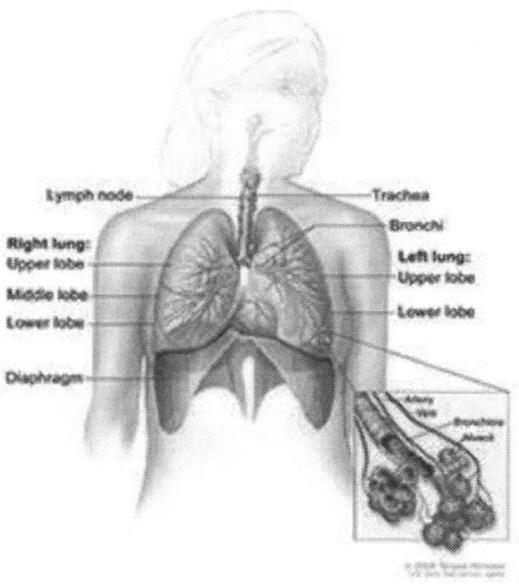

Anatomy of the respiratory system, showing the trachea and both lungs and their lobes and airways

Gene Theme Scene Synopsis of Gene Based Targeted Therapies

Advanced Non-Small Cell Lung Cancer Patients with the ErbB, EGFR Mutation Positive target

Afatinib (Gilotrif) Pharmaceutical Company Boehringer Ingelheim

Non Small Cell Lung Cancer Patients with the VEGF target

Bevacizumab (Avastin) Pharmaceutical Company Genentech

Kinase Positive Non Small Cell Lung Canceer Patients with the ALK, HGFR, C-MET targets

Crizotinib (Xalkori) Pharmaceutical Company Pfizer

Advanced Non-Small Cell Lung Cancer Patients with the EGFR

target Tarceva (Erlotinib) Pharmaceutical Company Genentech

Note:Tarceva (Erlotinib) is a Tyrosine Kinase Inhibitor that targets the Epidemal Growth Factor Receptor (EGFR)

Non Small Cell Lung Cancer Patients with the EGFR Mutation

Gefitinib (Iressa) Pharmaceutical Company AstraZeneca

Non Small Cell Lung Cancer Patients with ligand activation of PD-1 receptor on activated T cells (immunomodulator target)

Nivolumab (Opodivo) Pharmaceutical Company Bristol-Myers Squibb

NCI Breast Cancer Overview

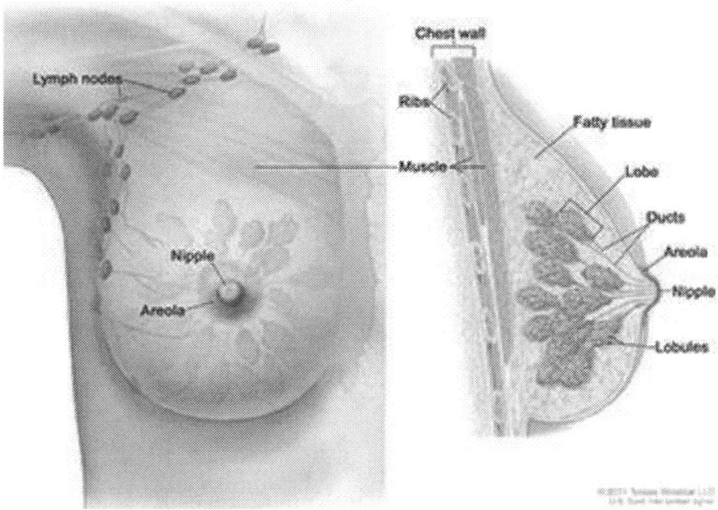

Anatomy of the female breast. The nipple, areola, lymph nodes, lobes, lobules, ducts, and other parts of the breast are shown.

Overview

The breast is made up of glands called lobules that can make milk and thin tubes called ducts that carry the milk from the lobules to the nipple. Breast tissue also contains fat and connective tissue, lymph nodes, and blood vessels.

The most common type of breast cancer is ductal carcinoma, which begins in the cells of the ducts. Breast cancer can also begin in the cells of the lobules and in other tissues in the breast. Invasive breast cancer is breast cancer that has spread from where it began in the ducts or lobules to surrounding tissue.

In the U.S., breast cancer is the second most common cancer in women after skin cancer. It can occur in both men and women, but it is very rare in men. Each year there are about 2,300 new cases of breast cancer in men and about 230,000 new cases in women.

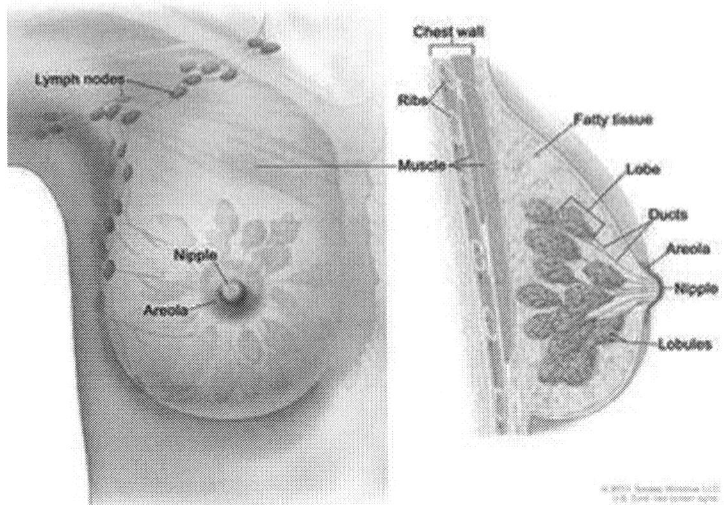

Anatomy of the female breast. The nipple, areola, lymph nodes, lobes, lobules, ducts, and other parts of the breast are shown.

Gene Theme Scene Synopsis of Gene Based Targeted Therapies

Breast Cancer Patients with the Receptor Tyrosine Protein Kinase erbB2 (CD340 proto-oncogene Neu which is a protein

Encoded by the ERBB2 gene abrviated HER2 or HER2/Neu (Human Epidermal Growth Factor Receptor 2)

Lapatinib (Tykerb) Pharmaceutical Company Glaxo Smith Kline

Breast Cancer Patients with the ERBB2 Gene located at Chromosome 17q12/HER2 target

Trastuzumab (Herceptin) Pharmaceutical Company Genentech

Stomach (Gastric) Cancer Overview

Gastric (stomach) cancer is a disease in which malignant (cancer) cells form in the lining of the stomach. The stomach is in the upper abdomen and helps digest food.

Almost all gastric cancers are adenocarcinomas (cancers that begin in cells that make and release mucus and other fluids). Other types of gastric cancer are gastrointestinal carcinoid tumors, gastrointestinal stromal tumors, and lymphomas.

Infection with bacteria called *H. pylori* is a common cause of gastric cancer.

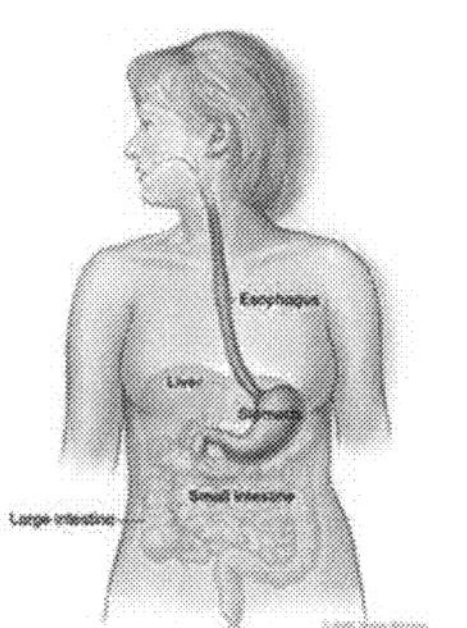

Gastric cancer is often diagnosed at an advanced stage because there are no early signs or symptoms.

The esophagus and stomach are part of the upper gastrointestinal system

Gene Theme Scene Synopsis of Gene Based Targeted Therapy

GIST Gastrointestinal Stromal Tumors in Patients with the RET, VEGFR, PDGFR targets

Regorafenib (Stivarga) Pharmaceutical Company Roche

GIST Gastrointestinal Stromal Tumor Patients with the VEGFR, PDGFR target

Sunitinib (Sutent) Pharmaceutical Company Pfizer

Metastatic Gastric Cancer Patients with the HER2/ ERBB2 Oncogene target

Trastuzumab (Herceptin) Pharmaceutical Company Genentech

NCI Pancreatic Cancer Overview

The pancreas lies behind the stomach and in front of the spine. There are two kinds of cells in the pancreas. Exocrine pancreas cells make enzymes that are released into the small intestine to help the body digest food. Neuroendocrine pancreas cells (such as islet cells) make several hormones, including insulin and glucagon, that help control sugar levels in the blood.

Most pancreatic cancers form in exocrine cells. These tumors do not secrete hormones and do not cause signs or symptoms. This makes it hard to diagnose this type of pancreatic cancer early. For most patients with exocrine pancreatic cancer, current treatments do not cure the cancer.

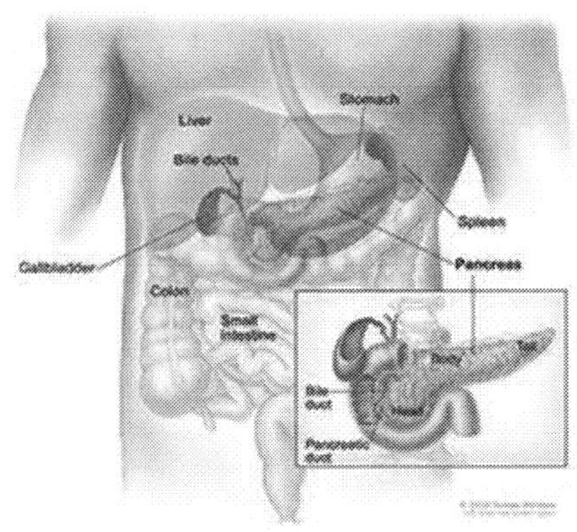

Some types of malignant pancreatic neuroendocrine tumors, such as islet cell tumors, have a better prognosis than pancreatic exocrine cancers.

Anatomy of the pancreas. The pancreas has three areas: head, body, and tail. It is found in the abdomen near the stomach, intestines, and other organs.

Gene Theme Scene Synopsis of Gene Based Targeted Therapy

Pancreatic Cancer Patients with the EGFR target

Erlotinib (Tarceva) Pharmaceutical Company Roche

Pancreatic Neuroendocrine Tumor Patients with the VEGFR, PDGFR targets

Sunitinib (Sutent) Pharmaceutical Company Pfizer

NCI Liver Cancer Overview

The liver has many important functions in the body. For example, it cleans toxins from the blood, makes bile that helps digest fat, makes substances that help blood clot, and makes, stores, and releases sugar for energy.

Primary liver cancer is cancer that starts in the liver. The most common type of primary liver cancer is hepatocellular carcinoma, which occurs in the tissue of the liver. When cancer starts in other parts of the body and spreads to the liver, it is called liver metastasis.

Liver cancer is rare in children and teenagers, but there are two types of liver cancer that can form in children. Hepatoblastoma occurs in younger children, and hepatocellular carcinoma occurs in older children and teenagers.

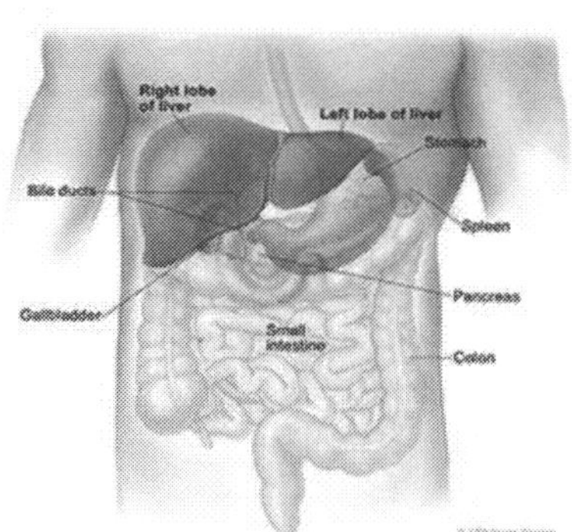

The bile ducts are tubes that carry bile between the liver and gallbladder and the intestine. Bile duct cancer is also called cholangiocarcinoma. When it begins in the bile ducts inside the liver, it is called intrahepatic cholangiocarcinoma. When it begins in the bile ducts outside the liver, it is called extrahepatic cholangiocarcinoma. Extrahepatic cholangiocarcinoma is much more common than intrahepatic cholangiocarcinoma.

Anatomy of the liver. The liver has four lobes. Two lobes are on the front and two small lobes (not shown) are on the back of the liver.

Gene Theme Scene Synopsis of Gene Based Targeted Therapy

Hepatocellular Carcinoma Patients with the VEGFR, PDGFR, BRAF, c-KIT targets

Sorafenib (Nexavar) Pharmaceutical Company Bayer

NCI KIDNEY CANCER OVERVIEW

There are two kidneys, one on each side of the spine, above the waist. The kidneys clean the blood to take out waste and make urine. Urine collects in the renal pelvis, the area at the center of the kidney, and then passes through the ureter, into the bladder, and out of the body. The kidneys also make hormones that help control blood pressure and signal the bone marrow to make red blood cells when needed.

There are three main types of kidney cancer. Renal cell cancer is the most common type in adults and Wilms tumors are the most common in children. These types form in the tissues of the kidney that make urine. Transitional cell cancer forms in the renal pelvis and ureter in adults.

Smoking and taking certain pain medicines for a long time can increase the risk of adult kidney cancer. Certain inherited disorders can increase the risk of kidney cancer in children and adults. These include von Hippel-Lindau syndrome, hereditary leiomyomatosis and renal cell cancer, Birt-Hogg-Dubé syndrome, and hereditary papillary renal cancer.

Kidney cancer is often diagnosed at an advanced stage because usually there are no early signs or symptoms.

Kidney tumors may be benign or malignant.

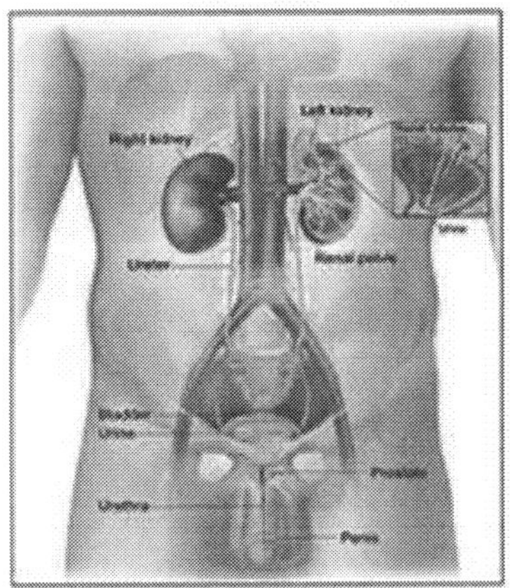

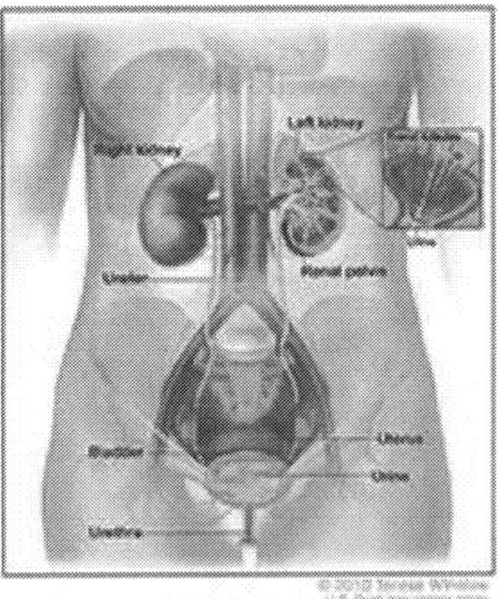

Anatomy of the male urinary system (left panel) and female urinary system (right panel).

Gene Theme Scene Synopsis of Gene Based Targeted Therapy

Renal Cell Carcinoma Patients with the VEGFR, PDGFR, c-KIT target

Axitinib (Inlyta) Pharmaceutical Company Pfizer

Renal Cell Carcinoma Patients with the VEGF target

Bevacizumab (Avastin) Pharmaceutical Company Genetech

Renal Cell Carcinoma Patients with the VEGFR, PDGFR, c-KIT targets

Pazopanib (Votrient) Pharmaceutical Company GlaxoSmithKline

Advanced Renal Cell Carcinoma Patients with the VEGFR, PDGFR, BRAF, c-KIT targets

Soraninib (Nexavar) Pharmaceutical Company Bayer

Renal Cell Carcinoma Patients with the VEGFR, PDGFR

target Sunitinib (Sutent) Pharmaceutical Company Pfizer

NCI Colon Cancer Overview

Colorectal cancer is cancer that starts in the colon or rectum. The colon and the rectum are parts of the large intestine, which is the lower part of the body's digestive system. During digestion, food moves through the stomach and small intestine into the colon. The colon absorbs water and nutrients from the food and stores waste matter (stool). Stool moves from the colon into the rectum before it leaves the body.

Most colorectal cancers are adenocarcinomas (cancers that begin in cells that make and release mucus and other fluids). Colorectal cancer often begins as a growth called a polyp, which may form on the inner wall of the colon or rectum. Some polyps become cancer over time. Finding and removing polyps can prevent colorectal cancer.

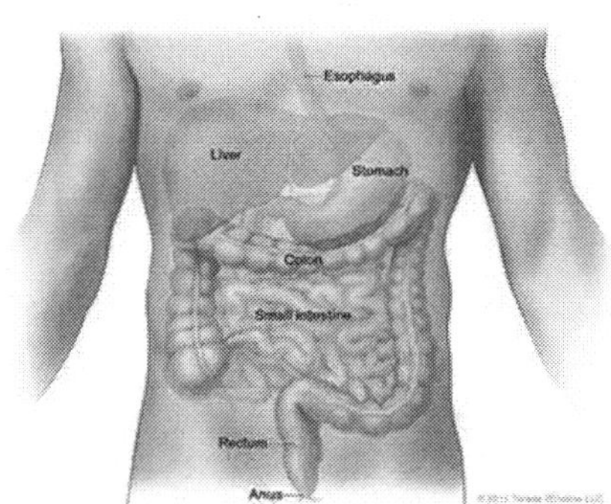

Colorectal cancer is the third most common type of cancer in men and women in the United States. Deaths from colorectal cancer have decreased with the use of colonoscopies and fecal occult blood tests, which check for blood in the stool.

Gene Theme Scene Synopsis of Gene Based Targeted Therapies

Metastatic Colorectal Cancer Patients with the VEGF target

Bevacizumab (Avastin) Pharmaceutical Company Genetech

Metastatic Colorectal Cancer in patients with the EGFR target

Cetuximab (Erbitux) Pharmaceutical Company Bristol-Myers Squibb

Metastatic Colorectal Cancer Patients with the EGFR target

Panitumumab (Vectibix) Pharmaceutical Company Amgen

Advanced Colorectal Cancer Patients with the VEGFR target

Aflibercept (Stivarga) Pharmaceutical Company Bayer

Prostate Cancer Overview

The prostate gland makes fluid that forms part of semen. The prostate lies just below the bladder in front of the rectum. It surrounds the urethra (the tube that carries urine and semen through the penis and out of the body).

Prostate cancer is the most common cancer in men in the United States, after skin cancer. It is the second leading cause of death from cancer in men. Prostate cancer occurs more often in African-American men than in white men. African-American men with prostate cancer are more likely to die from the disease than white men with prostate cancer.

Almost all prostate cancers are adenocarcinomas (cancers that begin in cells that make and release mucus and other fluids). Prostate cancer often has no early symptoms. Advanced prostate cancer can cause men to urinate more often or have a weaker flow of urine, but these symptoms can also be caused by benign prostate conditions.

Gene Theme Scene Synopsis of Gene Base Targeted Therapy

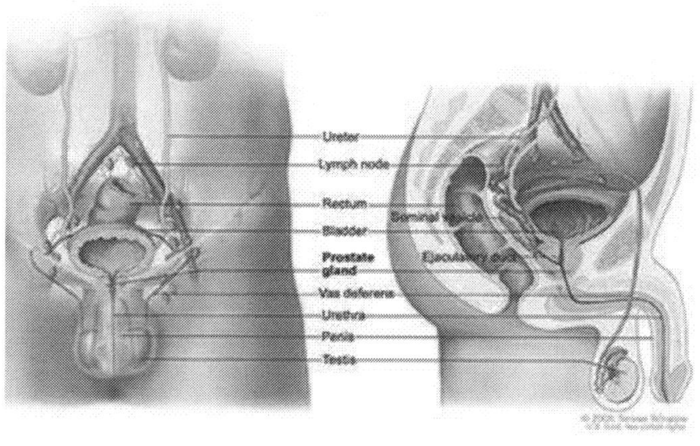

Prostate Cancer Patients with the Prostate Acid Phospatase (PAP) Antigen can Benefit from Immunotherapy with Sipuleucel-T (Provenge) Pharmaceutical Company Dendreon Corp

Anatomy of the male reproductive and urinary systems showing the prostate, testicles, bladder, and other organs.

NCI Cervical Cancer Overview

The cervix is the lower, narrow end of the uterus (the organ where a fetus grows). The cervix leads from the uterus to the vagina (birth canal).

The main types of cervical cancer are squamous cell carcinoma and adenocarcinoma. Squamous cell carcinoma begins in the thin, flat cells that line the cervix. Adenocarcinoma begins in cervical cells that make mucus and other fluids.

Long-lasting infections with certain types of human papillomavirus (HPV) cause almost all cases of cervical cancer. Vaccines that protect against infection with these types of HPV can greatly reduce the risk of cervical cancer. Having a Pap test to check for abnormal cells in the cervix or a test to check for HPV can find cells that may become cervical cancer. These cells can be treated before cancer forms.

Cervical cancer can usually be cured if it is found and treated in the early stages.

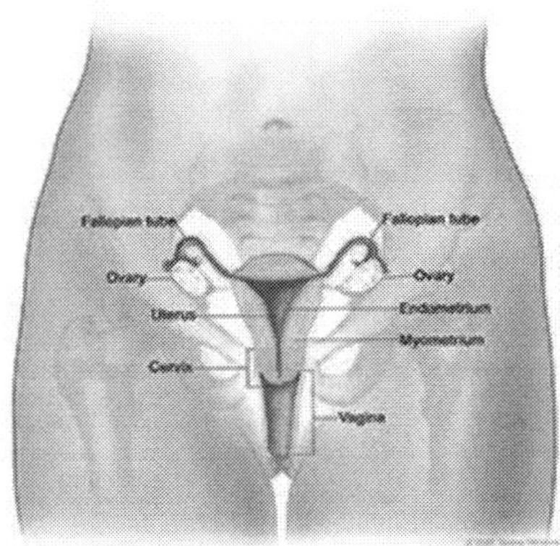

Anatomy of the female reproductive system. The organs in the female reproductive system include the uterus, ovaries, fallopian tubes, cervix, and vagina

Gene Theme Scene Cervical Cancer Statements

HPV Human Papilloma Virus Strains 16 and 18 cause 70% of Cervical Cancer and 80% of Anal Cancers

Two Vaccines Gardasil and Cervarix cover the HPV stains 16 and 18

Gardasil also covers HPV strains 6 and 11 that cause 90% of Genital Warts

NCI Uterine Cancer Overview

The uterus is a hollow, muscular organ where a fetus grows. Uterine cancer can start in different parts of the uterus. Most uterine cancers start in the endometrium (the inner lining of the uterus). This is called endometrial cancer. Most endometrial cancers are adenocarcinomas (cancers that begin in cells that make mucus and other fluids).

Uterine sarcoma is an uncommon form of uterine cancer that forms in the muscle and tissue that support the uterus.

Obesity, certain inherited conditions, and taking estrogen alone (without progesterone) can increase the risk of endometrial cancer. Radiation therapy to the pelvis can increase the risk of uterine sarcoma. Taking tamoxifen for breast cancer can increase the risk of both endometrial cancer and uterine sarcoma.

The most common sign of endometrial cancer is unusual vaginal bleeding. Endometrial cancer can usually be cured. Uterine sarcoma is harder to cure.

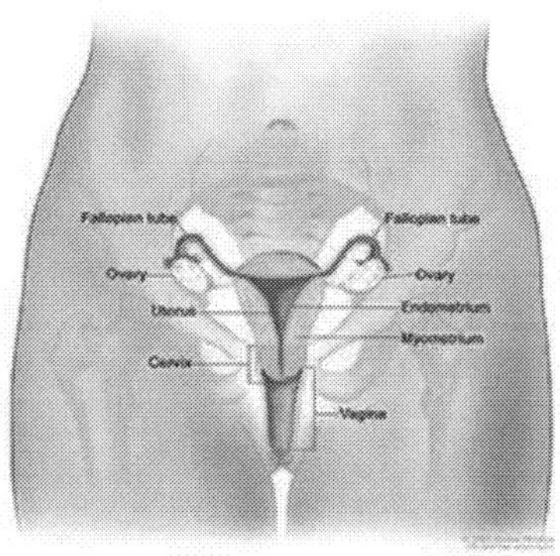

Anatomy of the female reproductive system. The organs in the female reproductive system include the uterus, ovaries, fallopian tubes, cervix, and vagina.

Gene Theme Scene Synopsis of Gene Based Targeted Therapy (Uterus)

Leiomyosarcoma Patients with the P53 Mutation

Thioureiodobutyronitrile (Kevetrin) (P53 Inducer) Pharmaceutical Company Cellceutix

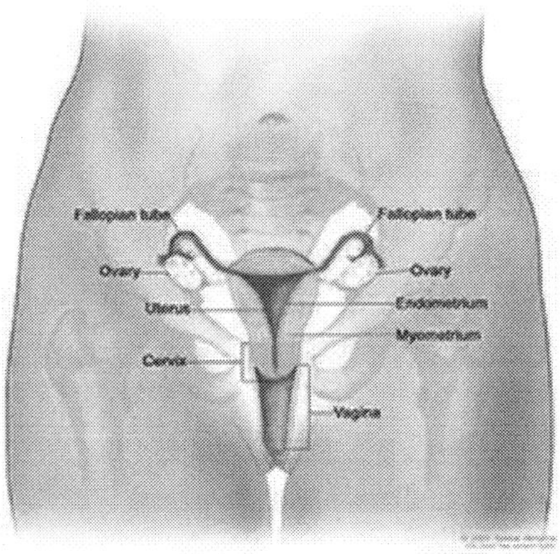

Anatomy of the female reproductive system. The organs in the female reproductive system include the uterus, ovaries, fallopian tubes, cervix, and vagina.

Gene Theme Scene Synopsis of Gene Based Targeted Therapy (Soft Tissue Sarcoma)

Soft Tissue Sarcoma Patients with the VEGFR, PDGFR, c-KIT targets

Pazopanib (Votrient) Pharmaceutical Company GlaxoSmithKline

NCI Bone Cancer Overview

There are several types of bone cancer.

Osteosarcoma is the most common bone cancer. It starts in bone cells that make new bone tissue. It usually forms at the end of long bones, such as the leg bones, but can form in any bone. It is most common in teenagers and in adults older than 65 years. Malignant fibrous histiocytoma of bone is a very rare bone cancer. It is treated like osteosarcoma.

Ewing sarcoma includes several types of bone tumors. Ewing sarcoma tumors usually form in the hip bones, the ribs, or in the middle of long bones. The disease occurs most often in teenagers and young adults. Ewing tumors are most common in bone but can also form in soft tissue.

Having past treatment with radiation can increase the risk of osteosarcoma. A small number of bone cancers are caused by inherited conditions. Signs and symptoms of bone tumors include a lump, swelling, and pain.

Bone cancer is rare. Most bone tumors are benign (not cancer).

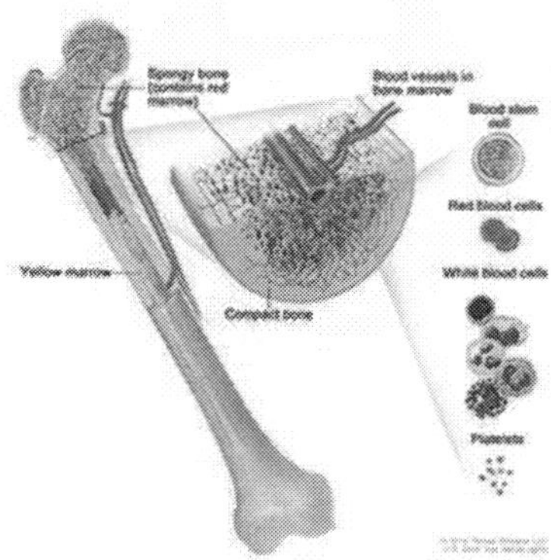

Anatomy of the bone. The bone is made up of compact bone, spongy bone, and bone marrow

NCI Skin Cancer Overview

The skin protects against heat, sunlight, injury, and infection. Skin also helps control body temperature and stores water and fat. Skin cancer is the most common type of cancer. It usually forms in skin that has been exposed to sunlight, but can occur anywhere on the body.

Skin has several layers. Skin cancer begins in the epidermis (outer layer), which is made up of squamous cells, basal cells, and melanocytes.

There are several different types of skin cancer. Squamous cell and basal cell skin cancers are sometimes called nonmelanoma skin cancers. Nonmelanoma skin cancer usually responds to treatment and rarely spreads to other parts of the body. Melanoma is more aggressive than most other types of skin cancer. If it isn't diagnosed early, it is likely to invade nearby tissues and spread to other parts of the body. The number of cases of melanoma is increasing each year. Only 2 percent of all skin cancers are melanoma, but it causes most deaths from skin cancer.

Rare types of skin cancer include Merkel cell carcinoma, skin lymphoma, and Kaposi sarcoma.

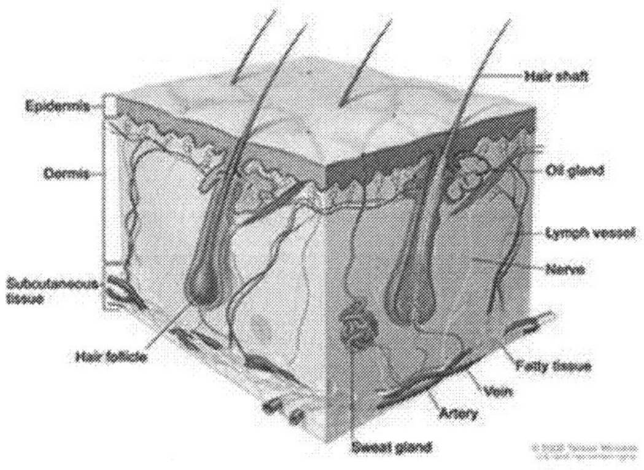

Anatomy of the skin, showing the epidermis, dermis, and subcutaneous tissue.

Gene Theme Scene Synopsis of Gene Based Targeted Therapy

Metastatic Malignant Melanoma Patients with the BRAF V600E mutation target

Vemurafenib (Zelboraf) Pharmaceutical Company Genentech

Metastatic Melanoma Patients with the CTLA4 target

Ipilimumab (Yervoy) Pharmaceutical Company Bristol-MyersSquibb

Metastatic Melanoma Patients with Ligand Activation of PD-1 Receptor on Activated T Cells

Nivolumab (Opdivo) Pharmaceutical Company Bristol-Myers

Squibb Metastatic Melanoma Patients with the PD-1 receptor target

Pembrolizumab (Keytruda) Pharmaceutical Company Merck

Leukemia Overview

Leukemia is cancer of the blood cells. Most blood cells form in the bone marrow. In leukemia, cancerous blood cells form and crowd out the healthy blood cells in the bone marrow.

The type of leukemia depends on the type of blood cell that has become cancerous. For example, acute lymphoblastic leukemia is a cancer of the lymphoblasts (white blood cells that fight infection). White blood cells are the most common type of blood cell to become cancer. But red blood cells (cells that carry oxygen from the lungs to the rest of the body) and platelets (cells that clot the blood) may also become cancer.

Leukemia occurs most often in adults older than 55 years, and it is the most common cancer in children younger than 15 years.

Leukemia is either acute or chronic. Acute leukemia is a fast-growing cancer that usually gets worse quickly. Chronic leukemia is a slower-growing cancer that gets worse slowly over time. The treatment and prognosis for leukemia depend on the type of blood cell affected and whether the leukemia is acute or chronic. Chemotherapy is often used to treat leukemia.

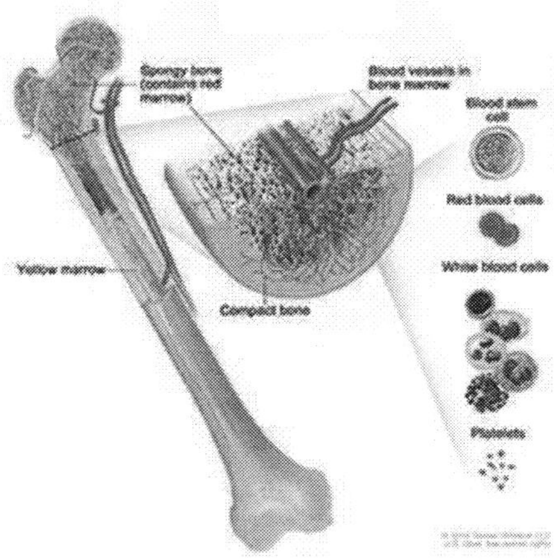

Anatomy of the bone. The bone is made up of compact bone, spongy bone, and bone marrow

Gene Theme Scene Synopsis of Gene Based Targeted Therapy

Chronic Myelogenous Leukemia Patients with Ph+ Bcr-Abl c-KIT ProtoOncogene encodes for the Tyrosine Kinase Receptor

Mast/Stem Cell Factor Receptor (SCFR) CD117 (KIT) Tyrosine Kinase Receptor

Imatinib (Gleevec) Pharmaceutical Company Novartis

Chronic Myelogenous Leukemia Patients with the Bcr-Abl, Src, c-KIT target

Dasatinib (Sprycel) Pharmaceutical Company Bristol-MyersSquibb

Chronic Myelogenous Leukemia Patients with the Bcr-Abl Ph+

target Nilotinib (Tasigna) Pharmaceutical Company Novartis

Chronic Myelogenous Leukemia Patients with the Bcr-Abl

target Bosutinib (Bosulif) Pharmaceutical Company Pfizer

T3151- Positive Chronic Myelogenous Leukemia and Acute Lymphoblastic Leukemia Patients with the Bcr-Abl, EGFR, PDGFR, SRC, c-KIT, RET, FLT3, TIE2 targets

Ponatinib (Inclusig) Pharmaceutical Company ARIAD

Non Hodgkin Lymphoma and CLL Chronic Lymphocytic Leukemia Patients with the CD20 target

Rituximab (Rituxan) Pharmaceutical Company Genentech

B Cell Non Hodgkin Lymphoma Patients with the CD20 target

As an Antibody Radioimmunotherapy Ibritumomab congugated to the metal Tiuxetan (Zevalin) Pharmaceutial Co. IDEC

Anaplastic Lymphoma Patients with the ALK, HGFR, c-MET

Crizotinib (Xalkori) Pharmaceutical Company Pfizer

Chronic Lymphocytic Leukemia Patients with the CD20 target

Ofatumumab (Arerra) Pharmaceutical Company Glaxo-SmithKline

Relapsed Anaplastic Large Cell Lymphoma and Relapsed Hodgkin Lymphoma Patients with the CD30 target

Brentuximab Vedotin (Adcetris) Pharmaceutical Companies Seattle Genetics (SGEN) and Immunogen (IMGN)

Chronic Lymphocytic Leukemia and Multiple Sclerosis Patients with the CD52 target

Alemtuzumab (Lemtrada) Pharmaceutical Company Bayer

Future of Gene Based Targeted Treatment

The field of Genetics Needs You!!!! Since my goal is to have Gene Based Targeted Treatments available in our lifetime, we must enhance our research efforts in the Cloning of Embryonic and Adult Stem Cells, which includes Cloning Neurologic Stem Cells for ALS Amyotropic Lateral Sclerosis and MS Multiple Sclerosis patients, as well the use of the washed extracellular matrix (scaffold) of a pig organ on which cells of the recipient are used to create an organ that will not be rejected (no immunosuppresant medication needed) as featured in the NOVA 1/26/11 Replacing Body Parts Video available for viewing on You Tube now! Yes I want to use my 31.02 acres in CT to reproduce this Clinical Trial by hiring Joseph Vacanti, MD from MGH to train our research team. Have we forgotten the man that got a kidney from a Woman who donated her kidney before she knew she had Uterine Cancer to her male donor who was immunosuppresed to prevent the rejection of her kidney, sadly, dying of her uterine cancer!!!!

I have also included articles on interventional radiological skills that need to be offered via cross training Neurosurgeons and our Vascular Surgeons. The widow of the actor John Ritter who died from an Aortic Dissection would agree that the best endovascular stent options should be available in all hospitals stating they have a vascular surgeon.

Articles on the Instrumentations for the Removal of DVT (Deep Vein Thrombosis), yes, those clots that when present in my Honorable Uncle Tommy, who served our Country for over 20 years including Viet Nam, whose 14cm DVT prompted his physician to anticoagulate him, until he bleed into his brain and died due to uncal herniation of his brainstem. With an IC Care Data accumulation in place, his physician could have had access to the fact that a CT scan performed at a Civilian Hospital six weeks prior, showed a contraindication to being anticoagulated, therefore my gentle statement to his Physician that we could Sign him out AMA to take him to a vascular surgeon should have been heeded. Recently his widow, my father's sister, my Precious Aunt Geraldine lost a friend of decades that had a successful coronary artery stent placement only to die recently after decades of anticoagulation, of uncal herniation of her brain stem also. Note: with in a few hours of an embolic type of stroke brain clots can be removed in order to preserve as much remaining brain function as possible, of course a Stroke Center with the proper instrumentation and a Neurosurgeon with the endovascular interventional radiological training must be available!!!

Thus the need to implement these latest technological advances is my life's work. You can contact me on my website Gene Dac.com and donate to Cancer Screening Centers.org a (501c3 non-profit organization). Stay tuned as Regional Gene Theme Scene Franchise Opportunities will be available soon. Respectfully, Clarisse Clemons Ferrara,MD also reachable via clarisseclemonsmd@outlook.com.

Endovascular Neurosurgery and Interventional Neuroradiology

Neurosurgery is a branch of surgery that treats conditions and diseases of the brain and nervous system. Radiology is a medical specialty that helps diagnose and treat conditions and diseases using various radiology techniques.

Endovascular neurosurgery is a subspecialty within neurosurgery that uses catheters and radiology to diagnose and treat various conditions and diseases of the central nervous system. The central nervous system is made up of the brain and the spinal cord. This medical specialty is also called neurointerventional surgery.

Interventional neuroradiology is a subspecialty within radiology that also involves catheters and radiology to diagnose and treat neurological conditions and diseases.

The term endovascular means "inside a blood vessel." Endovascular neurosurgery uses tools that pass through the blood vessels to diagnose and treat diseases and conditions rather than using open surgery. The surgeon often uses radiology images to help him or her to see the part of the body involved in the procedure.

Doctors in these specialties may also diagnose and treat conditions of the spinal cord using similar techniques, although not through a blood vessel.

These types of procedures are called "minimally invasive" because they generally require only a tiny incision instead of a larger incision necessary for open surgery.

Doctors who specialize in endovascular neurosurgery require training in both neurosurgery and radiology. After a neurosurgery residency, they complete a fellowship program in endovascular neurosurgery. A neurosurgeon may be board certified in neurosurgery through the American Board of Neurological Surgery. Currently, there is no board certified specialty in endovascular neurosurgery.

Doctors who specialize in interventional neuroradiology also require training in both radiology and neurology or neurosurgery. After a radiology residency, they complete a fellowship in interventional neuroradiology. A radiologist may be board certified in neuroradiology through the American Board of Radiology.

What kinds of procedures are performed by these specialists?

Endovascular neurosurgical procedures include:

Thrombolytic therapy. This procedure uses "clot-busting" medication to dissolve a clot in a blood vessel in the brain or elsewhere in the body.

Endovascular coiling. A surgeon inserts a very thin metal wire that forms a coil inside a brain aneurysm to block blood flow. A brain aneurysm is a bulging, weakened area in the wall of an artery in the brain, resulting in an abnormal widening or ballooning. Because the artery wall has a weakened spot, the aneurysm is at risk for bursting if blood flow isn't blocked. Aneurysm can be treated by coiling or clipping it closed.

Minimally invasive spine surgery. This procedure is used to treat spinal disorders, such as fractures, tumors, compressed nerves, and other conditions that put pressure on the spinal cord.

Cerebral angiography. This is a radiology procedure that looks at how blood is flowing in the brain.

Carotid artery angioplasty/stenting. This procedure uses a small balloon and/or a tiny metal scaffold called a *stent* to open a narrowed carotid artery. The carotid arteries supply blood to the brain.

Related Videos

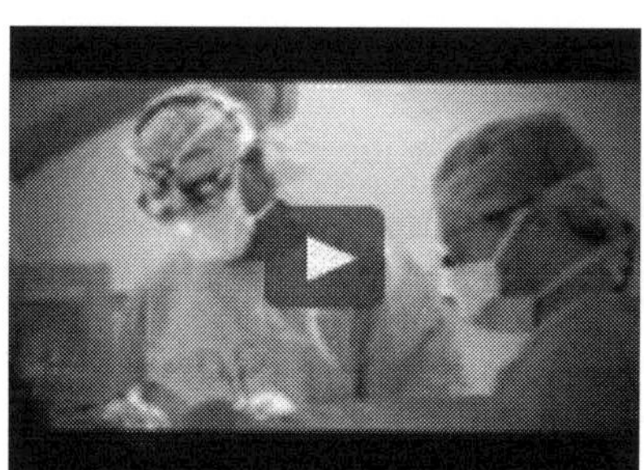

Visit www.hopkinsmedicine.org/neuro for more info. Dr Henry Brem, Director...

Ophthalmic Nerve Aneurysm

Appendix 1 Stent Graft Procedures

Technical features of the INCRAFT™ AAA Stent Graft System.

Bertoglio L1, Logaldo D, Marone EM, Rinaldi E, Chiesa R.

Author information

Vascular Surgery, "Vita-Salute" University, San Raffaele Hosp. Scientific Institute, Milan, Italy - bertoglio.luca@hsr.i.

Abstract

The INCRAFT® AAA Stent Graft System is the advanced endovascular aneurysm repair (EVAR) technology for the treatment of infrarenal abdominal aneurysms. This new system is designed to address the unmet needs of current endografts by combining unique features and adding new refinements compared to existing endografts delivered through a flexible 14-Fr ultra-low system. The INCRAFT® AAA Stent Graft System introduces innovative features without deviating from proven stent-graft design principles. It is a three-piece modular system, made of low porosity polyester and segmented nitinol stents. However, the introduction of cap-free delivery and partial proximal repositioning enhances the ability of the device to better match individual aortoiliac anatomy with a high deliverability and placement accuracy in a easy to use system. Moreover, the INCRAFT® System allows a "customization" of the implant during the procedure with bilateral in-situ length adjustment features. The present data from the ongoing clinical trials confirm excellent results with this system, but postmarket studies will be necessary to verify the effectiveness of this system in the real-world setting.

Endovascular abdominal aortic aneurysm repair

M G A Norwood, G M Lloyd, M J Bown, G Fishwick, N J London, and R D Sayers

M G A Norwood, G M Lloyd, M J Bown, N J London, R D Sayers, Department of Vascular Surgery, The Leicester Royal Infirmary, Leicester, UK

G Fishwick, Department of Radiology, The Leicester Royal Infirmary, Leicester, UK

Correspondence to: MrM G A Norwood Vascular Surgery Group, RKCSB, Leicester Royal Infirmary, Leicester LE2 7LX, UK; drmikenorwood@ hotmail.com

Author information - Article notes - Copyright and License information

Received 2006 Jul 6; Accepted 2006 Aug 7.

Abstract

The operative mortality following conventional abdominal aortic aneurysm (AAA) repair has not fallen significantly over the past two decades. Since its inception in 1991, endovascular aneurysm repair (EVAR) has provided an alternative to open AAA repair and perhaps an opportunity to improve operative mortality. Two recent large randomised trials have demonstrated the short and medium term benefit of EVAR over open AAA repair, although data on the long term efficacy of the technique are still lacking. This review aimed at providing an overview of EVAR and a discussion of the potential benefits and current limitations of the technique.

Abdominal aortic aneurysms (AAAs) occur in about 5% of men aged over 50 years. The incidence appears to be increasing, partly as a result of the increasing age of the population and also because of greater diagnostic awareness.[1,2,3] The natural history of an AAA is expansion at a rate of approximately 0.5 cm a year and eventual rupture, unless death from another cause supervenes. Most AAAs are asymptomatic unless rupture occurs, which presents with the classical triad of abdominal and back pain, hypovolaemic shock, and a pulsatile abdominal mass. The risk of rupture is related to size. AAAs greater than 5.5 cm in diameter have a significant risk of rupture, and elective repair is therefore indicated.[4,5]

Elective open surgical repair of an AAA is a complex procedure with a 30 day mortality of approximately 5% and a major complication rate of 15–30%, figures that have not changed significantly over the past 20 years.[6,7,8] The operative mortality increases significantly in medically unfit patients, and can be as high as 50%, resulting in many high risk patients being denied surgical repair.[8] The operative technique has changed very little since Dubost's original description in 1952, and involves exposure of the aneurysm through a trans- or retroperitoneal route, clamping of the aneurysm neck and iliac arteries, opening of the aneurysm sac, and insertion of a prosthetic graft.[9] The major risks during open elective repair are perioperative cardiac events (for example, myocardial infarction), but respiratory and renal failure are also common.[6] It is difficult to predict which patients are at risk despite techniques such as non-invasive cardiac risk assessment with echocardiography and physiological scoring systems such as POSSUM (Physiological and Operative Severity Score for the enUmeration of Mortality).[10] However, in surviving patients the long term durability of the technique is good; the annual rate of graft failure is in the region of 0.3% and in the vast majority of cases the graft will outlive the patient.[5,11,12]

A ruptured AAA (RAAA) accounts for about 7000 deaths a year in England and Wales, and the incidence is increasing.[1] The overall mortality is at least 80%, with near 50% mortality in patients who reach hospital alive. Despite advances in surgery, anaesthesia, and critical care, there has been little reduction in mortality over the past 40 years.[13] Surgical repair of an RAAA is a formidable undertaking. The technical problems during surgery include difficulties with vascular control, avoiding damage to surrounding structures, and achieving haemostasis in coagulopathic patients. Despite this, most patients (85%) survive surgery, but many die in the postoperative period from progressive multiple organ failure.[14] The cause of multiple organ failure is unknown, but probably related to widespread activation of inflammatory pathways stimulated by major surgery, hypovolaemic shock, acidosis, hypothermia, and massive blood transfusion.

At the present time, there are two strategies that might reduce the overall mortality from AAA. These are screening and endovascular aneurysm repair (EVAR). Ultrasound screening of a population at risk (usually men over 65 years of age) has been shown to reduce the incidence of RAAA in a recent multicentred trial, but as yet there is no national screening programme in the UK.[1] EVAR of elective (non-ruptured) AAAs has been performed in the UK since the early 1990s. More recently, several centres have also described their initial experience of repair of ruptured AAAs using EVAR. EVAR of elective AAAs is currently being evaluated by randomised trials; the early and intermediate term results of two of these—the UK Endovascular Aneurysm Repair (EVAR) and the Dutch Randomised Endovascular Aneurysm Management (DREAM) trials—have recently been reported

in the *Lancet* and the *New England Journal of Medicine*, respectively.[15,16,17,18] This article reviews the current role of EVAR in the treatment of this complex disease process.

Endovascular AAA repair

EVAR was first described in humans by Parodi *et al* in 1991.[19] This minimally invasive technique can be performed under general or regional anaesthesia and involves the placement of a stent-graft inside the aneurysm sac through the common femoral arteries. The stent-graft is a prosthetic vascular graft, (typically made of Dacron or polytetrafluoroethylene), which is reinforced by metallic struts and has metallic stents at the ends that fixate to the arterial wall. The stent-graft is usually delivered into the aneurysm inside a sheath, manoeuvred into position under x ray (fluoroscopic) guidance, and deployed by withdrawal of the sheath. Final fixation at the proximal and distal ends of the stent-grafts is usually achieved by inflating a balloon inside the stent-graft to ensure firm attachment to the arterial wall.

Over the past decade a number of UK vascular units have been performing EVAR, and extensive experience of patient selection, intraoperative technique, and postoperative problems has been acquired. Indeed, not all patients are suitable for EVAR, and certain anatomical criteria must be satisfied in order for EVAR to be performed successfully. Several studies have shown that about 50% of patients with AAA are suitable for EVAR based upon anatomical and technical criteria,[20,21,22] but this figure is likely to increase as stent-graft technology improves. Suitability is usually determined by a preoperative contrast enhanced computed tomography (CT) scan, or less frequently, angiography.[23,24,25] Modern spiral CT scanners and, more recently, multidetector (multislice) scanners, allow accurate assessment of aortic morphology and the size of the stent-graft required by the rapid acquisition of thin axial slice (1–3 mm cut) images and two- and three-dimensional angiographic reconstructions. The anatomical criteria required for EVAR have

Table 1 Favourable anatomical features for EVAR

Anatomical feature	Dimension
Neck length	>15 mm
Neck diameter	<30 mm
Neck angulation	<60°
Neck mural thrombus	<2 mm
External iliac artery diameter	>7 mm
Iliac angulation	<90°
Common iliac diameter	<18 mm

been developed and refined over the years (table), but the underlying principles remain: the arterial anatomy proximal and distal to the aneurysm must allow firm fixation of the stent-graft to the arterial wall, and the iliac artery configuration must permit the stent-graft to access the aorta through the femoral arteries.

Table 1 Favourable anatomical features for EVAR

The precise anatomical requirements for each design of stent-graft differ but, in general, the proximal aneurysm neck (non-dilated segment of aorta between the renal arteries and the aneurysm) should be free of thrombus, have adequate length, limited diameter, and limited angulation.[26] The iliac arteries should be of sufficient calibre to allow the stent-graft to be introduced, should not be tortuous, and have minimal calcification.[26] Attempted EVAR in the presence of unfavourable anatomy predisposes to intraoperative failure or postoperative complications.[27] Although not considered an absolute contraindication to EVAR, patients with larger AAA have been found to be more likely to experience adverse outcomes.[28,29,30]

Historically, the early stent-grafts were tubular, single body aorto-aortic devices that fixed to the aortic wall proximally, and to the aneurysm sac distally.31 However, a high incidence of complications, particularly attachment site failure and distal migration of the stent-graft, has rendered these devices almost obsolete.[32] Most stent-grafts currently in use are either bifurcated aorto-iliac grafts or straight aorto-monoiliac grafts. They are either unitary (one piece) or modular (composite) devices. If a monoiliac device is employed, the contralateral iliac artery is thrombosed (radiologically) and a femoro-femoral crossover graft is performed in order to perfuse the contralateral leg. Stent-graft design has rapidly evolved over the past decade, often in response to the identification of specific design related flaws and there has been a consistent trend towards newer devices outperforming their predecessors.[28,33] There is continuing research directed towards designing stent-grafts that will be able to contend with anatomical features that currently preclude or increase the risk of EVAR, with the aim of increasing the numbers of patients in whom EVAR can be performed. Fenestrated stent-grafts that extend above the renal arteries with holes placed over the ostia of the renal, superior mesenteric and coeliac arteries are being developed and will allow treatment of short necked or suprarenal AAAs.[34,35] Flexible devices that can follow the contour of the aneurysm neck may be able to treat AAA with marked neck angulation, and smaller devices may be able to navigate narrow or tortuous iliac arteries.

Existing research and continuing trials in EVAR

In the past 15 years, there has been a large amount of research dedicated to EVAR. This research has largely consisted of either small studies investigating specific aspects of EVAR (usually in direct comparison with conventional AAA repair) or large national multicentre trials comparing EVAR with conventional AAA repair. The smaller studies have provided some convincing evidence that the minimally invasive nature of EVAR in conjunction with obviation of the need for laparotomy and complex abdominal dissection, results in less physiological insult than open AAA repair. Biological markers of inflammatory pathways and stress responses such as proinflammatory cytokines, adrenaline, cortisol, and complement activity are reduced.[36,37,38] There is convincing evidence that EVAR is associated with reduced cardiac, respiratory, and renal complications, reduced need for blood transfusion and analgesia, shorter duration of surgery and postoperative hospital stay, lower need for intensive care, a faster return to normal function, and lower infection rates.[16,32,39,40,41,42,43]

Until earlier this year, in the absence of evidence from large national randomised controlled trials, information on the efficacy of EVAR relied upon observational case series and data from large voluntary registries, such as the UK Registry for Endovascular Treatment of Aneurysms (RETA), and in particular, the European Collaborators on Stent Graft Techniques for Abdominal Aortic Aneurysm Repair (EUROSTAR) Registry.[44,45] EUROSTAR is a voluntary registry that has been prospectively collecting data on EVAR procedures performed in Europe since 1999, and has provided much of the current EVAR outcome data.[45,46] These sources have suggested that the initial operative mortality of EVAR was comparable to that of open AAA repair but has consistently fallen and may now be as low as 1%.[33,46,47,48,49] However, early meaningful interpretation and comparison of the data from these registries and case series was difficult to make because of factors such as publication bias, lack of patient randomisation, the likelihood of a preponderance of patients with small aneurysms, the inclusion of patients considered unfit for open AAA repair, and patients with less challenging anatomy.[26,50,51,52]

A number of controlled trials of EVAR against open AAA repair are currently in progress. In the United Kingdom there are two multicentre trials; EVAR 1 and EVAR 2, both of which started in 1999. EVAR 1 has randomised patients with an AAA >5.5 cm who are medically and anatomically suitable for both open AAA repair and EVAR to undergo one or other procedure. The EVAR 2 trial (discussed later) randomised patients who are considered medically unfit for open AAA repair (but anatomically suitable for EVAR) to undergo EVAR or receive "best medical treatment". Between September 1999 and December 2003, EVAR 1 recruited 1082 patients from 41 centres. Also in Europe, the smaller Dutch DREAM trial randomised 345 patients to either open AAA repair or EVAR. The short term (30 day mortality) outcome of these trials was published in 2004, with the EVAR 1 trial reporting a significantly improved 30 day mortality in patients undergoing EVAR compared with patients undergoing open AAA repair (1.7% *v* 4.7%)[15]; the DREAM trial reported a similarly improved 30 day mortality in EVAR versus open AAA patients (1.2% *v* 4.6%).[16] The DREAM trial was, however, underpowered, with the result that improved mortality rates did not reach statistical significance. The DREAM trial made the point that when combining their results with EVAR 1, the resulting operative mortality of 5.8% for open AAA repair and 1.9% for EVAR, yields a risk ratio of 3.1.[16] Following on from the initial data, during 2005 EVAR 1 published its three year mortality data,[17] which demonstrated that all-cause mortality was similar in the two groups, but that there was a persistent reduction in aneurysm related deaths in the EVAR group (4% *v* 7%). The results of EVAR 2 were also published in 2005,[53] but no benefit was shown in patients undergoing EVAR compared with those receiving "best medical treatment". The underpowered DREAM trial has also published further survival data,[18] but initial (30 day) perioperative survival advantage of EVAR over open AAA repair was not sustained after the first year. World wide, in addition to the UK EVAR trial and the Dutch DREAM trial, the French Aneurisme de l'aorte abdominale: Chirurgie versus Endoprothese (ACE) study and the United States Open Versus Endovascular Repair (OVER) study are yet to report.

Despite the potential survival benefit and perioperative advantage of EVAR over open repair, enthusiasm for EVAR has been tempered by a number of procedure-specific problems that may result in further hospital admissions and secondary interventions. These problems are endoleaks and other graft-specific problems. Recent data from the

EUROSTAR Registry[54] have shown that secondary intervention was required in 8.7% of patients at 12 months. The annual cumulative rate for secondary intervention was 6%, 8.7%, 12%, and 14% at 1, 2, 3, and 4 years, respectively, resulting in an annual mean secondary intervention rate of 4.6%. This has meant that long term surveillance of EVAR patients is needed, with its associated cost and resource implications. This also has implications for the patient and needs to be thoroughly discussed when consenting for surgery and deciding between EVAR and an open procedure. Indeed, some younger patients opt for a conventional,

open operation if they do not want to return for long term follow-up.

Endoleak

The primary objective of AAA treatment is to prevent aneurysm rupture, and to achieve this, arterial perfusion of the aneurysm sac must be prevented. Endoleak is a phenomenon unique to EVAR and is defined as continued blood flow within the aneurysm sac after graft placement. It affects between 10 and 50% of patients.[40,51,55,56] The conventional classification of endoleak into types 1–4 (table 2) is based on the anatomical source of the blood

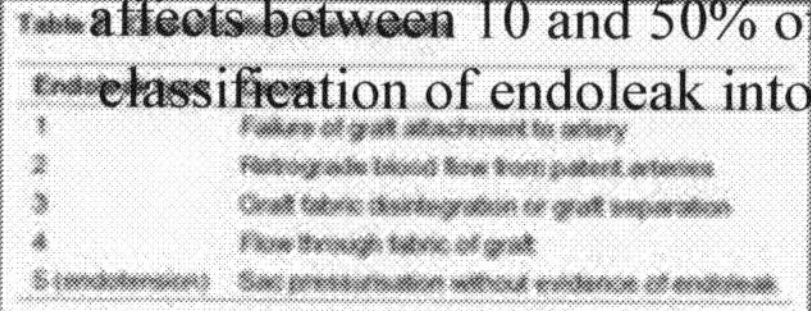

1	Failure of graft attachment to artery
2	Retrograde blood flow from patent arteries
3	Graft fabric disintegration or graft separation
4	Flow through fabric of graft
5 (endotension)	Sac pressurisation without evidence of endoleak

entering the aneurysm sac.[56,57]

Table 2 Classification of endoleaks

Type 1 endoleak is caused by blood flow into the aneurysm sac due to failure of the graft attachment to the aortic wall, either at the proximal or distal end of the graft. The principal causes appear to be adverse aneurysm morphology such as a short or angulated neck or gradual neck dilatation, probably as a consequence of progression of the aneurysmal process.[58] Type 2 endoleak, the most common variety, results from retrograde blood flow into the sac from patent lumbar, inferior mesenteric, hypogastric or accessory renal arteries. Type 3 endoleak results from blood flow through the graft due to fabric tear or through the junction sites between graft components. This was a particular problem with early grafts and appears to be less of a problem with the newer devices. Type 4 endoleak usually occurs in very thin grafts, the fabric of which becomes relatively porous and allows blood flow through the graft. In common with type 3 endoleaks, this appears to be less of a problem with the newer devices. Endotension, or type 5 endoleak is a curious phenomenon characterised by continued sac pressurisation without evidence of endoleak.[59,60] A number of theories to explain type 5 endoleaks have been proposed, including sac thrombus transmitting pressure or the presence of small undetectable endoleaks.[59,61]

Type 1 and type 3 endoleaks are of particular significance, as they frequently signify blood entering the aneurysm sac at systemic blood pressure, resulting in continued aneurysm expansion and risk of rupture.[59,62,63] Consequently, they require early intervention, either by balloon angioplasty to improve apposition to the arterial wall landing zone, the insertion of additional stent-grafts or cuffs to extend graft coverage in the landing zone or conversion to open repair.[33,64] The significance of type 2 and type 4 endoleaks is unclear.[40] In many cases they are probably of low pressure and are benign. However, management of this phenomenon is complicated by the finding that even thrombosed endoleaks can transmit systemic pressure and result in rupture.[65,66] There is currently no consensus about the management of type 2 endoleaks. In some centres, routine intervention, usually in the form of percutaneous coil embolisation or laparoscopic ligation of feeding vessels, is undertaken,[33,67] although recently there has been a trend towards intervention only if there is radiological evidence of continued expansion of the aneurysm sac, indicating a possible increased risk of rupture.[33,35,67,68] Current opinion is that the aneurysm sac size change is of greater significance than the presence of an endoleak. Type 4 endoleaks are usually treated conservatively.

Morphological changes

After EVAR, postoperative morphological changes in the aneurysm sac can occur. Sac shrinkage is seen as a reassuring sign of successful aneurysm exclusion from the circulation. However, this can be accompanied by shortening in the longitudinal axis,[69] which has been found to predispose to stent-graft limb kinking69,70 or separation of modular components, leading to graft occlusion43 or type 3 endoleak.[71] Continued expansion of the aneurysm neck after EVAR is a cause of late proximal attachment site failure,[72] graft migration, and risk of rupture.[46] The graft therefore has to cope with these morphological changes even after successful EVAR. It is expected that newer fenestrated grafts with suprarenal extension will be better able to survive these morphological changes.[73]

Graft migration

Distal migration of the stent-graft results from an attachment site

failure and is a cause of late type 1 endoleak, graft occlusion, and aneurysm rupture.[46] This was a significant problem with early aorto-aortic devices and the use of bifurcated or aorto-monoiliac

grafts has reduced the incidence of migration. It is predicted that grafts with suprarenal fixation or extension will further reduce the risk.

Structural failure

Failure of stent-graft components, particularly metal strut breakage and

fabric failure, has been reported, and is usually visible on a plain abdominal radiograph. Significant problems have been identified with early proprietary devices, including hook fractures with the EVT device,74 fabrication flaws with the Min Tec Stentor graft, loose ligatures on the Stentor endografts, and suture breakage in the Vanguard grafts,[75] leading to withdrawal or modification of the affected devices.

Rupture

Rupture is a recognised complication of EVAR, with an incidence of between 1 and 1.5% a year, equivalent to the rupture risk of untreated AAA <5.5 cm.[4,46,76] The EUROSTAR Registry reported rupture in 34 out of 4291 patients.[77] However, most of these ruptures occurred in patients with early generation grafts, although aneurysm related death after EVAR, excluding that for withdrawn devices, is still 0.6–0.7% annually.[78] Worryingly, although type 1 and type 3 endoleaks and graft migration are predictors of increased risk of rupture, rupture can still occur occasionally in patients with no evidence of endoleak or previously increasing sac size.[46,79]

Continued surveillance and need for further intervention

The risk of the postoperative complications detailed above necessitates continued radiological surveillance of patients after EVAR, with yearly abdominal CT and plain abdominal radiographs.[56] Complications such as endoleak can develop late in patients with normal early CT scan appearance,[43] and as no definitive long term follow up data are available it is unclear how long postoperatively endoleak remains a risk. In most

centres CT scans with contrast enhancement are performed before the patient is discharged from hospital, and at 6 and 12 months postoperatively, followed by lifelong annual scans. Clearly, patients need to be aware of this and accept the need for continued surveillance and possibly further procedures after EVAR, as poor compliance is associated with risk of rupture.[77] Although graft failure, para-anastomotic aneurysm, and AAA rupture can follow open AAA repair, these occurrences are sufficiently uncommon to justify no postoperative surveillance and in most cases open AAA repair can be regarded as a definitive procedure.

The incidence of postoperative complications currently necessitates further hospital admission and secondary intervention in approximately 14% of patients over four years after EVAR.[54] Despite this relatively frequent occurrence, approximately 80% can be treated with percutaneous radiological techniques.[42,80] The concept of emphasising the "second outcome success" has been advocated as the correct way for the surgeon and patient to regard EVAR. The primary objective of AAA repair is to prevent rupture, and the need for further minimally invasive procedures to achieve this should perhaps not be regarded as failure.[75] Along with aneurysm rupture, the need to convert to open AAA repair is a true indicator of failure of EVAR. Open procedures such as graft replacement after EVAR are technically difficult and carry a high mortality, in the region of 20%.[81,82,83] The EUROSTAR Registry reported that an early conversion to open repair is necessary in 2% of patients undergoing EVAR.[62]

Cost

The uncertainty surrounding long term outcome after EVAR means that the true cost of the procedure is unknown. It was hoped that the savings resulting from a shorter operating time, shorter length of hospital stay, reduced intensive care costs, fewer operative complications, and faster recovery and return to work might offset the initial cost of EVAR.

However, the high cost of stent-grafts, additional radiological imaging, readmissions to hospital, adjuvant procedures, and prolonged patient surveillance mean that EVAR is probably, at present, more expensive than open repair.[84,85,86] If, as envisaged, EVAR related complications become less common and competition among stent-graft manufacturers leads to cheaper devices the cost of EVAR may compare more favourably with that of open AAA repair.

Role of EVAR in elderly or medically unfit patients

Historically, many elderly or medically unfit patients with AAA have not been referred to a surgeon or have been denied open AAA repair because of a high predicted perioperative mortality.[68,87] However, even in patients with significant comorbidities who have large or symptomatic AAA this is a difficult decision to make as rupture leading to almost certain death is frequently imminent. The decreased physiological insult, lower operative morbidity and mortality associated with EVAR, feasibility of performing the procedure under local anaesthesia, and the lesser importance of long term stent-graft durability have resulted in enthusiasm for performing EVAR in this patient group. A number of centres offering EVAR have reported increasing numbers of patients with AAA referred and subsequently treated, a likely consequence of the growing awareness by physicians of the potential role of EVAR in patients at high risk for open AAA repair.[21,88] Published case series have reported encouraging morbidity and mortality rates in elderly patients and those considered high risk or unfit for open AAA repair.[89,90] However, despite EVAR appearing to be an attractive proposition in these patients, the results of the recently published EVAR 2 trial have failed to show any benefit of EVAR over "best medical treatment".[53] The trial demonstrated a 9% 30 day mortality in the EVAR arm of the trial compared with a rupture rate of 9.0 per 100 patient years in the "no intervention" group. The mean hospital cost over four years in the EVAR group was also high (£13 632) compared with the "no intervention" group (£4983).

Role of EVAR in RAAA

The less invasive nature and physiological advantages of EVAR over open AAA repair may be of particular benefit in patients with RAAA with deranged physiology and approaching extremis. Since the first report in 1994 that RAAA can be successfully treated with EVAR,[91] a number of centres have reported their early results, raising the possibility that the technique may provide the first real opportunity to improve operative mortality significantly in RAAA. Although experience is still limited, Ohki *et al* reported performing EVAR in 20/25 patients with RAAA with two deaths,[92] and Lee *et al* reported being able to perform EVAR in 13/17 consecutive patients with RAAA with a single death.[93]

This compares favourably with the near 50% in-hospital mortality of open AAA repair.[13]

A number of potential problems persist, however. To perform EVAR a CT scan is required to assess the suitability of the aneurysm for EVAR and to size the graft correctly. Initially, only patients with "stable" ruptures, who were haemodynamically stable underwent EVAR as it was felt that the delay in obtaining a CT scan, which would delay starting surgery, would be detrimental. However, it has been shown that almost 90% of patients with RAAA who reach hospital alive survive longer than 2 hours, and most centres performing EVAR have spiral CT scanners that allow a complete abdomen and chest CT scan to be performed within seconds and protocols to facilitate the transfer of patients with RAAA to theatre within about 25 minutes of admission.[94] As a result even haemodynamically unstable patients are currently undergoing successful EVAR in specialist units.

Key references

Greenhalgh RM, Brown LC, Kwong GP, et al, EVAR trial participants. Comparison of endovascular aneurysm repair with open repair in patients with abdominal aortic aneurysm (EVAR trial 1), 30-day operative mortality results: randomised controlled trial. Lancet 2004;364:843–8.

Prinssen M, Verhoeven EL, Buth J, et al, Dutch Randomized Endovascular Aneurysm Management (DREAM) Trial Group. A randomized trial comparing conventional and endovascular repair of abdominal aortic aneurysms. N Engl J Med 2004;351:1607–18.

EVAR trial participants. Endovascular aneurysm repair versus open repair in patients with abdominal aortic aneurysm (EVAR trial 1): randomised controlled trial. Lancet 2005;365:2179–86.

Blankensteijn JD, de Jong, SECA, Prinssen M, et al. Two year outcomes after conventional or endovascular repair of abdominal aortic aneurysms. N Engl J Med 2005;352:2398–405.

EVAR trial participants. Endovascular aneurysm repair and outcome in patients unfit for open repair of abdominal aortic aneurysm (EVAR trial 2): randomised controlled trial. Lancet 2005;365:2187–92.

Conclusion

EVAR is a new technique, and represents a major advance in the treatment of AAA. It has undergone considerable refinement since its advent, and confers a number of advantages over open AAA repair, including no need for laparotomy, less postoperative pain, no need for routine intensive care, and shorter hospital stay. The initial results look promising, and although there is no difference in all-cause mortality between patients undergoing EVAR and open repair, there is now evidence that EVAR is associated with a persistent reduction in aneurysm related deaths over three years (EVAR 1). Longer term follow-up data are still awaited, and in common with all new techniques EVAR has its problems—namely, secondary stent related problems that require intervention, the need for long term follow-up and the associated cost. However, stent-graft technology is rapidly advancing, and many of the initially identified problems are now being overcome. We think that EVAR is here to stay and that patients who are anatomically suitable for EVAR should be offered the opportunity to make an informed choice between EVAR and open AAA repair.

Multiple choice questions (true (T)/false (F)); answers at the end of references

1. The mortality from AAA repair has fallen significantly over the past 20 years.
2. EVAR can be performed under regional anaesthesia.
3. The long term benefit of EVAR over open AAA repair has been proved.
4. Type 1 endoleak is frequently caused by a patent inferior mesenteric artery.
5. EVAR can be used to treat RAAA.

Abbreviations

AAA - abdominal aortic aneurysm
CT - computed tomography
DREAM - Dutch Randomised Endovascular Aneurysm Management (trial)
EUROSTAR - European Collaborators on Stent Graft Techniques for Abdominal Aortic Aneurysm Repair
EVAR - endovascular aneurysm repair
RAAA - ruptured AAA

Answers

(1) F, (2) T, (3) F, (4) F, (5) T

References

1. Ashton H A, Buxton M J, Day N E. *et al* The Multicentre Aneurysm Screening Study (MASS) into the effect of abdominal aortic aneurysm screening on mortality in men: a randomised controlled trial. *Lancet* 20023601531– 1539.1539

2. Naylor A R, Webb J, Fowkes F G. *et al* Trends in abdominal aortic aneurysm surgery in Scotland (1971–1984). *Eur J Vasc Surg* 19882217–221.221 [PubMed]

3. Samy A K, Whyte B, MacBain G. Abdominal aortic aneurysm in Scotland. *Br J Surg* 1994811104–1106.1106

4. The UK Small Aneurysm Trial participants Mortality results for randomised controlled trial of early elective surgery or ultrasonographic surveillance for small abdominal aortic aneurysms. *Lancet* 19983521649–1655.1655

5. Lederle F A, Wilson S E, Johnson G R, Aneurysm Detection and Management Veterans Affairs Cooperative Study Group *et al* Immediate repair compared with surveillance of small abdominal aortic aneurysms. *N Engl J Med* 20023461437–1444.1444

6. Blankensteijn J D, Lindenburg F P, Van der Graaf Y. *et al* Influence of study design on reported mortality and morbidity rates after abdominal aortic aneurysm repair. *Br J Surg* 1998851624–1630.1630

7. Ernst C B. Abdominal aortic aneurysm. *N Engl J Med* 19933281167–1172.1172

8. Johnston K W. Multicenter prospective study of nonruptured abdominal aortic aneurysm. Part II. Variables predicting morbidity and mortality. *J Vasc Surg* 19899437–447.447

9. Dubost C, Allary M, Oeconomos N. Resection of an aneurysm of the abdominal aorta: reestablishment of the continuity by a preserved human arterial graft, with result after five months. *AMA Arch Surg* 195264405–408.408

10. Heller J A, Weinberg A, Arons R. *et al* Two decades of abdominal aortic aneurysm repair: have we made any progress? *J Vasc Surg* 2000321091–1100.1100

11. Hallett J W, Jr, Marshall D M, Petterson T M. et al Graft-related complications after abdominal aortic aneurysm repair: reassurance from a 36-year population-based experience. J Vasc Surg 199725277–284.284

12. Johnston K W. Nonruptured abdominal aortic aneurysm: six-year follow-up results from the multicenter prospective Canadian aneurysm study. Canadian Society for Vascular Surgery Aneurysm Study Group. *J Vasc Surg*

199420163–170.170 [PubMed]

13. Bown M J, Sutton A J, Bell P R. et al A meta-analysis of 50 years of ruptured abdominal aortic aneurysm repair. *Br J Surg* 200289714–730.730 [PubMed]

14. Sayers R D, Thompson M M, Nasim A. *et al* Surgical management of 671 abdominal aortic aneurysms: a 13 year review from a single centre. *Eur J Vasc Endovasc Surg* 199713322–327.327

15. Greenhalgh RM, Brown LC, Kwong GP, *et al*, Comparison of endovascular aneurysm repair with open repair in patients with abdominal aortic aneurysm (EVAR trial 1), 30-day operative mortality results: randomised controlled trial. *Lancet* 2004364843–848.848

16. Prinssen M, Verhoeven E L, Buth J, Dutch Randomized Endovascular Aneurysm Management (DREAM) Trial Group *et al* A randomized trial comparing conventional and endovascular repair of abdominal aortic aneurysms. *N Engl J Med* 20043511607–1618.1618

17. EVAR trial participants Endovascular aneurysm repair versus open repair in patients with abdominal aortic aneurysm (EVAR trial 1): randomised controlled trial. *Lancet* 20053652179–2186.2186

18. Blankensteijn J D, de Jong S E C A, Prinssen M. *et al* Two year outcomes after conventional or endovascular repair of abdominal aortic aneurysms. *N Engl J Med* 20053522398–2405.2405

19. Parodi J C, Palmaz J C, Barone H D. Transfemoral intraluminal graft implantation for abdominal aortic aneurysms. *Ann Vasc Surg* 19915491– 499.499

20. Brewster D C, Geller S C, Kaufman J A. *et al* Initial experience with endovascular aneurysm repair: comparison of early results with outcome of conventional open repair. *J Vasc Surg* 199827992–1003.1003

21. Wolf Y G, Fogarty T J, Olcott C I V. *et al* Endovascular repair of abdominal aortic aneurysms: eligibility rate and impact on the rate of open repair. *J Vasc Surg* 200032519–523.523

22. Armon M P, Yusuf S W, Latief K. *et al* Anatomical suitability of abdominal aortic aneurysms for endovascular repair. *Br J Surg* 199784178–180.180

23. Beebe H G, Kritpracha B, Serres S. *et al* Endograft planning without preoperative arteriography: a clinical feasibility study. *J Endovasc Ther* 200078–15.15

24. Broeders I A, Blankensteijn J D, Olree M. *et al* Preoperative sizing of grafts for transfemoral endovascular aneurysm management: a prospective comparative study of spiral CT angiography, arteriography, and conventional CT imaging. *J Endovasc Surg* 19974252–261.261

25. Gorham T J, Taylor J, Raptis S. Endovascular treatment of abdominal aortic aneurysm. *Br J Surg* 200491815–827.827 [PubMed]

26. Carpenter J P, Baum R A, Barker C F. *et al* Impact of exclusion criteria on patient selection for endovascular abdominal aortic aneurysm repair. *J Vasc Surg* 2001341050–1054.1054 [PubMed]

27. Stanley B M, Semmens J B, Mai Q. *et al* Evaluation of patient selection guidelines for endoluminal AAA repair with the Zenith Stent-Graft: the Australasian experience. *J Endovasc Ther* 20018457–464.464 [PubMed]

28. Torella F. Effect of improved endograft design on outcome of endovascular aneurysm repair. *J Vasc Surg* 200440216–221.221

29. Ouriel K, Srivastava S D, Sarac T P. *et al* Disparate outcome after endovascular treatment of small versus large abdominal aortic aneurysm. *J Vasc Surg* 20033712O6–1212.1212 [PubMed]

30. Peppelenbosch N, Buth J, Harris P L, EUROSTAR Collaborators *et al* Diameter of abdominal aortic aneurysm and outcome of endovascular aneurysm repair: does size matter? A report from EUROSTAR. *J Vasc Surg* 200439288–297.297 [PubMed]

31. Uflacker R, Robison J. Endovascular treatment of abdominal aortic aneurysms: a review. *Eur Radiol* 200111739–753.753 [PubMed]

32. Faries P L, Briggs V L, Rhee J Y. *et al* Failure of endovascular aortoaortic tube grafts: a plea for preferential use of bifurcated grafts. *J Vasc Surg* 200235868–873.873 [PubMed]

33. Holzenbein T J, Kretschmer G, Thurnher S. *et al* Midterm durability of abdominal aortic aneurysm endograft repair: a word of caution. *J Vasc Surg* 200133(suppl)S46–S54.S54 [PubMed]

34. Greenberg R K, Haulon S, O'Neill S. *et al* Primary endovascular repair of juxtarenal aneurysms with fenestrated endovascular grafting. *Eur J Vasc Endovasc Surg* 200427484–491.491 [PubMed]

35. Verhoeven E L, Tielliu I F, Prins T R. *et al* Frequency and outcome of reinterventions after endovascular repair for abdominal aortic aneurysm: a prospective cohort study. *Eur J Vasc Endovasc Surg* 200428357–364.364 [PubMed]

36. Thompson J P, Boyle J R, Thompson M M. *et al* Cardiovascular and catecholamine responses during endovascular and conventional abdominal aortic aneurysm repair. *Eur J Vasc Endovasc Surg* 199917326–333.333 [PubMed]

37. Boyle J R, Goodall S, Thompson J P. *et al* Endovascular AAA repair attenuates the inflammatory and renal responses associated with conventional surgery. *J Endovasc Ther* 20007359–371.371 [PubMed]

38. Thompson M M, Nasim A, Sayers R D. *et al* Oxygen free radical and cytokine generation during endovascular and conventional aneurysm repair. *Eur J Vasc Endovasc Surg* 19961270–75.75 [PubMed]

39. Boyle J R, Thompson J P, Thompson M M. *et al* Improved respiratory function and analgesia control after endovascular AAA repair. *J Endovasc Surg* 1997462–65.65 [PubMed]

40. Zarins C K, White R A, Schwarten D. *et al* AneuRx stent graft versus open surgical repair of abdominal aortic aneurysms: multicenter prospective clinical trial. *J Vasc Surg* 199929292–305.305 [PubMed]

41. Zeebregts C J, Geelkerken R H, van der Palen J. *et al* Outcome of abdominal aortic aneurysm repair in the era of endovascular treatment. *Br J Surg* 200491563–568.568 [PubMed]

42. Elkouri S, Gloviczki P, McKusick M A. *et al* Perioperative complications and early outcome after endovascular and open surgical repair of abdominal aortic aneurysms. *J Vasc Surg* 200439497–505.505 [PubMed]

43. Carpenter J P, Baum R A, Barker C F. *et al* Durability of benefits of endovascular versus conventional abdominal aortic aneurysm repair. *J Vasc Surg* 200235222–228.228 [PubMed]

44. Thomas S M, Gaines P A, Beard J D. Vascular surgical society of Great Britain and Ireland: RETA: the registry of endovascular treatment of abdominal aortic aneurysms. *Br J Surg* 199986711 [PubMed]

45. Harris P L, Buth J, Mialhe C. *et al* The need for clinical trials of endovascular abdominal aortic aneurysm stentgraft repair: the EUROSTAR Project. EUROpean collaborators on Stentgraft Techniques for abdominal aortic Aneurysm Repair. *J Endovasc Surg* 1997472–77.77 [PubMed]

46. Harris P L, Vallabhaneni S R, Desgranges P. *et al* Incidence and risk factors of late rupture, conversion, and death after endovascular repair of infrarenal aortic aneurysms: the EUROSTAR experience. European Collaborators on Stent/graft techniques for aortic aneurysm repair. *J Vasc Surg* 200032739– 749.749 [PubMed]

47. Zarins C K, Shaver D M, Arko F R. *et al* Introduction of endovascular aneurysm repair into community practice: initial results with a new Food and Drug Administration approved device. *J Vasc Surg* 200236226–232.232 [PubMed]

48. Pfammatter T, Lachat M L, Kunzli A. *et al* Shortterm results of endovascular AAA repair with the Excluder bifurcated stentgraft. *J Endovasc Ther* 20029474–480.480 [PubMed]

49. Blum U, Voshage G, Lammer J. *et al* Endoluminal stent - grafts for infrarenal abdominal aortic aneurysms. *N Engl J Med* 199733613–20.20 [PubMed]

50. Collin J, Murie J A. Endovascular treatment of abdominal aortic aneurysm: a failed experiment. *Br J Surg* 2001881281–1282.1282 [PubMed]

51. Buth J, Laheij R J. Early complications and endoleaks after endovascular abdominal aortic aneurysm repair: report of a multicenter study. *J Vasc Surg* 200031134–146.146 [PubMed]

52. May J, White G H, Waugh R. *et al* Improved survival after endoluminal repair with second - generation prostheses compared with open repair in the treatment of abdominal aortic aneurysms: a 5 - year concurrent comparison using life table method. *J Vasc Surg* 200133(suppl)S21–S26.S26 [PubMed]

53. EVAR trial participants Endovascular aneurysm repair and outcome in patients unfit for open repair of abdominal aortic aneurysm (EVAR trial 2): randomised controlled trial. *Lancet* 20053652187–2192.2192 [PubMed]

54. Hobo R, Bath J, EUROSTAR collaborators Secondary interventions following EVAR using current endografts. A EUROSTAR report. *J Vasc Surg* 200643896–902.902 [PubMed]

55. White G H, Yu W, May J. Endoleak - a proposed new terminology to describe incomplete aneurysm exclusion by an endoluminal graft. *J Endovasc Surg* 19963124–125.125 [PubMed]

56. Wain R A, Marin M L, Ohki T. *et al* Endoleaks after endovascular graft treatment of aortic aneurysms: classification, risk factors, and outcome. *J Vasc Surg* 19982769–78.78 [PubMed]

57. White G H, Yu W, May J. *et al* Endoleak as a complication of endoluminal grafting of abdominal aortic aneurysms: classification, incidence, diagnosis, and management. *J Endovasc Surg* 19974152–168.168 [PubMed]

58. Hinchliffe R J, Hopkinson B R. Current concepts and controversies in endovascular repair of abdominal aortic aneurysms. *J Cardiovasc Surg (Torino)* 200344481–502.502 [PubMed]

59. Veith F J, Baum R A, Ohki T. *et al* Nature and significance of endoleaks and endotension: summary of opinions expressed at an international conference. *J Vasc Surg* 2002351029–1035.1035 [PubMed]

60. White G H, May J, Petrasek P. *et al* Endotension: an explanation for continued AAA growth after successful endoluminal repair. *J Endovasc Surg* 19996308–315.315 [PubMed]

61. Dalal S, Donlon M, Beard J D. Thrombosed abdominal aortic aneurysms. Do they need surveillance to prevent late rupture? *Eur J Vasc Endovasc Surg* 200122570–572.572 [PubMed]

62. van Marrewijk C, Buth J, Harris P L. *et al* Significance of endoleaks after endovascular repair of abdominal aortic aneurysms: the EUROSTAR experience. *J Vasc Surg* 200235461–473.473 [PubMed]

63. Chuter T A, Faruqi R M, Sawhney R. *et al* Endoleak after endovascular

repair of abdominal aortic aneurysm. *J Vasc Surg* 20013498–105.105 [PubMed]

64. Faries P L, Cadot H, Agarwal G. *et al* Management of endoleak after endovascular aneurysm repair: cuffs, coils, and conversion. *J Vasc Surg* 2003371155–1161.1161 [PubMed]

65. Velazquez O C, Baum R A, Carpenter J P. *et al* Relationship between preoperative patency of the inferior mesenteric artery and subsequent occurrence of type II endoleak in patients undergoing endovascular repair of abdominal aortic aneurysms. *J Vasc Surg* 200032777–788.788 [PubMed]

66. Schurink G W, van Baalen J M, Visser M J. *et al* Thrombus within an aortic aneurysm does not reduce pressure on the aneurysmal wall. *J Vasc Surg* 200031501–506.506 [PubMed]

67. Gorich J, Rilinger N, Sokiranski R. *et al* Embolization of type II endoleaks fed by the inferior mesenteric artery: using the superior mesenteric artery approach. *J Endovasc Ther* 20007297–301.301 [PubMed]

68. Dattilo J B, Brewster D C, Fan C M. *et al* Clinical failures of endovascular abdominal aortic aneurysm repair: incidence, causes, and management. *J Vasc Surg* 2002351137–1144.1144 [PubMed]

69. Harris P, Brennan J, Martin J. *et al* Longitudinal aneurysm shrinkage following endovascular aortic aneurysm repair: a source of intermediate and late complications. *J Endovasc Surg* 1999611–16.16 [PubMed]

70. Gould D A, Edwards R D, McWilliams R G. *et al* Graft distortion after endovascular repair of abdominal aortic aneurysm: association with sac morphology and mid - term complications. *Cardiovasc Intervent Radiol* 200023358–363.363 [PubMed]

71. Beebe H G, Cronenwett J L, Katzen B T, Vanguard Endograft Trial investigators *et al* Results of an aortic endograft trial: impact of device failure beyond 12 months. *J Vasc Surg* 200133(suppl)S55–S63.S63 [PubMed]

72. Matsumura J S, Chaikof E L. Continued expansion of aortic necks after endovascular repair of abdominal aortic aneurysms. EVT investigators. EndoVascular Technologies, Inc. *J Vasc Surg* 199828422–430.430 [PubMed]

73. Anderson J L, Berce M, Hartley D E. Endoluminal aortic grafting with renal and superior mesenteric artery incorporation by graft fenestration. *J Endovasc Ther* 200183–15.15 [PubMed]

74. Rutherford R B. Problems with the dissemination of up - to - date information on the results of endograft repair for abdominal aortic aneurysm. *J Vasc Surg* 1999291167–1169.1169 [PubMed]

75. Zarins C K. The limits of endovascular aortic aneurysm repair. *J Vasc Surg* 1999291164–1166.1166 [PubMed]

76. Bernhard V M, Mitchell R S, Matsumura J S. *et al* Ruptured abdominal aortic aneurysm after endovascular repair. *J Vasc Surg* 2002351155–1162.1162

[PubMed]

77. Fransen G A, Vallabhaneni S R, Sr, van Marrewijk C J, EUROSTAR. *et al* Rupture of infra - renal aortic aneurysm after endovascular repair: a series from EUROSTAR registry. *Eur J Vasc Endovasc Surg* 200326487–493.493 [PubMed]

78. Lindholt J S. Endovascular aneurysm repair. *Lancet* 2004364818–820.820 [PubMed]

79. Alimi Y S, Chakfe N, Rivoal E. *et al* Rupture of an abdominal aortic aneurysm after endovascular graft placement and aneurysm size reduction. *J Vasc Surg* 199828178–183.183 [PubMed]

80. Makaroun M S, Chaikof E, Naslund T. *et al* Efficacy of a bifurcated endograft versus open repair of abdominal aortic aneurysms: a reappraisal. *J Vasc Surg* 200235203–210.210 [PubMed]

81. Bockler D, Probst T, Weber H. *et al* Surgical conversion after endovascular grafting for abdominal aortic aneurysms. *J Endovasc Ther* 20029111–118.118 [PubMed]

82. Cuypers P W, Laheij R J, Buth J. Which factors increase the risk of conversion to open surgery following endovascular abdominal aortic aneurysm repair? The EUROSTAR collaborators. *Eur J Vasc Endovasc Surg* 200020183–189.189 [PubMed]

83. May J, White G H, Yu W. *et al* Conversion from endoluminal to open repair of abdominal aortic aneurysms: a hazardous procedure. *Eur J Vasc Endovasc Surg* 1997144–11.11 [PubMed]

84. Bertges D J, Zwolak R M, Deaton D H. *et al* Current hospital costs and Medicare reimbursement for endovascular abdominal aortic aneurysm repair. *J Vasc Surg* 200337272–279.279 [PubMed]

85. Clair D G, Gray B, O'Hara P J. *et al* An evaluation of the costs to health care institutions of endovascular aortic aneurysm repair. *J Vasc Surg* 200032148–152.152 [PubMed]

86. Patel S T, Haser P B, Bush H L., Jr *et al* The cost - ffectiveness of endovascular repair versus open surgical repair of abdominal aortic aneurysms: a decision analysis model. *J Vasc Surg* 199929958–972.972 [PubMed]

87. Lederle F A. Risk of rupture of large abdominal aortic aneurysms. Disagreement among vascular surgeons. *Arch Intern Med* 19961561007–1009.1009 [PubMed]

88. Zarins C K, Wolf Y G, Lee W A. *et al* Will endovascular repair replace open surgery for abdominal aortic aneurysm repair? *Ann Surg* 2000232501–507.507 [PMC free article] [PubMed]

89. Jordan W D, Alcocer F, Wirthlin D J. *et al* Abdominal aortic aneurysms in “high - risk” surgical patients: comparison of open and endovascular repair. *Ann*

Surg 2003237623–629.629 [PMC free article]

90. Biebl M, Lau L L, Hakaim A G. *et al* Midterm outcome of endovascular abdominal aortic aneurysm repair in octogenarians: a single institution's experience. *J Vasc Surg* 200440435–442.442 [PubMed]

91. Yusuf S W, Whitaker S C, Chuter T A. *et al* Emergency endovascular repair of leaking aneurysm. *Lancet* 19943441645 [PubMed]

92. Ohki T, Veith F J, Sanchez L A. *et al* Endovascular grafts and other image - guided catheter - based adjuncts to improve the treatment of ruptured aortoiliac aneurysms. *Ann Surg* 2000232466–479.479 [PMC free article] [PubMed]

93. Lee W A, Hirneise C M, Tayyarah M. *et al* Impact of endovascular repair on early outcomes of ruptured abdominal aortic aneurysms. *J Vasc Surg* 200440211–215.215 [PubMed]

94. Lloyd G M, Bown M J, Norwood M G. *et al* Feasibility of preoperative computer tomography in patients with ruptured abdominal aortic aneurysm: a time - to - eath study in patients without operation. *J Vasc Surg* 200439788– 791.791 [PubMed]

Related citations in PubMed

Randomized clinical trials of endovascular repair versus surveillance for treatment of small abdominal aortic aneurysms.[J Endovasc Ther. 2009]

Ouriel K. J Endovasc Ther. 2009 Feb; 16 Suppl 1:I94-105.

Paradigm shifts in the treatment of abdominal aortic aneurysm: trends in 721 patients between 1996 and 2008.[J Vasc Surg. 2010]

Albuquerque FC Jr, Tonnessen BH, Noll RE Jr, Cires G, Kim JK, Sternbergh WC 3rd. J Vasc Surg. 2010 Jun; 51(6):1348-52; discussion 1352-3.

The UK EndoVascular Aneurysm Repair (EVAR) trials: randomised trials of EVAR versus standard therapy.[Health Technol Assess. 2012]

Brown LC, Powell JT, Thompson SG, Epstein DM, Sculpher MJ, Greenhalgh RM. Health Technol Assess. 2012; 16(9):1-218.

Systematic review of recent evidence for the safety and efficacy of elective endovascular repair in the management of infrarenal abdominal aortic aneurysm.[Br J Surg. 2005]

Drury D, Michaels JA, Jones L, Ayiku L. Br J Surg. 2005 Aug; 92(8):937-46.

Treatment options for delayed AAA rupture following endovascular repair.[J Vasc Surg. 2011]

Mehta M, Paty PS, Roddy SP, Taggert JB, Sternbach Y, Kreienberg PB, Chang BB, Darling RC 3rd. J Vasc Surg. 2011 Jan; 53(1):14-20. Epub 2010 Sep 26.

Cited by other articles in PMC

Polymeric Endoaortic Paving (PEAP): Mechanical, Thermoforming, and Degradation Properties of Polycaprolactone/Polyurethane Blends for Cardiovascular Applications[Acta biomaterialia. 2011]

Ashton JH, Mertz JA, Harper JL, Slepian MJ, Mills JL, McGrath DV, Vande Geest JP. Acta biomaterialia. 2011/01/01 00:00; 7(1)287-294

Recent Activity

Endovascular abdominal aortic aneurysm repair

Endovascular abdominal aortic aneurysm repair
Postgraduate Medical Journal. 2007 Jan; 83(975)21

Technical features of the INCRAFT™ AAA Stent Graft System.

Technical features of the INCRAFT™ AAA Stent Graft System.
J Cardiovasc Surg (Torino). 2014 Oct ;55(5):705-15. Epub 2014 Jul 16 .

Interventional approaches to deep vein thrombosis.

Interventional approaches to deep vein thrombosis.
Am J Hematol. 2012 May ;87 Suppl 1:S113-8. doi: 10.1002/ajh.23145. Epub 2012 Mar 3 .

New Tools and Techniques in Interventional Radiology: Ultrasound-Enhanced Thromb...

New Tools and Techniques in Interventional Radiology: Ultrasound-Enhanced Thrombolysis: EKOS EndoWave Infusion Catheter System
Seminars in Interventional Radiology. 2008 Mar; 25(1)37

The Multicentre Aneurysm Screening Study (MASS) into the effect of abdominal aortic aneurysm screening on mortality in men: a randomised controlled trial.

[Lancet. 2002]
Ashton HA, Buxton MJ, Day NE, Kim LG, Marteau TM, Scott RA, Thompson SG, Walker NM, Multicentre Aneurysm Screening Study Group Lancet. 2002 Nov 16; 360(9345):1531-9.

Trends in abdominal aortic aneurysm surgery in Scotland (1971-1984).

[Eur J Vasc Surg. 1988]

Naylor AR, Webb J, Fowkes FG, Ruckley
CV Eur J Vasc Surg. 1988 Aug; 2(4):217-21.

Abdominal aortic aneurysm in Scotland.
[Br J Surg. 1994]
Samy AK, Whyte B, MacBain G
Br J Surg. 1994 Aug; 81(8):1104-6.

Mortality results for randomised controlled trial of early elective surgery or ultrasonographic surveillance for small abdominal aortic aneurysms. The UK Small Aneurysm Trial Participants.
[Lancet. 1998]
Lancet. 1998 Nov 21; 352(9141):1649-55.

Immediate repair compared with surveillance of small abdominal aortic aneurysms.
[N Engl J Med. 2002]
Lederle FA, Wilson SE, Johnson GR, Reinke DB, Littooy FN, Acher CW, Ballard DJ, Messina LM, Gordon IL, Chute EP, Krupski WC, Busuttil SJ, Barone GW, Sparks S, Graham LM, Rapp JH, Makaroun MS, Moneta GL, Cambria RA, Makhoul RG, Eton D, Ansel HJ, Freischlag JA, Bandyk D, Aneurysm Detection and Management Veterans Affairs Cooperative Study Group
N Engl J Med. 2002 May 9; 346(19):1437-44.

Review Influence of study design on reported mortality and morbidity rates after abdominal aortic aneurysm repair.
[Br J Surg. 1998]
Blankensteijn JD, Lindenburg FP, Van der Graaf Y, Eikelboom
BC Br J Surg. 1998 Dec; 85(12):1624-30.

Review Abdominal aortic aneurysm.
[N Engl J Med. 1993]
Ernst CB
N Engl J Med. 1993 Apr 22; 328(16):1167-72.

Multicenter prospective study of nonruptured abdominal aortic aneurysm. Part II. Variables predicting morbidity and mortality.
[J Vasc Surg. 1989]
Johnston KW
J Vasc Surg. 1989 Mar; 9(3):437-47.

Resection of an aneurysm of the abdominal aorta: reestablishment of the continuity by a preserved human arterial graft, with result after five months.
[AMA Arch Surg. 1952]

DUBOST C, ALLARY M, OECONOMOS
N AMA Arch Surg. 1952 Mar; 64(3):405-8.

Two decades of abdominal aortic aneurysm repair: have we made any progress?
[J Vasc Surg. 2000]
Heller JA, Weinberg A, Arons R, Krishnasastry KV, Lyon RT, Deitch JS, Schulick AH, Bush HL Jr, Kent KC
J Vasc Surg. 2000 Dec; 32(6):1091-100.

Immediate repair compared with surveillance of small abdominal aortic aneurysms.
[N Engl J Med. 2002]
Lederle FA, Wilson SE, Johnson GR, Reinke DB, Littooy FN, Acher CW, Ballard DJ, Messina LM, Gordon IL, Chute EP, Krupski WC, Busuttil SJ, Barone GW, Sparks S, Graham LM, Rapp JH, Makaroun MS, Moneta GL, Cambria RA, Makhoul RG, Eton D, Ansel HJ, Freischlag JA, Bandyk D, Aneurysm Detection and Management Veterans Affairs Cooperative Study Group
N Engl J Med. 2002 May 9; 346(19):1437-44.

Graft-related complications after abdominal aortic aneurysm repair: reassurance from a 36-year population-based experience.
[J Vasc Surg. 1997]
Hallett JW Jr, Marshall DM, Petterson TM, Gray DT, Bower TC, Cherry KJ Jr, Gloviczki P, Pairolero PC
J Vasc Surg. 1997 Feb; 25(2):277-84; discussion 285-6.

Nonruptured abdominal aortic aneurysm: six-year follow-up results from the multicenter prospective Canadian aneurysm study. Canadian Society for Vascular Surgery Aneurysm Study Group.
[J Vasc Surg. 1994]
Johnston KW
J Vasc Surg. 1994 Aug; 20(2):163-70.

The Multicentre Aneurysm Screening Study (MASS) into the effect of abdominal aortic aneurysm screening on mortality in men: a randomised controlled trial.
[Lancet. 2002]
Ashton HA, Buxton MJ, Day NE, Kim LG, Marteau TM, Scott RA, Thompson SG, Walker NM, Multicentre Aneurysm Screening Study Group Lancet. 2002 Nov 16; 360(9345):1531-9.

Review A meta -analysis of 50 years of ruptured abdominal aortic aneurysm repair.
[Br J Surg. 2002]

Bown MJ, Sutton AJ, Bell PR, Sayers
RD Br J Surg. 2002 Jun; 89(6):714-30.

Surgical management of 671 abdominal aortic aneurysms: a 13 year review from a single centre.
[Eur J Vasc Endovasc Surg. 1997]
Sayers RD, Thompson MM, Nasim A, Healey P, Taub N, Bell PR
Eur J Vasc Endovasc Surg. 1997 Mar; 13(3):322-7.

The Multicentre Aneurysm Screening Study (MASS) into the effect of abdominal aortic aneurysm screening on mortality in men: a randomised controlled trial.
[Lancet. 2002]
Ashton HA, Buxton MJ, Day NE, Kim LG, Marteau TM, Scott RA, Thompson SG, Walker NM, Multicentre Aneurysm Screening Study Group Lancet. 2002 Nov 16; 360(9345):1531-9.

Comparison of endovascular aneurysm repair with open repair in patients with abdominal aortic aneurysm (EVAR trial 1), 30-day operative mortality results: randomised controlled trial.
[Lancet. 2004]
Greenhalgh RM, Brown LC, Kwong GP, Powell JT, Thompson SG, EVAR trial participants
Lancet. 2004 Sep 4-10; 364(9437):843-8.

A randomized trial comparing conventional and endovascular repair of abdominal aortic aneurysms.
[N Engl J Med. 2004]
Prinssen M, Verhoeven EL, Buth J, Cuypers PW, van Sambeek MR, Balm R, Buskens E, Grobbee DE, Blankensteijn JD, Dutch Randomized Endovascular Aneurysm Management (DREAM)Trial Group
N Engl J Med. 2004 Oct 14; 351(16):1607-18.

Endovascular aneurysm repair versus open repair in patients with abdominal aortic aneurysm (EVAR trial 1): randomised controlled trial.
[Lancet. 2005]
EVAR trial participants
Lancet. 2005 Jun 25-Jul 1; 365(9478):2179-86.

Two-year outcomes after conventional or endovascular repair of abdominal aortic aneurysms.
[N Engl J Med. 2005]
Blankensteijn JD, de Jong SE, Prinssen M, van der Ham AC, Buth J, van Sterkenburg SM, Verhagen HJ, Buskens E, Grobbee DE, Dutch Randomized Endovascular Aneurysm Management (DREAM) Trial Group
N Engl J Med. 2005 Jun 9; 352(23):2398-405.

Transfemoral intraluminal graft implantation for abdominal aortic aneurysms.

[Ann Vasc Surg. 1991]
Parodi JC, Palmaz JC, Barone HD Ann
Vasc Surg. 1991 Nov; 5(6):491-9.

Initial experience with endovascular aneurysm repair: comparison of early results with outcome of conventional open repair.
[J Vasc Surg. 1998]
Brewster DC, Geller SC, Kaufman JA, Cambria RP, Gertler JP, LaMuraglia GM, Atamian S, Abbott WM
J Vasc Surg. 1998 Jun; 27(6):992-1003; discussion 1004-5.

Endovascular repair of abdominal aortic aneurysms: eligibility rate and impact on the rate of open repair.
[J Vasc Surg. 2000]
Wolf YG, Fogarty TJ, Olcott C IV, Hill BB, Harris EJ, Mitchell RS, Miller DC, Dalman RL, Zarins CK
J Vasc Surg. 2000 Sep; 32(3):519-23.

Anatomical suitability of abdominal aortic aneurysms for endovascular repair.
[Br J Surg. 1997]
Armon MP, Yusuf SW, Latief K, Whitaker SC, Gregson RH, Wenham PW, Hopkinson BR
Br J Surg. 1997 Feb; 84(2):178-80.

Endograft planning without preoperative arteriography: a clinical feasibility study.
[J Endovasc Ther. 2000]
Beebe HG, Kritpracha B, Serres S, Pigott JP, Price CI, Williams
DM J Endovasc Ther. 2000 Feb; 7(1):8-15.

Preoperative sizing of grafts for transfemoral endovascular aneurysm management: a prospective comparative study of spiral CT angiography, arteriography, and conventional CT imaging.
[J Endovasc Surg. 1997]
Broeders IA, Blankensteijn JD, Olree M, Mali W, Eikelboom BC
J Endovasc Surg. 1997 Aug; 4(3):252-61.

Review Endovascular treatment of abdominal aortic aneurysm.
[Br J Surg. 2004]
Gorham TJ, Taylor J, Raptis S
Br J Surg. 2004 Jul; 91(7):815-27.

Impact of exclusion criteria on patient selection for endovascular abdominal aortic aneurysm repair.
[J Vasc Surg. 2001]
Carpenter JP, Baum RA, Barker CF, Golden MA, Mitchell ME, Velazquez OC, Fairman RM
J Vasc Surg. 2001 Dec; 34(6):1050-4.

Evaluation of patient selection guidelines for endoluminal AAA repair with the Zenith Stent-Graft: the Australasian experience.

[J Endovasc Ther. 2001]
Stanley BM, Semmens JB, Mai Q, Goodman MA, Hartley DE, Wilkinson C, Lawrence-Brown MD
J Endovasc Ther. 2001 Oct; 8(5):457-64.

Effect of improved endograft design on outcome of endovascular aneurysm repair.

[J Vasc Surg.
2004] Torella F
J Vasc Surg. 2004 Aug; 40(2):216-21.

Disparate outcome after endovascular treatment of small versus large abdominal aortic aneurysm.

[J Vasc Surg. 2003]
Ouriel K, Srivastava SD, Sarac TP, O'hara PJ, Lyden SP, Greenberg RK, Clair DG, Sampram E, Butler B
J Vasc Surg. 2003 Jun; 37(6):1206-12.

Diameter of abdominal aortic aneurysm and outcome of endovascular aneurysm repair: does size matter? A report from EUROSTAR.

[J Vasc Surg. 2004]
Peppelenbosch N, Buth J, Harris PL, van Marrewijk C, Fransen G, EUROSTAR Collaborators
J Vasc Surg. 2004 Feb; 39(2):288-97.

Review Endovascular treatment of abdominal aortic aneurysms: a review.

[Eur Radiol. 2001]
Uflacker R, Robison J
Eur Radiol. 2001; 11(5):739-53.

Failure of endovascular aortoaortic tube grafts: a plea for preferential use of bifurcated grafts.

[J Vasc Surg. 2002]
Faries PL, Briggs VL, Rhee JY, Burks JA Jr, Gravereaux EC, Carroccio A, Morrissey NJ, Teodorescu V, Hollier LH, Marin ML
J Vasc Surg. 2002 May; 35(5):868-73.

Effect of improved endograft design on outcome of endovascular aneurysm repair.

[J Vasc Surg.
2004] Torella F
J Vasc Surg. 2004 Aug; 40(2):216-21.

Midterm durability of abdominal aortic aneurysm endograft repair: a word of caution.

[J Vasc Surg. 2001]
Hölzenbein TJ, Kretschmer G, Thurnher S, Schoder M, Aslim E, Lammer J, Polterauer P
J Vasc Surg. 2001 Feb; 33(2 Suppl):S46-54.

Primary endovascular repair of juxtarenal aneurysms with fenestrated endovascular grafting.
[Eur J Vasc Endovasc Surg. 2004]
Greenberg RK, Haulon S, O'Neill S, Lyden S, Ouriel K
Eur J Vasc Endovasc Surg. 2004 May; 27(5):484-91.

Frequency and outcome of re- interventions after endovascular repair for abdominal aortic aneurysm: a prospective cohort study.
[Eur J Vasc Endovasc Surg. 2004]
Verhoeven EL, Tielliu IF, Prins TR, Zeebregts CJ, van Andringa de Kempenaer MG, Cinà CS, van den Dungen JJ
Eur J Vasc Endovasc Surg. 2004 Oct; 28(4):357-64.

Cardiovascular and catecholamine responses during endovascular and conventional abdominal aortic aneurysm repair.
[Eur J Vasc Endovasc Surg. 1999]
Thompson JP, Boyle JR, Thompson MM, Strupish J, Bell PR, Smith G
Eur J Vasc Endovasc Surg. 1999 Apr; 17(4):326-33.

Endovascular AAA repair attenuates the inflammatory and renal responses associated with conventional surgery.
[J Endovasc Ther. 2000]
Boyle JR, Goodall S, Thompson JP, Bell PR, Thompson MM
J Endovasc Ther. 2000 Oct; 7(5):359-71.

Oxygen free radical and cytokine generation during endovascular and conventional aneurysm repair.
[Eur J Vasc Endovasc Surg. 1996]
Thompson MM, Nasim A, Sayers RD, Thompson J, Smith G, Lunec J, Bell PR
Eur J Vasc Endovasc Surg. 1996 Jul; 12(1):70-5.

A randomized trial comparing conventional and endovascular repair of abdominal aortic aneurysms.
[N Engl J Med. 2004]
Prinssen M, Verhoeven EL, Buth J, Cuypers PW, van Sambeek MR, Balm R, Buskens E, Grobbee DE, Blankensteijn JD, Dutch Randomized Endovascular Aneurysm Management (DREAM)Trial Group
N Engl J Med. 2004 Oct 14; 351(16):1607-18.

Failure of endovascular aortoaortic tube grafts: a plea for preferential use of bifurcated grafts.
[J Vasc Surg. 2002]

Faries PL, Briggs VL, Rhee JY, Burks JA Jr, Gravereaux EC, Carroccio A, Morrissey NJ, Teodorescu V, Hollier LH, Marin ML
J Vasc Surg. 2002 May; 35(5):868-73.

Improved respiratory function and analgesia control after endovascular AAA repair.
[J Endovasc Surg. 1997]
Boyle JR, Thompson JP, Thompson MM, Sayers RD, Smith G, Bell PR
J Endovasc Surg. 1997 Feb; 4(1):62-5.

AneuRx stent graft versus open surgical repair of abdominal aortic aneurysms: multicenter prospective clinical trial.
[J Vasc Surg. 1999]
Zarins CK, White RA, Schwarten D, Kinney E, Diethrich EB, Hodgson KJ, Fogarty TJ
J Vasc Surg. 1999 Feb; 29(2):292-305; discussion 306-8.

Outcome of abdominal aortic aneurysm repair in the era of endovascular treatment.
[Br J Surg. 2004]
Zeebregts CJ, Geelkerken RH, van der Palen J, Huisman AB, de Smit P, van Det RJ
Br J Surg. 2004 May; 91(5):563-8.

Perioperative complications and early outcome after endovascular and open surgical repair of abdominal aortic aneurysms.
[J Vasc Surg. 2004]
Elkouri S, Gloviczki P, McKusick MA, Panneton JM, Andrews J, Bower TC, Noel AA, Harmsen WS, Hoskin TL, Cherry K
J Vasc Surg. 2004 Mar; 39(3):497-505.

Durability of benefits of endovascular versus conventional abdominal aortic aneurysm repair.
[J Vasc Surg. 2002]
Carpenter JP, Baum RA, Barker CF, Golden MA, Velazquez OC, Mitchell ME, Fairman RM
J Vasc Surg. 2002 Feb; 35(2):222-8.

Vascular surgical society of great britain and ireland: RETA: the registry of endovascular treatment of abdominal aortic aneurysms
[Br J Surg. 1999]
Thomas SM, Gaines PA, Beard JD
Br J Surg. 1999 May; 86(5):711.

Review The need for clinical trials of endovascular abdominal aortic aneurysm stent -graft repair: The EUROSTAR Project. EUROpean collaborators on Stent-graft Techniques for abdominal aortic Aneurysm Repair.

[J Endovasc Surg. 1997]
Harris PL, Buth J, Mialhe C, Myhre HO, Norgren L
J Endovasc Surg. 1997 Feb; 4(1):72-7; discussion 78-9.

Incidence and risk factors of late rupture, conversion, and death after endovascular repair of infrarenal aortic aneurysms: the EUROSTAR experience. European Collaborators on Stent/graft techniques for aortic aneurysm repair.
[J Vasc Surg. 2000]
Harris PL, Vallabhaneni SR, Desgranges P, Becquemin JP, van Marrewijk C, Laheij RJ
J Vasc Surg. 2000 Oct; 32(4):739-49.

Midterm durability of abdominal aortic aneurysm endograft repair: a word of caution.
[J Vasc Surg. 2001]
Hölzenbein TJ, Kretschmer G, Thurnher S, Schoder M, Aslim E, Lammer J, Polterauer P
J Vasc Surg. 2001 Feb; 33(2 Suppl):S46-54.

Introduction of endovascular aneurysm repair into community practice: initial results with a new Food and Drug Administration-approved device.
[J Vasc Surg. 2002]
Zarins CK, Shaver DM, Arko FR, Schubart PJ, Lengle SJ, Dixon SM
J Vasc Surg. 2002 Aug; 36(2):226-32; discussion 232-3.

Short-term results of endovascular AAA repair with the Excluder bifurcated stent-graft.
[J Endovasc Ther. 2002]
Pfammatter T, Lachat ML, Künzli A, Baur DR, Koppensteiner R, Turina M, Blum U
J Endovasc Ther. 2002 Aug; 9(4):474-80.

Endoluminal stent-grafts for infrarenal abdominal aortic aneurysms.
[N Engl J Med. 1997]
Blum U, Voshage G, Lammer J, Beyersdorf F, Töllner D, Kretschmer G, Spillner G, Polterauer P, Nagel G, Hölzenbein T
N Engl J Med. 1997 Jan 2; 336(1):13-20.

Impact of exclusion criteria on patient selection for endovascular abdominal aortic aneurysm repair.
[J Vasc Surg. 2001]
Carpenter JP, Baum RA, Barker CF, Golden MA, Mitchell ME, Velazquez OC, Fairman RM
J Vasc Surg. 2001 Dec; 34(6):1050-4.

Endovascular treatment of abdominal aortic aneurysm: a failed experiment.

[Br J Surg. 2001]
Collin J, Murie JA
Br J Surg. 2001 Oct; 88(10):1281-2.

Early complications and endoleaks after endovascular abdominal aortic aneurysm repair: report of a multicenter study.
[J Vasc Surg. 2000]
Buth J, Laheij RJ
J Vasc Surg. 2000 Jan; 31(1 Pt 1):134-46.

Improved survival after endoluminal repair with second-generation prostheses compared with open repair in the treatment of abdominal aortic aneurysms: a 5-year concurrent comparison using life table method.
[J Vasc Surg. 2001]
May J, White GH, Waugh R, Ly CN, Stephen MS, Jones MA, Harris
JP J Vasc Surg. 2001 Feb; 33(2 Suppl):S21-6.

Comparison of endovascular aneurysm repair with open repair in patients with abdominal aortic aneurysm (EVAR trial 1), 30-day operative mortality results: randomised controlled trial.
[Lancet. 2004]
Greenhalgh RM, Brown LC, Kwong GP, Powell JT, Thompson SG, EVAR trial participants
Lancet. 2004 Sep 4-10; 364(9437):843-8.

A randomized trial comparing conventional and endovascular repair of abdominal aortic aneurysms.
[N Engl J Med. 2004]
Prinssen M, Verhoeven EL, Buth J, Cuypers PW, van Sambeek MR, Balm R, Buskens E, Grobbee DE, Blankensteijn JD, Dutch Randomized Endovascular Aneurysm Management (DREAM)Trial Group
N Engl J Med. 2004 Oct 14; 351(16):1607-18.

Endovascular aneurysm repair versus open repair in patients with abdominal aortic aneurysm (EVAR trial 1): randomised controlled trial.
[Lancet. 2005]
EVAR trial participants
Lancet. 2005 Jun 25-Jul 1; 365(9478):2179-86.

Endovascular aneurysm repair and outcome in patients unfit for open repair of abdominal aortic aneurysm (EVAR trial 2): randomised controlled trial.
[Lancet. 2005]
EVAR trial participants
Lancet. 2005 Jun 25-Jul 1; 365(9478):2187-92.

Two-year outcomes after conventional or endovascular repair of abdominal aortic aneurysms.

[N Engl J Med. 2005]
Blankensteijn JD, de Jong SE, Prinssen M, van der Ham AC, Buth J, van Sterkenburg SM, Verhagen HJ, Buskens E, Grobbee DE, Dutch Randomized Endovascular Aneurysm Management (DREAM) Trial Group
N Engl J Med. 2005 Jun 9; 352(23):2398-405.

Secondary interventions following endovascular abdominal aortic aneurysm repair using current endografts. A EUROSTAR report.
[J Vasc Surg. 2006]
Hobo R, Buth J, EUROSTAR collaborators
J Vasc Surg. 2006 May; 43(5):896-902.

AneuRx stent graft versus open surgical repair of abdominal aortic aneurysms: multicenter prospective clinical trial.
[J Vasc Surg. 1999]
Zarins CK, White RA, Schwarten D, Kinney E, Diethrich EB, Hodgson KJ, Fogarty TJ
J Vasc Surg. 1999 Feb; 29(2):292-305; discussion 306-8.

Early complications and endoleaks after endovascular abdominal aortic aneurysm repair: report of a multicenter study.
[J Vasc Surg. 2000]
Buth J, Laheij RJ
J Vasc Surg. 2000 Jan; 31(1 Pt 1):134-46.

Endoleak--a proposed new terminology to describe incomplete aneurysm exclusion by an endoluminal graft.
[J Endovasc Surg. 1996]
White GH, Yu W, May J
J Endovasc Surg. 1996 Feb; 3(1):124-5.

Endoleaks after endovascular graft treatment of aortic aneurysms: classification, risk factors, and outcome.
[J Vasc Surg. 1998]
Wain RA, Marin ML, Ohki T, Sanchez LA, Lyon RT, Rozenblit A, Suggs WD, Yuan JG, Veith FJ
J Vasc Surg. 1998 Jan; 27(1):69-78; discussion 78-80.

Review Endoleak as a complication of endoluminal grafting of abdominal aortic aneurysms: classification, incidence, diagnosis, and management.
[J Endovasc Surg. 1997]
White GH, Yu W, May J, Chaufour X, Stephen MS
J Endovasc Surg. 1997 May; 4(2):152-68.

Review Current concepts and controversies in endovascular repair of abdominal aortic aneurysms.

[J Cardiovasc Surg (Torino). 2003]
Hinchliffe RJ, Hopkinson BR
J Cardiovasc Surg (Torino). 2003 Aug; 44(4):481-502.

Review Nature and significance of endoleaks and endotension: summary of opinions expressed at an international conference.

[J Vasc Surg. 2002]
Veith FJ, Baum RA, Ohki T, Amor M, Adiseshiah M, Blankensteijn JD, Buth J, Chuter TA, Fairman RM, Gilling- Smith G, Harris PL, Hodgson KJ, Hopkinson BR, Ivancev K, Katzen BT, Lawrence -Brown M, Meier GH, Malina M, Makaroun MS, Parodi JC, Richter GM, Rubin GD, Stelter WJ, White GH, White RA, Wisselink W, Zarins CK
J Vasc Surg. 2002 May; 35(5):1029-35.

Endotension: an explanation for continued AAA growth after successful endoluminal repair.

[J Endovasc Surg. 1999]
White GH, May J, Petrasek P, Waugh R, Stephen M, Harris
J J Endovasc Surg. 1999 Nov; 6(4):308-15.

Thrombosed abdominal aortic aneurysms. Do they need surveillance to prevent late rupture?

[Eur J Vasc Endovasc Surg. 2001]
Dalal S, Donlon M, Beard JD
Eur J Vasc Endovasc Surg. 2001 Dec; 22(6):570-2.

Review Nature and significance of endoleaks and endotension: summary of opinions expressed at an international conference.

[J Vasc Surg. 2002]
Veith FJ, Baum RA, Ohki T, Amor M, Adiseshiah M, Blankensteijn JD, Buth J, Chuter TA, Fairman RM, Gilling- Smith G, Harris PL, Hodgson KJ, Hopkinson BR, Ivancev K, Katzen BT, Lawrence -Brown M, Meier GH, Malina M, Makaroun MS, Parodi JC, Richter GM, Rubin GD, Stelter WJ, White GH, White RA, Wisselink W, Zarins CK
J Vasc Surg. 2002 May; 35(5):1029-35.

Significance of endoleaks after endovascular repair of abdominal aortic aneurysms: The EUROSTAR experience.

[J Vasc Surg. 2002]
van Marrewijk C, Buth J, Harris PL, Norgren L, Nevelsteen A, Wyatt
MG J Vasc Surg. 2002 Mar; 35(3):461-73.

Endoleak after endovascular repair of abdominal aortic

aneurysm. [J Vasc Surg. 2001]

Chuter TA, Faruqi RM, Sawhney R, Reilly LM, Kerlan RB, Canto CJ, Lukaszewicz GC, Laberge JM, Wilson MW, Gordon RL, Wall SD, Rapp J, Messina LM
J Vasc Surg. 2001 Jul; 34(1):98-105.

Midterm durability of abdominal aortic aneurysm endograft repair: a word of caution.
[J Vasc Surg. 2001]
Hölzenbein TJ, Kretschmer G, Thurnher S, Schoder M, Aslim E, Lammer J, Polterauer P
J Vasc Surg. 2001 Feb; 33(2 Suppl):S46-54.

Management of endoleak after endovascular aneurysm repair: cuffs, coils, and conversion.
[J Vasc Surg. 2003]
Faries PL, Cadot H, Agarwal G, Kent KC, Hollier LH, Marin ML
J Vasc Surg. 2003 Jun; 37(6):1155-61.

AneuRx stent graft versus open surgical repair of abdominal aortic aneurysms: multicenter prospective clinical trial.
[J Vasc Surg. 1999]
Zarins CK, White RA, Schwarten D, Kinney E, Diethrich EB, Hodgson KJ, Fogarty TJ
J Vasc Surg. 1999 Feb; 29(2):292-305; discussion 306-8.

Relationship between preoperative patency of the inferior mesenteric artery and subsequent occurrence of type II endoleak in patients undergoing endovascular repair of abdominal aortic aneurysms.
[J Vasc Surg. 2000]
Velazquez OC, Baum RA, Carpenter JP, Golden MA, Cohn M, Pyeron A, Barker CF, Criado FJ, Fairman RM
J Vasc Surg. 2000 Oct; 32(4):777-88.

Thrombus within an aortic aneurysm does not reduce pressure on the aneurysmal wall.
[J Vasc Surg. 2000]
Schurink GW, van Baalen JM, Visser MJ, van Bockel JH
J Vasc Surg. 2000 Mar; 31(3):501-6.

Embolization of type II endoleaks fed by the inferior mesenteric artery: using the superior mesenteric artery approach.
[J Endovasc Ther. 2000]
Görich J, Rilinger N, Sokiranski R, Krämer S, Schütz A, Sunder-Plassmann L, Pamler R
J Endovasc Ther. 2000 Aug; 7(4):297-301.

Frequency and outcome of re-interventions after endovascular repair for abdominal aortic aneurysm: a prospective cohort study.

[Eur J Vasc Endovasc Surg. 2004]
Verhoeven EL, Tielliu IF, Prins TR, Zeebregts CJ, van Andringa de Kempenaer MG, Cinà CS, van den Dungen JJ
Eur J Vasc Endovasc Surg. 2004 Oct; 28(4):357-64.

Clinical failures of endovascular abdominal aortic aneurysm repair: incidence, causes, and management.

[J Vasc Surg. 2002]
Dattilo JB, Brewster DC, Fan CM, Geller SC, Cambria RP, Lamuraglia GM, Greenfield AJ, Lauterbach SR, Abbott WM
J Vasc Surg. 2002 Jun; 35(6):1137-44.

Longitudinal aneurysm shrinkage following endovascular aortic aneurysm repair: a source of intermediate and late complications.

[J Endovasc Surg. 1999]
Harris P, Brennan J, Martin J, Gould D, Bakran A, Gilling-Smith G, Buth J, Gevers E, White D
J Endovasc Surg. 1999 Feb; 6(1):11-6.

Graft distortion after endovascular repair of abdominal aortic aneurysm: association with sac morphology and mid-term complications.

[Cardiovasc Intervent Radiol. 2000]
Gould DA, Edwards RD, McWilliams RG, Rowlands PC, Martin J, White D, Fear S, Bakran A, Brennan J, Gilling-Smith G, Harris PL
Cardiovasc Intervent Radiol. 2000 Sep-Oct; 23(5):358-63.

Durability of benefits of endovascular versus conventional abdominal aortic aneurysm repair.

[J Vasc Surg. 2002]
Carpenter JP, Baum RA, Barker CF, Golden MA, Velazquez OC, Mitchell ME, Fairman RM
J Vasc Surg. 2002 Feb; 35(2):222-8.

Results of an aortic endograft trial: impact of device failure beyond 12 months.

[J Vasc Surg. 2001]
Beebe HG, Cronenwett JL, Katzen BT, Brewster DC, Green RM, Vanguard Endograft Trial Investigators
J Vasc Surg. 2001 Feb; 33(2 Suppl):S55-63.

Continued expansion of aortic necks after endovascular repair of abdominal aortic aneurysms. EVT Investigators. EndoVascular Technologies, Inc.

Matsumura JS, Chaikof EL

J Vasc Surg. 1998 Sep; 28(3):422-30; discussion 430-1.

Incidence and risk factors of late rupture, conversion, and death after endovascular repair of infrarenal aortic aneurysms: the EUROSTAR experience. European Collaborators on Stent/graft techniques for aortic aneurysm repair.

[J Vasc Surg. 2000]
Harris PL, Vallabhaneni SR, Desgranges P, Becquemin JP, van Marrewijk C, Laheij RJ
J Vasc Surg. 2000 Oct; 32(4):739-49.

Endoluminal aortic grafting with renal and superior mesenteric artery incorporation by graft fenestration.

[J Endovasc Ther. 2001]
Anderson JL, Berce M, Hartley DE
J Endovasc Ther. 2001 Feb; 8(1):3-15.

Incidence and risk factors of late rupture, conversion, and death after endovascular repair of infrarenal aortic aneurysms: the EUROSTAR experience. European Collaborators on Stent/graft techniques for aortic aneurysm repair.

[J Vasc Surg. 2000]
Harris PL, Vallabhaneni SR, Desgranges P, Becquemin JP, van Marrewijk C, Laheij RJ
J Vasc Surg. 2000 Oct; 32(4):739-49.

Problems with the dissemination of up-to-date information on the results of endograft repair for abdominal aortic aneurysm.

[J Vasc Surg. 1999]
Rutherford RB
J Vasc Surg. 1999 Jun; 29(6):1167-9.

The limits of endovascular aortic aneurysm repair.

[J Vasc Surg. 1999]
Zarins CK
J Vasc Surg. 1999 Jun; 29(6):1164-6.

Mortality results for randomised controlled trial of early elective surgery or ultrasonographic surveillance for small abdominal aortic aneurysms. The UK Small Aneurysm Trial Participants.

[Lancet. 1998]
Lancet. 1998 Nov 21; 352(9141):1649-55.

Incidence and risk factors of late rupture, conversion, and death after endovascular repair of infrarenal aortic aneurysms: the EUROSTAR experience. European Collaborators on Stent/graft techniques for aortic aneurysm repair.

[J Vasc Surg. 2000]

Harris PL, Vallabhaneni SR, Desgranges P, Becquemin JP, van Marrewijk C, Laheij RJ
J Vasc Surg. 2000 Oct; 32(4):739-49.

Review Ruptured abdominal aortic aneurysm after endovascular repair.
[J Vasc Surg. 2002]
Bernhard VM, Mitchell RS, Matsumura JS, Brewster DC, Decker M, Lamparello P, Raithel D, Collin J
J Vasc Surg. 2002 Jun; 35(6):1155-62.

Rupture of infra-renal aortic aneurysm after endovascular repair: a series from EUROSTAR registry.
[Eur J Vasc Endovasc Surg. 2003]
Fransen GA, Vallabhaneni SR Sr, van Marrewijk CJ, Laheij RJ, Harris PL, Buth J, EUROSTAR
Eur J Vasc Endovasc Surg. 2003 Nov; 26(5):487-93.

Lindholt JS
Lancet. 2004 Sep 4-10; 364(9437):818-20.

Rupture of an abdominal aortic aneurysm after endovascular graft placement and aneurysm size reduction.

Alimi YS, Chakfe N, Rivoal E, Slimane KK, Valerio N, Riepe G, Kretz JG, Juhan C
J Vasc Surg. 1998 Jul; 28(1):178-83.

Endoleaks after endovascular graft treatment of aortic aneurysms: classification, risk factors, and outcome.
[J Vasc Surg. 1998]
Wain RA, Marin ML, Ohki T, Sanchez LA, Lyon RT, Rozenblit A, Suggs WD, Yuan JG, Veith FJ
J Vasc Surg. 1998 Jan; 27(1):69-78; discussion 78-80.

Durability of benefits of endovascular versus conventional abdominal aortic aneurysm repair.
[J Vasc Surg. 2002]
Carpenter JP, Baum RA, Barker CF, Golden MA, Velazquez OC, Mitchell ME, Fairman RM
J Vasc Surg. 2002 Feb; 35(2):222-8.

Rupture of infra-renal aortic aneurysm after endovascular repair: a series from EUROSTAR registry.
[Eur J Vasc Endovasc Surg. 2003]
Fransen GA, Vallabhaneni SR Sr, van Marrewijk CJ, Laheij RJ, Harris PL, Buth J, EUROSTAR
Eur J Vasc Endovasc Surg. 2003 Nov; 26(5):487-93.

Secondary interventions following endovascular abdominal aortic aneurysm repair using current endografts. A EUROSTAR report.
[J Vasc Surg. 2006]
Hobo R, Buth J, EUROSTAR collaborators
J Vasc Surg. 2006 May; 43(5):896-902.

Perioperative complications and early outcome after endovascular and open surgical repair of abdominal aortic aneurysms.
[J Vasc Surg. 2004]
Elkouri S, Gloviczki P, McKusick MA, Panneton JM, Andrews J, Bower TC, Noel AA, Harmsen WS, Hoskin TL, Cherry K
J Vasc Surg. 2004 Mar; 39(3):497-505.

Efficacy of a bifurcated endograft versus open repair of abdominal aortic aneurysms: a reappraisal.
[J Vasc Surg. 2002]
Makaroun MS, Chaikof E, Naslund T, Matsumura JS
J Vasc Surg. 2002 Feb; 35(2):203-10.

The limits of endovascular aortic aneurysm repair.
[J Vasc Surg. 1999]
Zarins CK
J Vasc Surg. 1999 Jun; 29(6):1164-6.

Surgical conversion after endovascular grafting for abdominal aortic aneurysms.
[J Endovasc Ther. 2002]
Böckler D, Probst T, Weber H, Raithel D
J Endovasc Ther. 2002 Feb; 9(1):111-8.

Which factors increase the risk of conversion to open surgery following endovascular abdominal aortic aneurysm repair? The EUROSTAR collaborators.
[Eur J Vasc Endovasc Surg. 2000]
Cuypers PW, Laheij RJ, Buth J
Eur J Vasc Endovasc Surg. 2000 Aug; 20(2):183-9.

Conversion from endoluminal to open repair of abdominal aortic aneurysms: a hazardous procedure.
[Eur J Vasc Endovasc Surg. 1997]
May J, White GH, Yu W, Waugh R, Stephen M, Sieunarine K, Harris JP
Eur J Vasc Endovasc Surg. 1997 Jul; 14(1):4-11.

Significance of endoleaks after endovascular repair of abdominal aortic aneurysms: The EUROSTAR experience.
[J Vasc Surg. 2002]

van Marrewijk C, Buth J, Harris PL, Norgren L, Nevelsteen A, Wyatt MG J Vasc Surg. 2002 Mar; 35(3):461-73.

Current hospital costs and medicare reimbursement for endovascular abdominal aortic aneurysm repair.
[J Vasc Surg. 2003]
Bertges DJ, Zwolak RM, Deaton DH, Teigen C, Tapper S, Koslow AR, Makaroun MS
J Vasc Surg. 2003 Feb; 37(2):272-9.

An evaluation of the costs to health care institutions of endovascular aortic aneurysm repair.
[J Vasc Surg. 2000]
Clair DG, Gray B, O'hara PJ, Ouriel K
J Vasc Surg. 2000 Jul; 32(1):148-52.

The cost-effectiveness of endovascular repair versus open surgical repair of abdominal aortic aneurysms: A decision analysis model.
[J Vasc Surg. 1999]
Patel ST, Haser PB, Bush HL Jr, Kent KC J Vasc Surg. 1999 Jun; 29(6):958-72.

Clinical failures of endovascular abdominal aortic aneurysm repair: incidence, causes, and management.
[J Vasc Surg. 2002]
Dattilo JB, Brewster DC, Fan CM, Geller SC, Cambria RP, Lamuraglia GM, Greenfield AJ, Lauterbach SR, Abbott WM
J Vasc Surg. 2002 Jun; 35(6):1137-44.

Risk of rupture of large abdominal aortic aneurysms. Disagreement among vascular surgeons.
[Arch Intern Med. 1996]
Lederle FA
Arch Intern Med. 1996 May 13; 156(9):1007-9.

Endovascular repair of abdominal aortic aneurysms: eligibility rate and impact on the rate of open repair.
[J Vasc Surg. 2000]
Wolf YG, Fogarty TJ, Olcott C IV, Hill BB, Harris EJ, Mitchell RS, Miller DC, Dalman RL, Zarins CK
J Vasc Surg. 2000 Sep; 32(3):519-23.

Will endovascular repair replace open surgery for abdominal aortic aneurysm repair?
[Ann Surg. 2000]
Zarins CK, Wolf YG, Lee WA, Hill BB, Olcott C IV, Harris EJ, Dalman RL, Fogarty TJ
Ann Surg. 2000 Oct; 232(4):501-7.

Abdominal aortic aneurysms in "high-risk" surgical patients: comparison of open and endovascular repair.

[Ann Surg. 2003]
Jordan WD, Alcocer F, Wirthlin DJ, Westfall AO, Whitley D
Ann Surg. 2003 May; 237(5):623-9; discussion 629-30.

Midterm outcome of endovascular abdominal aortic aneurysm repair in octogenarians: a single institution's experience.

[J Vasc Surg. 2004]
Biebl M, Lau LL, Hakaim AG, Oldenburg WA, Klocker J, Neuhauser B, McKinney JM, Paz-Fumagalli R
J Vasc Surg. 2004 Sep; 40(3):435-42.

Endovascular aneurysm repair and outcome in patients unfit for open repair of abdominal aortic aneurysm (EVAR trial 2): randomised controlled trial.

[Lancet. 2005]
EVAR trial participants
Lancet. 2005 Jun 25-Jul 1; 365(9478):2187-92.

Emergency endovascular repair of leaking aortic aneurysm.

[Lancet. 1994]
Yusuf SW, Whitaker SC, Chuter TA, Wenham PW, Hopkinson
BR Lancet. 1994 Dec 10; 344(8937):1645.

Endovascular grafts and other image-guided catheter-based adjuncts to improve the treatment of ruptured aortoiliac aneurysms.

[Ann Surg. 2000]
Ohki T, Veith FJ
Ann Surg. 2000 Oct; 232(4):466-79.

Impact of endovascular repair on early outcomes of ruptured abdominal aortic aneurysms.

[J Vasc Surg. 2004]
Lee WA, Hirneise CM, Tayyarah M, Huber TS, Seeger JM
J Vasc Surg. 2004 Aug; 40(2):211-5.

Review A meta-analysis of 50 years of ruptured abdominal aortic aneurysm repair.

[Br J Surg. 2002]
Bown MJ, Sutton AJ, Bell PR, Sayers
RD Br J Surg. 2002 Jun; 89(6):714-30.

Feasibility of preoperative computer tomography in patients with ruptured abdominal aortic aneurysm: a time-to-death study in patients without operation.

[J Vasc Surg. 2004]
Lloyd GM, Bown MJ, Norwood MG, Deb R, Fishwick G, Bell PR, Sayers RD
J Vasc Surg. 2004 Apr; 39(4):788-91.

The Multicentre Aneurysm Screening Study (MASS) into the effect of abdominal aortic aneurysm screening on mortality in men: a randomised controlled trial.

Ashton HA, Buxton MJ, Day NE, Kim LG, Marteau TM, Scott RA, Thompson SG, Walker NM, Multicentre Aneurysm Screening Study Group Lancet. 2002 Nov 16; 360(9345):1531-9.

Trends in abdominal aortic aneurysm surgery in Scotland (1971-1984).
Naylor AR, Webb J, Fowkes FG, Ruckley CV
Eur J Vasc Surg. 1988 Aug; 2(4):217-21.

Abdominal aortic aneurysm in Scotland.
Samy AK, Whyte B, MacBain G
Br J Surg. 1994 Aug; 81(8):1104-6.

Mortality results for randomised controlled trial of early elective surgery or ultrasonographic surveillance for small abdominal aortic aneurysms. The UK Small Aneurysm Trial Participants. Lancet. 1998 Nov 21; 352(9141):1649-55.

Immediate repair compared with surveillance of small abdominal aortic aneurysms.
Lederle FA, Wilson SE, Johnson GR, Reinke DB, Littooy FN, Acher CW, Ballard DJ, Messina LM, Gordon IL, Chute EP, Krupski WC, Busuttil SJ, Barone GW, Sparks S, Graham LM, Rapp JH, Makaroun MS, Moneta GL, Cambria RA, Makhoul RG, Eton D, Ansel HJ, Freischlag JA, Bandyk D, Aneurysm Detection and Management Veterans Affairs Cooperative Study Group
N Engl J Med. 2002 May 9; 346(19):1437-44.

Review Influence of study design on reported mortality and morbidity rates after abdominal aortic aneurysm repair.
Blankensteijn JD, Lindenburg FP, Van der Graaf Y, Eikelboom BC
Br J Surg. 1998 Dec; 85(12):1624-30.

Review Abdominal aortic aneurysm. Ernst CB
N Engl J Med. 1993 Apr 22; 328(16):1167-72.

Multicenter prospective study of nonruptured abdominal aortic aneurysm. Part II. Variables predicting morbidity and mortality.
Johnston KW
J Vasc Surg. 1989 Mar; 9(3):437-47.

Resection of an aneurysm of the abdominal aorta: reestablishment of the continuity by a preserved human arterial graft, with result after five months. DUBOST C, ALLARY M, OECONOMOS N
AMA Arch Surg. 1952 Mar; 64(3):405-8.

Two decades of abdominal aortic aneurysm repair: have we made any progress?

Heller JA, Weinberg A, Arons R, Krishnasastry KV, Lyon RT, Deitch JS, Schulick AH, Bush HL Jr, Kent KC
J Vasc Surg. 2000 Dec; 32(6):1091-100.

Graft -related complications after abdominal aortic aneurysm repair: reassurance from a 36-year population-based experience.
Hallett JW Jr, Marshall DM, Petterson TM, Gray DT, Bower TC, Cherry KJ Jr, Gloviczki P, Pairolero PC
J Vasc Surg. 1997 Feb; 25(2):277-84; discussion 285-6.

Nonruptured abdominal aortic aneurysm: six-year follow-up results from the multicenter prospective Canadian aneurysm study. Canadian Society for Vascular Surgery Aneurysm Study Group.
Johnston KW
J Vasc Surg. 1994 Aug; 20(2):163-70.

Review A meta-analysis of 50 years of ruptured abdominal aortic aneurysm repair.
Bown MJ, Sutton AJ, Bell PR, Sayers
RD Br J Surg. 2002 Jun; 89(6):714-30.

Surgical management of 671 abdominal aortic aneurysms: a 13 year review from a single centre.
Sayers RD, Thompson MM, Nasim A, Healey P, Taub N, Bell PR
Eur J Vasc Endovasc Surg. 1997 Mar; 13(3):322-7.

Comparison of endovascular aneurysm repair with open repair in patients with abdominal aortic aneurysm (EVAR trial 1), 30-day operative mortality results: randomised controlled trial.
Greenhalgh RM, Brown LC, Kwong GP, Powell JT, Thompson SG, EVAR trial participants
Lancet. 2004 Sep 4-10; 364(9437):843-8.

A randomized trial comparing conventional and endovascular repair of abdominal aortic aneurysms.
Prinssen M, Verhoeven EL, Buth J, Cuypers PW, van Sambeek MR, Balm R, Buskens E, Grobbee DE, Blankensteijn JD, Dutch Randomized Endovascular Aneurysm Management (DREAM)Trial Group
N Engl J Med. 2004 Oct 14; 351(16):1607-18.

Endovascular aneurysm repair versus open repair in patients with abdominal aortic aneurysm (EVAR trial 1): randomised controlled trial.
EVAR trial participants
Lancet. 2005 Jun 25-Jul 1; 365(9478):2179-86.

Two-year outcomes after conventional or endovascular repair of abdominal aortic aneurysms.
Blankensteijn JD, de Jong SE, Prinssen M, van der Ham AC, Buth J, van Sterkenburg SM, Verhagen HJ, Buskens E, Grobbee DE, Dutch Randomized Endovascular Aneurysm Management (DREAM) Trial Group

N Engl J Med. 2005 Jun 9; 352(23):2398-405.

Transfemoral intraluminal graft implantation for abdominal aortic aneurysms.
Parodi JC, Palmaz JC, Barone HD
Ann Vasc Surg. 1991 Nov; 5(6):491-9.

Initial experience with endovascular aneurysm repair: comparison of early results with outcome of conventional open repair.
Brewster DC, Geller SC, Kaufman JA, Cambria RP, Gertler JP, LaMuraglia GM, Atamian S, Abbott WM
J Vasc Surg. 1998 Jun; 27(6):992-1003; discussion 1004-5.

Endovascular repair of abdominal aortic aneurysms: eligibility rate and impact on the rate of open repair.
Wolf YG, Fogarty TJ, Olcott C IV, Hill BB, Harris EJ, Mitchell RS, Miller DC, Dalman RL, Zarins CK
J Vasc Surg. 2000 Sep; 32(3):519-23.

Anatomical suitability of abdominal aortic aneurysms for endovascular repair.
Armon MP, Yusuf SW, Latief K, Whitaker SC, Gregson RH, Wenham PW, Hopkinson BR
Br J Surg. 1997 Feb; 84(2):178-80.

Endograft planning without preoperative arteriography: a clinical feasibility study.
Beebe HG, Kritpracha B, Serres S, Pigott JP, Price CI, Williams
DM J Endovasc Ther. 2000 Feb; 7(1):8-15.

Preoperative sizing of grafts for transfemoral endovascular aneurysm management: a prospective comparative study of spiral CT angiography, arteriography, and conventional CT imaging.
Broeders IA, Blankensteijn JD, Olree M, Mali W, Eikelboom BC
J Endovasc Surg. 1997 Aug; 4(3):252-61.

Review Endovascular treatment of abdominal aortic aneurysm.
Gorham TJ, Taylor J, Raptis S
Br J Surg. 2004 Jul; 91(7):815-27.

Impact of exclusion criteria on patient selection for endovascular abdominal aortic aneurysm repair.
Carpenter JP, Baum RA, Barker CF, Golden MA, Mitchell ME, Velazquez OC, Fairman RM
J Vasc Surg. 2001 Dec; 34(6):1050-4.

Evaluation of patient selection guidelines for endoluminal AAA repair with the Zenith Stent-Graft: the Australasian experience.
Stanley BM, Semmens JB, Mai Q, Goodman MA, Hartley DE, Wilkinson C, Lawrence-Brown MD
J Endovasc Ther. 2001 Oct; 8(5):457-64.

Effect of improved endograft design on outcome of endovascular aneurysm

repair.
Torella F
J Vasc Surg. 2004 Aug; 40(2):216-21.

Disparate outcome after endovascular treatment of small versus large abdominal aortic aneurysm.
Ouriel K, Srivastava SD, Sarac TP, O'hara PJ, Lyden SP, Greenberg RK, Clair DG, Sampram E, Butler B
J Vasc Surg. 2003 Jun; 37(6):1206-12.

Diameter of abdominal aortic aneurysm and outcome of endovascular aneurysm repair: does size matter? A report from EUROSTAR.
Peppelenbosch N, Buth J, Harris PL, van Marrewijk C, Fransen G, EUROSTAR Collaborators
J Vasc Surg. 2004 Feb; 39(2):288-97.

Review Endovascular treatment of abdominal aortic aneurysms: a review. Uflacker R, Robison J
Eur Radiol. 2001; 11(5):739-53.

Failure of endovascular aortoaortic tube grafts: a plea for preferential use of bifurcated grafts.
Faries PL, Briggs VL, Rhee JY, Burks JA Jr, Gravereaux EC, Carroccio A, Morrissey NJ, Teodorescu V, Hollier LH, Marin ML
J Vasc Surg. 2002 May; 35(5):868-73.

Midterm durability of abdominal aortic aneurysm endograft repair: a word of caution.
Hölzenbein TJ, Kretschmer G, Thurnher S, Schoder M, Aslim E, Lammer J, Polterauer P
J Vasc Surg. 2001 Feb; 33(2 Suppl):S46-54.

Primary endovascular repair of juxtarenal aneurysms with fenestrated endovascular grafting.
Greenberg RK, Haulon S, O'Neill S, Lyden S, Ouriel K
Eur J Vasc Endovasc Surg. 2004 May; 27(5):484-91.

Frequency and outcome of re- interventions after endovascular repair for abdominal aortic aneurysm: a prospective cohort study.
Verhoeven EL, Tielliu IF, Prins TR, Zeebregts CJ, van Andringa de Kempenaer MG, Cinà CS, van den Dungen J
Eur J Vasc Endovasc Surg. 2004 Oct; 28(4):357-64.

Cardiovascular and catecholamine responses during endovascular and conventional abdominal aortic aneurysm repair.
Thompson JP, Boyle JR, Thompson MM, Strupish J, Bell PR, Smith G Eur J Vasc Endovasc Surg. 1999 Apr; 17(4):326-33.

Endovascular AAA repair attenuates the inflammatory and renal responses

associated with conventional surgery.
Boyle JR, Goodall S, Thompson JP, Bell PR, Thompson MM
J Endovasc Ther. 2000 Oct; 7(5):359-71.

Oxygen free radical and cytokine generation during endovascular and conventional aneurysm repair.
Thompson MM, Nasim A, Sayers RD, Thompson J, Smith G, Lunec J, Bell PR
Eur J Vasc Endovasc Surg. 1996 Jul; 12(1):70-5.

Improved respiratory function and analgesia control after endovascular AAA repair.
Boyle JR, Thompson JP, Thompson MM, Sayers RD, Smith G, Bell PR
J Endovasc Surg. 1997 Feb; 4(1):62-5.

AneuRx stent graft versus open surgical repair of abdominal aortic aneurysms: multicenter prospective clinical trial.
Zarins CK, White RA, Schwarten D, Kinney E, Diethrich EB, Hodgson KJ, Fogarty TJ
J Vasc Surg. 1999 Feb; 29(2):292-305; discussion 306-8.

Outcome of abdominal aortic aneurysm repair in the era of endovascular treatment.
Zeebregts CJ, Geelkerken RH, van der Palen J, Huisman AB, de Smit P, van Det RJ
Br J Surg. 2004 May; 91(5):563-8.

Perioperative complications and early outcome after endovascular and open surgical repair of abdominal aortic aneurysms.
Elkouri S, Gloviczki P, McKusick MA, Panneton JM, Andrews J, Bower TC, Noel AA, Harmsen WS, Hoskin TL, Cherry K
J Vasc Surg. 2004 Mar; 39(3):497-505.

Durability of benefits of endovascular versus conventional abdominal aortic aneurysm repair.
Carpenter JP, Baum RA, Barker CF, Golden MA, Velazquez OC, Mitchell ME, Fairman RM
J Vasc Surg. 2002 Feb; 35(2):222-8.

Vascular surgical society of great britain and ireland: RETA: the registry of endovascular treatment of abdominal aortic aneurysms
Thomas SM, Gaines PA, Beard JD
Br J Surg. 1999 May; 86(5):711.

Review The need for clinical trials of endovascular abdominal aortic aneurysm stent -graft repair: The EUROSTAR Project. EUROpean collaborators on Stent-graft Techniques for abdominal aortic Aneurysm Repair.
Harris PL, Buth J, Mialhe C, Myhre HO, Norgren L
J Endovasc Surg. 1997 Feb; 4(1):72-7; discussion 78-9.

Incidence and risk factors of late rupture, conversion, and death after endovascular repair of infrarenal aortic aneurysms: the EUROSTAR experience.

European Collaborators on Stent/graft techniques for aortic aneurysm repair.
Harris PL, Vallabhaneni SR, Desgranges P, Becquemin JP, van Marrewijk C, Laheij RJ
J Vasc Surg. 2000 Oct; 32(4):739-49.

Introduction of endovascular aneurysm repair into community practice: initial results with a new Food and Drug Administration-approved device.
Zarins CK, Shaver DM, Arko FR, Schubart PJ, Lengle SJ, Dixon SM J Vasc Surg. 2002 Aug; 36(2):226-32; discussion 232-3.

Short-term results of endovascular AAA repair with the Excluder bifurcated stent-graft.
Pfammatter T, Lachat ML, Künzli A, Baur DR, Koppensteiner R, Turina M, Blum U
J Endovasc Ther. 2002 Aug; 9(4):474-80.

Endoluminal stent -grafts for infrarenal abdominal aortic aneurysms.
Blum U, Voshage G, Lammer J, Beyersdorf F, Töllner D, Kretschmer G, Spillner G, Polterauer P, Nagel G, Hölzenbein T
N Engl J Med. 1997 Jan 2; 336(1):13-20.

Endovascular treatment of abdominal aortic aneurysm: a failed experiment.
Collin J, Murie JA
Br J Surg. 2001 Oct; 88(10):1281-2.

Early complications and endoleaks after endovascular abdominal aortic aneurysm repair: report of a multicenter study.
Buth J, Laheij RJ
J Vasc Surg. 2000 Jan; 31(1 Pt 1):134-46.

Improved survival after endoluminal repair with second-generation prostheses compared with open repair in the treatment of abdominal aortic aneurysms: a 5-year concurrent comparison using life table method.
May J, White GH, Waugh R, Ly CN, Stephen MS, Jones MA, Harris JP J Vasc Surg. 2001 Feb; 33(2 Suppl):S21-6.

Endovascular aneurysm repair and outcome in patients unfit for open repair of abdominal aortic aneurysm (EVAR trial 2): randomised controlled trial.
EVAR trial participants
Lancet. 2005 Jun 25-Jul 1; 365(9478):2187-92.

Secondary interventions following endovascular abdominal aortic aneurysm repair using current endografts. A EUROSTAR report.
Hobo R, Buth J, EUROSTAR collaborators
J Vasc Surg. 2006 May; 43(5):896-902.

Endoleak--a proposed new terminology to describe incomplete aneurysm exclusion by an endoluminal graft.
White GH, Yu W, May J
J Endovasc Surg. 1996 Feb; 3(1):124-5.

Endoleaks after endovascular graft treatment of aortic aneurysms: classification, risk factors, and outcome.
Wain RA, Marin ML, Ohki T, Sanchez LA, Lyon RT, Rozenblit A, Suggs WD, Yuan JG, Veith FJ
J Vasc Surg. 1998 Jan; 27(1):69-78; discussion 78-80.

Review Endoleak as a complication of endoluminal grafting of abdominal aortic aneurysms: classification, incidence, diagnosis, and management.
White GH, Yu W, May J, Chaufour X, Stephen MS
J Endovasc Surg. 1997 May; 4(2):152-68.

Review Current concepts and controversies in endovascular repair of abdominal aortic aneurysms.
Hinchliffe RJ, Hopkinson BR
J Cardiovasc Surg (Torino). 2003 Aug; 44(4):481-502.

Review Nature and significance of endoleaks and endotension: summary of opinions expressed at an international conference.Veith FJ, Baum RA, Ohki T, Amor M, Adiseshiah M, Blankensteijn JD, Buth J, Chuter TA, Fairman RM, Gilling-Smith G, Harris PL, Hodgson KJ, Hopkinson BR, Ivancev K, Katzen BT, Lawrence-Brown M, Meier GH, Malina M, Makaroun MS, Parodi JC, Richter GM, Rubin GD, Stelter WJ, White GH, White RA, Wisselink W, Zarins CK
J Vasc Surg. 2002 May; 35(5):1029-35.

Endotension: an explanation for continued AAA growth after successful endoluminal repair.
White GH, May J, Petrasek P, Waugh R, Stephen M, Harris
J J Endovasc Surg. 1999 Nov; 6(4):308-15.

Thrombosed abdominal aortic aneurysms. Do they need surveillance to prevent late rupture?
Dalal S, Donlon M, Beard JD
Eur J Vasc Endovasc Surg. 2001 Dec; 22(6):570-2.

Significance of endoleaks after endovascular repair of abdominal aortic aneurysms: The EUROSTAR experience.
van Marrewijk C, Buth J, Harris PL, Norgren L, Nevelsteen A, Wyatt MG
J Vasc Surg. 2002 Mar; 35(3):461-73.

Endoleak after endovascular repair of abdominal aortic aneurysm.
Chuter TA, Faruqi RM, Sawhney R, Reilly LM, Kerlan RB, Canto CJ, Lukaszewicz GC, Laberge JM, Wilson MW, Gordon RL, Wall SD, Rapp J, Messina LM
J Vasc Surg. 2001 Jul; 34(1):98-105.

Management of endoleak after endovascular aneurysm repair: cuffs, coils, and conversion.
Faries PL, Cadot H, Agarwal G, Kent KC, Hollier LH, Marin ML
J Vasc Surg. 2003 Jun; 37(6):1155-61.

Relationship between preoperative patency of the inferior mesenteric artery and subsequent occurrence of type II endoleak in patients undergoing endovascular repair of abdominal aortic aneurysms.
Velazquez OC, Baum RA, Carpenter JP, Golden MA, Cohn M, Pyeron A, Barker CF, Criado FJ, Fairman RM
J Vasc Surg. 2000 Oct; 32(4):777-88.

Thrombus within an aortic aneurysm does not reduce pressure on the aneurysmal wall.
Schurink GW, van Baalen JM, Visser MJ, van Bockel JH
J Vasc Surg. 2000 Mar; 31(3):501-6.

Embolization of type II endoleaks fed by the inferior mesenteric artery: using the superior mesenteric artery approach.
Görich J, Rilinger N, Sokiranski R, Krämer S, Schütz A, Sunder-Plassmann L, Pamler R
J Endovasc Ther. 2000 Aug; 7(4):297-301.

Clinical failures of endovascular abdominal aortic aneurysm repair: incidence, causes, and management.
Dattilo JB, Brewster DC, Fan CM, Geller SC, Cambria RP, Lamuraglia GM, Greenfield AJ, Lauterbach SR, Abbott WM
J Vasc Surg. 2002 Jun; 35(6):1137-44.

Longitudinal aneurysm shrinkage following endovascular aortic aneurysm repair: a source of intermediate and late complications.
Harris P, Brennan J, Martin J, Gould D, Bakran A, Gilling-Smith G, Buth J, Gevers E, White D
J Endovasc Surg. 1999 Feb; 6(1):11-6.

Graft distortion after endovascular repair of abdominal aortic aneurysm: association with sac morphology and mid-term complications.
Gould DA, Edwards RD, McWilliams RG, Rowlands PC, Martin J, White D, Fear S, Bakran A, Brennan J, Gilling-Smith G, Harris PL
Cardiovasc Intervent Radiol. 2000 Sep-Oct; 23(5):358-63.

Results of an aortic endograft trial: impact of device failure beyond 12 months. Beebe HG, Cronenwett JL, Katzen BT, Brewster DC, Green RM, Vanguard Endograft Trial Investigators
J Vasc Surg. 2001 Feb; 33(2 Suppl):S55-63.

Continued expansion of aortic necks after endovascular repair of abdominal aortic aneurysms. EVT Investigators. EndoVascular Technologies, Inc. Matsumura JS, Chaikof EL
J Vasc Surg. 1998 Sep; 28(3):422-30; discussion 430-1.

Endoluminal aortic grafting with renal and superior mesenteric artery incorporation by graft fenestration.
Anderson JL, Berce M, Hartley DE
J Endovasc Ther. 2001 Feb; 8(1):3-15.

Problems with the dissemination of up-to-date information on the results of endograft repair for abdominal aortic aneurysm.
Rutherford RB
J Vasc Surg. 1999 Jun; 29(6):1167-9.

The limits of endovascular aortic aneurysm repair.
Zarins CK
J Vasc Surg. 1999 Jun; 29(6):1164-6.

Review Ruptured abdominal aortic aneurysm after endovascular repair. Bernhard VM, Mitchell RS, Matsumura JS, Brewster DC, Decker M, Lamparello P, Raithel D, Collin J
J Vasc Surg. 2002 Jun; 35(6):1155-62.

upture of infra-renal aortic aneurysm after endovascular repair: a series from EUROSTAR registry.
Fransen GA, Vallabhaneni SR Sr, van Marrewijk CJ, Laheij RJ, Harris PL, Buth J, EUROSTAR
Eur J Vasc Endovasc Surg. 2003 Nov; 26(5):487-93.

Endovascular aneurysm repair. Lindholt JS
Lancet. 2004 Sep 4-10; 364(9437):818-20.

Rupture of an abdominal aortic aneurysm after endovascular graft placement and aneurysm size reduction.
Alimi YS, Chakfe N, Rivoal E, Slimane KK, Valerio N, Riepe G, Kretz JG, Juhan C
J Vasc Surg. 1998 Jul; 28(1):178-83.

Efficacy of a bifurcated endograft versus open repair of abdominal aortic aneurysms: a reappraisal.
Makaroun MS, Chaikof E, Naslund T, Matsumura JS
J Vasc Surg. 2002 Feb; 35(2):203-10.

Surgical conversion after endovascular grafting for abdominal aortic aneurysms.
Böckler D, Probst T, Weber H, Raithel D
J Endovasc Ther. 2002 Feb; 9(1):111-8.

Which factors increase the risk of conversion to open surgery following endovascular abdominal aortic aneurysm repair? The EUROSTAR collaborators.
Cuypers PW, Laheij RJ, Buth J
Eur J Vasc Endovasc Surg. 2000 Aug; 20(2):183-9.

Conversion from endoluminal to open repair of abdominal aortic aneurysms: a hazardous procedure.
May J, White GH, Yu W, Waugh R, Stephen M, Sieunarine K, Harris JP
Eur J Vasc Endovasc Surg. 1997 Jul; 14(1):4-11.

Current hospital costs and medicare reimbursement for endovascular abdominal

aortic aneurysm repair.
Bertges DJ, Zwolak RM, Deaton DH, Teigen C, Tapper S, Koslow AR, Makaroun MS
J Vasc Surg. 2003 Feb; 37(2):272-9.

An evaluation of the costs to health care institutions of endovascular aortic aneurysm repair.
Clair DG, Gray B, O'hara PJ, Ouriel K
J Vasc Surg. 2000 Jul; 32(1):148-52.

The cost-effectiveness of endovascular repair versus open surgical repair of abdominal aortic aneurysms: A decision analysis model.
Patel ST, Haser PB, Bush HL Jr, Kent
KC J Vasc Surg. 1999 Jun; 29(6):958-72.

Risk of rupture of large abdominal aortic aneurysms. Disagreement among vascular surgeons.
Lederle FA
Arch Intern Med. 1996 May 13; 156(9):1007-9.

Will endovascular repair replace open surgery for abdominal aortic aneurysm repair?
Zarins CK, Wolf YG, Lee WA, Hill BB, Olcott C IV, Harris EJ, Dalman RL, Fogarty TJ
Ann Surg. 2000 Oct; 232(4):501-7.

Abdominal aortic aneurysms in "high-risk" surgical patients: comparison of open and endovascular repair.
Jordan WD, Alcocer F, Wirthlin DJ, Westfall AO, Whitley D
Ann Surg. 2003 May; 237(5):623-9; discussion 629-30.

Midterm outcome of endovascular abdominal aortic aneurysm repair in octogenarians: a single institution's experience.
Biebl M, Lau LL, Hakaim AG, Oldenburg WA, Klocker J, Neuhauser B, McKinney JM, Paz-Fumagalli R
J Vasc Surg. 2004 Sep; 40(3):435-42.

Emergency endovascular repair of leaking aortic aneurysm.
Yusuf SW, Whitaker SC, Chuter TA, Wenham PW, Hopkinson
BR Lancet. 1994 Dec 10; 344(8937):1645.

Endovascular grafts and other image-guided catheter-based adjuncts to improve the treatment of ruptured aortoiliac aneurysms.
Ohki T, Veith FJ
Ann Surg. 2000 Oct; 232(4):466-79.

Impact of endovascular repair on early outcomes of ruptured abdominal aortic aneurysms.
Lee WA, Hirneise CM, Tayyarah M, Huber TS, Seeger JM
J Vasc Surg. 2004 Aug; 40(2):211-5.

Feasibility of preoperative computer tomography in patients with ruptured abdominal aortic aneurysm: a time-to-death study in patients without operation. Lloyd GM, Bown MJ, Norwood MG, Deb R, Fishwick G, Bell PR, Sayers RD J Vasc Surg. 2004 Apr; 39(4):788-91.

Appendix 2 Stroke Intervention Procedures

Front Neurol. 2011; 2: 9.

Published online 2011 Mar 8. Prepublished online 2011 Jan 31. doi: 10.3389/fneur.2011.00009

PMCID: PMC3079955

Intra-Arterial Treatment Methods in Acute Stroke Therapy

Thanh N. Nguyen,1,2,* Viken L. Babikian,1 Rafael Romero,1 Aleksandra Pikula,1 Carlos S. Kase,1 Tudor G. Jovin,3 and Alexander M. Norbash2

1Department of Neurology, Boston Medical Center, Boston University School of Medicine, Boston, MA, USA

2Department of Radiology, Boston Medical Center, Boston University School of Medicine, Boston, MA, USA

3Stroke Institute, University of Pittsburgh Medical Center, Pittsburgh, PA, USA

Edited by: Alex Abou-Chebl, University of Louisville School of Medicine, USA

Reviewed by: Osama O. Zaidat, Medical College of Wisconsin/Froedtert Hospital, USA; Randall Edgell, Saint Louis University, USA

*Correspondence: Thanh N. Nguyen, Department of Neurology, Boston Medical Center, Boston University School of Medicine, 720 Harrison Avenue, Suite 707, Boston, MA 02118, USA e-mail: gro.cmb@neyugn.hnaht

This article was submitted to Frontiers in Endovascular and Interventional Neurology, a specialty of Frontiers in Neurology.

Author information Article notes Copyright and License information

Received 2010 Nov 10; Accepted 2011 Feb 7.

Abstract

Acute revascularization is associated with improved outcomes in ischemic stroke patients. It is unclear which method of intra-arterial intervention, if any, is ideal. Promising approaches in acute stroke treatment are likely a combination of intravenous and endovascular revascularization efforts, combining early treatment initiation with direct clot manipulation and/or PTA/stenting. In this review, we will discuss available thrombolytic therapies and endovascular recanalization techniques, beginning with chemical thrombolytic agents, followed by mechanical devices, and a review of ongoing trials. Further randomized studies comparing medical therapy, intravenous and endovascular treatments are essential, and their implementation will require the wide support and enthusiasm from the neurologic, neuroradiologic, and neurosurgical stroke communities.

Keywords: intra-arterial therapy, intra-arterial thrombolysis, mechanical embolectomy, retrieval devices, aspiration devices, angioplasty, stent

Introduction

The National Institute of Neurological Disorders and Stroke (NINDS, 1995) trial initiated a new era for recanalization therapies for acute ischemic stroke patients. In 1996, after review of the results of the NINDS study demonstrating better clinical outcome in patients receiving intravenous (IV) recombinant tissue plasminogen activator (rt-PA) compared to placebo at 3 months, the US Food and Drug Administration (FDA) approved IV rt-PA for the treatment of ischemic stroke within 3 h from symptom onset (NINDS, 1995). Recently, the time window for IV thrombolysis was further extended to 4.5 h following the ECASS 3 trial which demonstrated benefit in a more select group of patients (Hacke et al., 2008). Following the ECASS 3 trial findings, the European Stroke Organization endorsed IV thrombolysis in their 2009 Stroke Guidelines for ischemic stroke patients presenting up to the 4.5-h time window. In the United States, many stroke centers have adopted administration of IV rt-PA in this time window using similar entry criteria as the ECASS 3 trial (Davis and Donnan, 2009).

Despite extension of the time window for acute stroke therapy, IV thrombolysis presents numerous challenges; early recanalization after IV rt-PA occurs in only 30–50% of patients, it has less favorable results for proximal occlusions (del Zoppo et al., 1992; NINDS, 1995; Lee et al., 2007), and re-occlusion is frequent after IV rt-PA (Grotta et al., 2001; Alexandrov and Grotta, 2002). While many patients do not present in early time windows (Kleindorfer et al., 2008), the number needed to treat (NNT) to get one added favorable outcome increases from 2 during the first 90 min to 7 within 3 h, and reaches 14 between 3 and 4.5 h (Hacke et al., 2008).

Endovascular therapy includes intra-arterial (IA) thrombolytic therapy and mechanical thrombectomy. It is an alternative treatment for acute ischemic stroke in patients who are ineligible for IV thrombolysis and who present beyond the thrombolysis time window. It is also an option for patients who do not improve after IV thrombolysis (Wolpert et al., 1993) . The main advantage of IA thrombolytic therapy is that it allows direct delivery of a highly concentrated and locally delivered thrombolytic drug to the thrombus within the end-organ distribution, permitting lower total amounts of systemic thrombolytics to achieve recanalization. The other advantage is that mechanical thrombectomy may spare usage of a chemical thrombolytic entirely. Disadvantages of IA approaches include the need for a neurointerventional team, the requirement to rapidly mobilize such a team, and the added time required to obtain subselective IA access in order to reopen the occluded vessel.

The principal goal of IA stroke therapy is to safely and rapidly restore flow. There are two components required to achieve this goal: recanalizing the original or primary arterial occlusive lesion (AOL) in order to reperfuse the distal arterial bed and brain parenchyma (Tomsick, 2007). Successful recanalization has been associated with improved outcomes (Rha and Saver, 2007; Nogueira et al., 2009a). Three types of reperfusion strategies have been proposed (Nogueira et al., 2009a):

1. recanalization or antegrade reperfusion
2. global reperfusion (flow augmentation by maximizing collateral circulation, or transarterial retrograde reperfusion),

and 3. transvenous retrograde reperfusion

Most endovascular therapies use recanalization or antegrade reperfusion as the mechanism for restoration of flow. In this review, we will discuss available thrombolytic therapies and endovascular recanalization techniques, beginning with chemical thrombolytic agents, followed by mechanical devices (Table (Table1),1), and a review of ongoing trials (Table (Table2),2). While not all chemical thrombolytic agents have been used or tested via IA routes, a review of chemical thrombolytics is important in the grand spectrum of acute ischemic stroke treatment.

Table 1

Selected summary of thrombolytic and endovascular ischemic stroke trials.

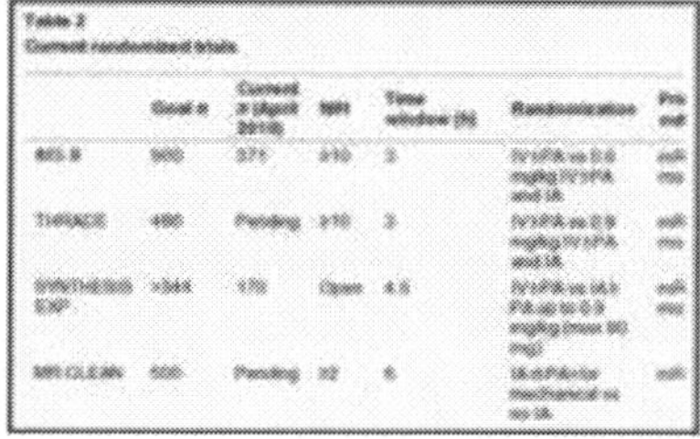

Table 2

Current randomized trials.

Intra-Arterial and/or Intravenous Chemical Thrombolysis

A number of agents have been used for thrombolytic treatment of stroke including: streptokinase, urokinase, alteplase, pro-urokinase, reteplase, tenecteplase, and desmoteplase. In principle, these agents convert the proenzyme plasminogen to plasmin, which when activated subsequently breaks down fibrin into fibrin degradation products. The main advantage of IA over IV chemical thrombolysis is that it allows direct delivery of drug to the thrombus, and compared to mechanical techniques it requires less vessel instrumentation to reopen occluded arteries. Chemical thrombolysis also can work on distal clots, in M2, M3, and higher order branches which may be inaccessible to most mechanical devices. Despite very low systemic doses of these drugs when administered IA, the occurrence of intracranial hemorrhage (ICH) is the most feared complication. Of additional concern often a relatively long interval can elapse between initiation of drug infusion and time to vessel reopening, and such therapy failure is not uncommon.

First generation agents such as *streptokinase* and *urokinase* (withdrawn from the market) are plasminogen activators and are mainly of historical interest. Both agents have been tested in IV and IA thrombolytic trials respectively. This class of drug is not fibrin specific, thus systemic hypofibrinogenemia may occur. Streptokinase is immunogenic and may lead to drug resistance, fever, and allergic reactions. The rate of ICH is increased with streptokinase, hence it is no longer used for IV or IA stroke therapy (Nogueira et al., 2009a).

In contrast to first generation agents, second generation agents, such as alteplase and pro-urokinase are fibrin specific and are not antigenic. Alteplase or rt-PA has a similar profile to native plasminogen activator and acts by converting the proenzyme plasminogen to the active enzyme plasmin. The plasma half life of alteplase is 4–6 min, and the dose used for IA lysis can range in practice from 1 to 60 mg. One possibly concerning feature of alteplase is that it may have neurotoxic properties including activation of metalloproteinases, which may result in increased blood–brain barrier permeability leading to ICH and edema (Nogueira et al., 2009a). Pro-urokinase is a second generation agent which is a precursor of urokinase and has a half life of 7 min.

Third generation thrombolytics, such as reteplase and tenecteplase, are modified forms of alteplase, have greater thrombolytic potency and longer half-lives compared to alteplase (Nogueira et al., 2009b). Compared with the second generation thrombolytics, reteplase does not bind as strongly with fibrin. In a phase I reteplase safety study, IA reteplase (0.5, 1, 1.5, 2 units) with IV abciximab (platelet glycoprotein IIb/IIIa inhibitor) at a dose of 0.25 mg/kg bolus followed by 0.125 μg/kg/min was administered to acute ischemic stroke patients presenting between 3 and 6 h after symptom onset. The goal was to achieve higher rates of recanalization with a medication combination using two different mechanisms of action – IA reteplase to lyse fibrin and IV abciximab to prevent platelet aggregation. Of 20 patients in this study, one symptomatic ICH was observed (1-unit tier); partial or complete recanalization was observed in 13 of 20 (65%) patients (Qureshi et al., 2006).

Compared to alteplase, tenecteplase has a longer half life (17 min), higher fibrin specificity, and better resistance to plasminogen activator inhibitor-1 (Nogueira et al., 2009b). In a pilot study of patients presenting 3–6 h after ischemic stroke onset with perfusion deficit,
0.1 mg/kg IV tenecteplase was given to 15 patients and compared with 29 patients treated with alteplase (Parsons et al., 2009). The primary outcome was reperfusion as assessed by interval reduction in the mean transit time lesion (MTT) at 24 h on MRI perfusion imaging, in addition to major vessel recanalization. Greater reperfusion volumes (mean 74 vs 44% for the alteplase control group) and major vessel recanalization rates (10/15 tenecteplase vs 7/29 alteplase, p = 0.01) were seen. Despite delayed time to treatment in the tenecteplase arm, more tenecteplase patients had major neurologic improvement at 24 h compared to patients treated with alteplase. Although the number of stroke events was small, there were no parenchymal hematomas in the tenecteplase patients.

There are additional new generation agents which include desmoteplase, a genetically engineered version of the clot-dissolving factor found in the saliva of the vampire bat desmodus rotundus (Nogueira et al., 2009b). Desmoteplase is more selective for fibrin-bound plasminogen than any other known plasminogen activator. In the Desmoteplase in Acute Ischemic Stroke (DIAS) 2 trial, MR or CT perfusion imaging was used to randomize 193 patients presenting within 3–9 h of symptom onset to two evaluated doses of desmoteplase (90 μg/g, 125 μg/kg) or a placebo arm. While the study did not show clinical benefit, and although the rates of symptomatic ICH were high (3.5% for 90 μg/kg desmoteplase, 4.5% for 125 μg/kg desmoteplase, and 0% for placebo), a phase III randomized trial (DIAS 3) comparing IV desmoteplase 90 μg/kg to placebo is currently ongoing with a planned sample size of 320 patients triaged with MRI or CT-angiography with vessel occlusion or high-grade stenosis.

IA Fibrinolytic Randomized Trials

The Prolyse in Acute Cerebral Thromboembolism (PROACT II) trial is a landmark study that randomized patients presenting within 6 h of a middle cerebral artery (MCA) occlusion to either IA pro-urokinase or IV heparin. Higher rates of recanalization were seen in the IA pro-urokinase treated cohort compared to the IV heparin treated group (66 vs 18%) and

40% of IA patients received an mRS score < 2 compared to 25% of controls (Furlan et al., 1999). Risk of symptomatic hemorrhage was 11% in patients receiving IA pro-urokinase compared to 3% of patients receiving heparin alone (Kase et al., 2001). An absolute risk improvement of 15% or NNT of 7 was achieved. PROACT II was pivotal as the first randomized trial to demonstrate clinical efficacy of IA therapies, extending the therapeutic window to 6 h for acute ischemic stroke patients. Although the success of PROACT II was insufficient to gain FDA approval for pro-urokinase, this led to a new era in IA therapy for acute ischemic stroke patients.

The Middle Cerebral Artery Embolism Local Fibrinolytic Intervention Trial (MELT), a trial performed in Japan, randomized patients presenting with MCA occlusion within 6 h of symptom onset to either an IA urokinase (UK) arm or a conventional treatment arm (Ogawa et al., 2007). The trial was stopped after approval of rt-PA in Japan in October 2005. While the primary endpoint of a modified Rankin Scale (mRS) ≤ 1 did not reach statistical significance, secondary analyses suggested that IA UK may offer increased likelihood of excellent functional outcomes at 90 days.

In patients with posterior circulation stroke presenting up to 24 h, the Australian Urokinase Stroke Trial randomized 16 patients to IA urokinase (increments of 100,000 IU to maximum dose of 1,000,000 IU) or control (no thrombolysis). All patients were anticoagulated acutely (5000 IU heparin IA followed by IV heparin to target ptt of 60–80 for 2 days minimum), and then received oral warfarin to target INR of 1.5–2.5 for 6 months. There was some imbalance between the groups, with more severe strokes occurring in the IA arm. A good outcome was observed in four of eight patients who received IA urokinase compared with one of eight patients in the control group (MacLeod et al., 2005).

A meta-analysis comprising five randomized controlled trials (PROACT I, II, MELT, Australian Urokinase Stroke Trial) found that IA fibrinolysis was associated with increased good clinical outcomes defined as mRS 0–2 (43 vs 28%; OR 2, 95% CI 1.3–3.1; NNT 6.8) and excellent clinical outcomes defined as mRS 0–1 (31 vs 18%; OR 2.1; 95% CI 1.3–3.5; NNT 7.7; Lee et al., 2007). The rate of any radiological or symptomatic hemorrhage was increased, although mortality was not higher.

Intra-Arterial Mechanical Thrombolysis

The principal goal of mechanical thrombolysis is to restore cerebral blood flow by either removing or fragmenting obstructing thrombus. Advantages of fragmenting thrombus may include increasing surface area exposure to thrombolysis (Levy et al., 2004; Nogueira et al., 2009a). This approach may minimize or obviate the use of chemical thrombolytics, thereby decreasing the risk of ICH while extending the time window of intervention up to an 8-h window for the anterior circulation. Faster recanalization may be achieved with mechanical techniques as compared with infusion lysis, especially with large clot burdens and with clots containing large amounts of calcium and cholesterol (Halloran and Bekavac, 2004). Mechanical embolectomy is particularly helpful for patients who are ineligible for chemical thrombolysis, patients who are recently post-operative, coagulopathic, or who are presenting late or with an unclear time of onset. The disadvantage of mechanical devices is that almost all mechanical methods are less flexible than the smallest simple microcatheters through which thrombolytics can be introduced, especially since access pathways to occlusive lesions are tortuous. Aggressive instrumentation can lead to unfortunate complications such as dissection, perforation, vessel rupture (Nguyen et al., 2008; Malik et al., 2010), and conversion of distal emboli to more proximal larger-artery occlusions. Robust clinical outcomes demonstrating benefits of mechanical embolectomy for acute ischemic stroke have yet to be generated from randomized trials.

Retrieval Devices

Several mechanical thrombectomy devices have been developed over the last decades. The MERCI (Mechanical Embolus Removal in Cerebral Ischemia) retriever (Concentric Medical, Mountain View, CA, USA) was the first stroke thrombectomy device approved by the FDA in 2004 and increased the practicality of mechanical thrombectomy beyond that formerly possible with snaring. More than 10,000 MERCI devices have since been used worldwide (Concentric Medical, personal communication). The MERCI device is an extremely flexible corkscrew-shaped device made of nitinol wire, designed to remove blood clots from the brain in patients experiencing ischemic stroke. An 8- or 9-F balloon guide catheter is placed in the internal carotid artery. The retrieval device is introduced into a 2.4-F Merci microcatheter and is advanced beyond the thrombus. The general principle of its use is that upon retrieval of the device, the balloon in the guide catheter is inflated to occlude the carotid artery with simultaneous guide catheter aspiration to prevent distal embolization of thrombus with retrieval. Recently, a 4.3-F distal access catheter (DAC) was added to navigate more distally and closer to the clot, and allow for better leverage in clot retrieval. The first generation MERCI X5 and X6 devices consisted of simple helical nitinol wire loops. The second generation L4, L5, and L6 MERCI devices added filaments to the helical nitinol coil to improve clot engagement and capture. The third generation devices (V series) have variable tension in the loops (Figure (Figure11).

Figure 1

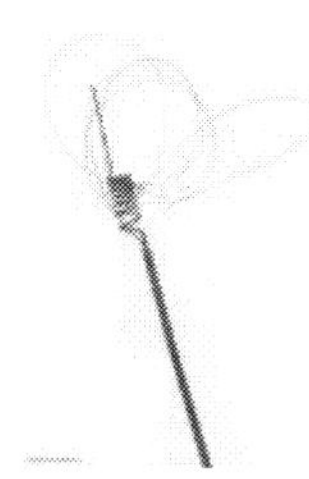

Merci Retriever V 2.5 series (with permission, Concentrics Medical).

The Multi MERCI trial was an international, multicenter prospective study that evaluated the combined safety of IV t-PA and mechanical thrombectomy when used together. Favorable clinical outcome (mRS 0–2) occurred in 36% of patients and mortality was 34%. Both outcomes were related to vascular recanalization. Symptomatic ICH and procedural complications occurred in 10 and 5.5% of patients, respectively (Smith et al., 2008).

Although the MERCI device has had the largest impact to date regarding sales volume, a number of other specifically directed thrombectomy devices with various designs have also been developed. The Catch Thromboembolectomy System (Balt Extrusion, Montmorency, France) is a self -expanding basket-like device with a maximum diameter of 4 mm, fixed to a pusher wire. This device has been tested and favorably compared with the MERCI device in animal models.

The Phenox clot retriever (Phenox, Bochum, Germany) is a flexible nitinol/platinum compound core wire resembling a bottle-brush (Henkes et al., 2006). The wire compound firmly holds perpendicularly-oriented polyamide microfilaments and carries radiodense markers on its proximal and distal end. The contour of this brush is trimmed in a conical shape with an increasing diameter distally. Animal model studies comparing the MERCI, Catch, and Phenox retriever system suggest that the Phenox retriever may be superior in preventing distal embolization because it can capture and filter the largest percentage of clot fragments

(Liebig et al., 2008). Preliminary case reports in patients suggest this device holds promise as a useful supplement for thrombectomy (Henkes et al., 2006).

The Alligator Retriever (Chestnut Medical Technologies, Menlo Park, CA, USA) is a four-pronged retrieval device on a stainless-steel wire which resembles a clawlike hand, and which closes on partial retraction into the tip of the carrying microcatheter. The Alligator Retriever is FDA approved for the retrieval of intravascular foreign bodies in the peripheral circulation and in the neurovasculature. Its use has been reported in six patients with acute MCA occlusion, with all patients achieving successful clot retrieval and improvement in NIH stroke scale (Kerber et al., 2007). Once a microcatheter is delivered to the thrombus, the retriever is pushed a few millimeters beyond the catheter tip, allowing the jaws to open. The microcatheter is then advanced to partially close the jaws and trap the clot. The whole unit is then withdrawn along with concomitant suction via a guide catheter in the internal carotid artery.

The Trevo Device (Concentric Medical) is a new thrombus retrieval system, involving a stentriever, a retrievable and non-deployable stent-like device that is designed to act as retriever. This device has received regulatory approval in Europe and Canada as of February 2010. The Solitaire device is another retrievable stent-on -a-wire (EV3) designed for neurovascular disease and may be effective for acute ischemic stroke revascularization as a retrievable open-ended basket, but which has the additional feature of controlled deployment permitting the Solitaire to be left in the vessel as a conventional stent (Henkes et al., 2006).

Aspiration Devices

Several aspiration techniques and fragmentation devices have been developed for acute ischemic stroke treatment and include the Penumbra System (Bose et al., 2008; Penumbra, Alameda, CA, USA), Angiojet and manual aspiration. In contrast to retrieval devices, which are deployed distal to a thrombus for clot extraction, most of the aspiration techniques and fragmentation devices are placed proximal to the thrombus. The advantage to aspiration technique and devices is a lower rate of embolic events. The theory of fragmentation assumes that fragmentation methods serve two purposes; the first, as mentioned earlier, is to increase the surface area of the clot volume thereby permitting more effective thrombolysis, and the second purpose is to send the clot fragments into the more distal circulation. This distal fragmentation recognizes that leptomeningeal collateral supply is rich, serving a protective and potentially redundant blood supply to more distal circulations, and also recognizes the relatively small surface area of true clinically eloquent cortex when compared with the total surface area of the brain. Fragmentation approaches have evolved from initial wire-maceration of clot to angioplasty of clot (thromboplasty), to specialized methods for thrombofragmentation using mechanical, ultrasound, and even low-energy laser-assistance.

Manual suction or aspiration with a syringe can be used in the presence of a large vessel thrombus, such as in the internal carotid artery or basilar artery. A 10 or 20 cc syringe can be placed at the end of a catheter or large microcatheter. This technique is inexpensive, rapid, but infrequently curative as solo technique. However, it can be used as a bridge to using another device to reopen the occluded vessel. A specific technical consideration is selecting an appropriate microcatheter which is sufficiently reinforced to present collapse or coaptation once suction is applied, in order to convey adequate suction to the tip of the microcatheter for clot retrieval.

The Penumbra System (Penumbra, Alameda, CA, USA), was FDA approved in January 2008, and designed to remove large vessel thrombi in acute ischemic stroke patients via aspiration and extraction. There are two main components of this system: a reperfusion catheter and a specially designed microwire with a distal olive known as a "separator" (Figure (Figure2).2). The reperfusion catheter is connected to a suction device, and as the separator is gently pulled back and forth into the tip of the catheter there is resultant fragmentation of the clot surface with fragment aspiration into the catheter. There are now four

different sizes of microcatheter, consisting of 0.054′, 0.041′, 0.032′, and 0.026′ sized to the vessel affected. Larger microcatheter sizes are much more effective in generating higher suction. Both the Penumbra Pivotal Stroke Trial (125 patients) and the Penumbra POST study (157 patients) evaluated the safety and effectiveness of the Penumbra system in patients with stroke symptom onset within 8 h, and NIHSS $\geq$ 8 (Penumbra POST, 2009a,b). Mean NIHSS was above 16 in both studies. Complete or partial recanalization (TIMI 2–3) was very high, achieved in over 80% of patients in both studies. Symptomatic ICH occurred in 11% in the Pivotal Trial and 6.4% in the POST study. In the POST study, good clinical outcome defined as mRS $\leq$ 2 was achieved in 41% of patients and a greater than 10 point improvement in the NIHSS was seen in 37% of patients.

Figure 2

Penumbra microcatheter and separator system (with permission, Penumbra Inc.).

Angioplasty

Angioplasty and stenting may be particularly advantageous when an acute vessel occlusion is related to local atherothrombotic disease. The diameter of the balloon should be slightly undersized in reference to the diameter of the occluded vessel due to the fragility of the intracranial vessels which distinctly and significantly lack an outer elastic lamina. Balloons commonly used include compliant (Hyperglide, EV3, Irvine, CA, USA) and semicompliant balloons (Gateway, Boston Scientific, Fremont, CA, USA). Balloon-assisted thromboplasty can rapidly re-establish flow, but its most feared risk is vessel rupture. Several studies have shown the high efficacy of percutaneous transluminal angioplasty (PTA) with recanalization rates of up to 90% (Ueda et al., 1998; Nakano et al., 2002; Nogueira et al., 2009a). However, its results can also be disappointing for acute cerebral occlusions (Leary et al., 2003). Re-occlusion is a common problem with balloon angioplasty, and adjunctive thrombolytics or GPIIb/IIa inhibitors are often administered in conjunction (Abou-Chebl et al., 2005).

Stents

Initially, the use of balloon mounted stents for acute ischemic stroke revascularization resulted in high morbidity and high rates of ICH. Development of self-expandable stents for the intracranial circulation gained increasing attention and usage with the introduction of the Wingspan and Enterprise stents. The Enterprise stent (Cordis, Raynham, MA, USA) is a closed cell partially retrievable stent, which was originally developed for adjunctive use in intracranial aneurysm treatment, and which was subsequently used off-label for acute stroke therapy in several reports (Levy et al., 2004; Kelly et al., 2008; Zaidat et al., 2008; Brekenfeld et al., 2009; Mocco et al., 2010). A retrospective multicenter series of 20 patients demonstrated revascularization in all patients (75% with TIMI score of 3, 25% TIMI score of 2) and improvement in NIH stroke scale in 75% of patients by four or more stroke scale points (Mocco et al., 2010). An FDA-approved pilot study of 20 patients evaluated the safety and efficacy of primary stent deployment for revascularization in acute stroke patients presenting within 8 h. Recanalization to TIMI 3 (60% of patients) or 2 (40% of patients) was achieved, while there was one (5%) symptomatic and 2 (10%) asymptomatic ICH. The mRS was ≤1 in 9 of 20 patients at 1 month. In the extracranial circulation, emergent stenting for acute carotid occlusion was shown to have very high recanalization rates, with rate of symptomatic ICH in 10% (Jovin et al., 2005). Good outcome (mRS ≤ 2) was achieved in 13/25 (52%) patients and mortality was 12%.

Despite the high success rates and fast times achieved for recanalization with acute stenting, one of the limitations of acute stenting remains the difficulty of navigating a stent through the tortuous cerebrovascular anatomy, and as second concern is the act of leaving behind a possibly unnecessarily implanted foreign body in the patient. Retrievable and partially retrievable stents such as Solitaire (EV3), the Trevo device (Concentric), and Enterprise (Cordis) may circumvent the latter problem. Thrombosis of an acutely placed stent, obligating the patient to clopidogrel and aspirin, and sometimes GPIIbIIIa inhibitors is another disadvantage of stents. These chemical agents can increase the risk of hemorrhagic conversion of infarcts, or the risk of reperfusion hemorrhage.

The need for Randomized Trials

Despite the multitude of medications and devices developed for acute ischemic stroke intervention, it is disappointing to recognize that only two large randomized trials have been completed; the PROACT I and II trials. The MELT trial was interrupted due to approval of IV rt -PA in Japan. The resistance to planning and initiating randomized trials lies in the neurointerventional community which questions whether the correct or optimal devices to randomize patients are available at present, and whether the optimal formula for patient selection using perfusion mapping have been established. The ethical concerns voiced by proceduralists who are asked to randomize patients is another significant challenge. Is it ethical to randomize a young patient with a large MCA stroke, large mismatch, and severe neurological deficit? For such a patient, a clinician may feel no clinical equipoise, and may exclude this patient from a trial. However, a 72-year-old patient presenting at 4.5 h from stroke onset with right MCA stroke and NIHSS of 14, ASPECTS score of 7, could be considered to have clinical equipoise for randomization. Where equipoise exists, randomizing subjects to clinical trials is a rational and appropriate approach (Tomsick, 2007).

There is an important need for randomized controlled trials of acute stroke intervention to demonstrate the clinical benefit of our therapies. The most important measure for efficacy of any acute stroke treatment is not recanalization, but ultimately is clinical outcome (Wechsler, 2006). Belief in our therapies is not knowledge and conviction is not proof (Saver, 2006). Our initial hypotheses have been proven wrong many times in our therapies for cerebrovascular disease. Two examples of flawed hypotheses include the EC/IC bypass study which showed no benefit from performing MCA-STA bypass in patients with carotid occlusion, and the WARRS study which showed absence of benefit with warfarin treatment compared to aspirin for secondary prevention of stroke.

Ongoing and Future Trials

IMS III

The Interventional Management Stroke Trial III is a NIH funded phase III randomized multicenter trial evaluating acute ischemic stroke patients presenting within 3 h, comparing IV rt -PA (0.9 mg/kg) to reduced dose IV rt-PA (0.6 mg/kg) followed by IA therapy, with primary outcome measure of mRS at 3 months (Khatri et al., 2008). Despite the recent results of the ECASS 3 trial (Hacke et al., 2008), there time window for IV t-PA was study inclusion was not extended to the 4.5-h window. As of April 2010, 371 of 900 patients have been recruited from 50 sites. The trial plans to expand to Europe in the summer of 2010. The target for trial completion is 2014 (Khatri, personal communication).

MR RESCUE

The Magnetic Resonance and Recanalization of Stroke Clots Using Embolectomy (MR RESCUE) study is a NIH funded trial evaluating whether mechanical embolectomy with the MERCI or Penumbra device is superior to standard medical management of acute ischemic stroke presenting within 8 h of stroke onset. Patients treated with IV t -PA up to 4.5 h from symptom onset with persistent vessel occlusion on post-treatment MRI may also be included. The study aims to identify patients who may benefit from intervention using MRI perfusion imaging.

SYNTHESIS

SYNTHESIS is the first randomized controlled trial comparing IAT to IVT with alteplase in acute ischemic stroke within 4.5 h of ischemic stroke onset. The study aims to evaluate the proportion of independent survivors at 3 months. Preliminary data suggests that rapid initiation of IA thrombolysis is a safe and feasible alternative to IV thrombolysis in acute ischemic stroke (Ciccone et al., 2010). The study began enrollment February 2008 and will end September 2011. As of March 2010 almost 50% of the target patient sample has been recruited (goal is at least 172 patients per arm; A. Ciccone, personal communication).

MR CLEAN

MR CLEAN is a multicenter randomized clinical trial of endovascular treatment of acute ischemic stroke in the Netherlands. The null hypothesis for this study is that endovascular treatment for acute ischemic stroke with onset less than 6 h in patients with a symptomatic proximal arterial occlusion leads to a similar distribution of functional outcomes as standard treatment. The primary outcome studied is the modified Rankin scale at 90 days. A goal of 500 patients is intended for recruitment, to be completed over 4 years. As of April 2010, the study awaits medical ethics approval (Dippel, personal communication).

RETRIEVE

Randomized Trial of Endovascular Treatment of Acute Ischemic Stroke vs Medical Management was meant to be a multicenter international prospective randomized controlled trial comparing mechanical thrombectomy using the MERCI retriever with best medical therapy (with or without IV thrombolysis) within 8 h of stroke symptom onset.

This study was funded by Concentric Medical, but unfortunately has been placed on hold (Concentric Medical, personal communication) for two reasons. The FDA objected to including IA thrombolytic in the intervention arm and four new national-sponsored randomized controlled trials in Europe emerged, which would directly compete with RETRIEVE (Concentrics, personal communication).

SWIFT

The Solitaire With the Intention for Thrombectomy (SWIFT) Study is a randomized study which aims to demonstrate equivalence of the SOLITAIRE™ FR Revascularization Device with the MERCI Retrieval System in patients with acute ischemic stroke presenting within 8 h of symptom onset and NIHSS 8–30. The primary objective is to measure arterial recanalization of the occluded target vessel as measured by TIMI score of 2 or 3 without any symptomatic ICH[1].

DAWN

In patients with unclear-onset or wake-up strokes, preliminary data suggest that IV or IA thrombolysis may be safely applied based on MRI criteria (positive perfusion–diffusion mismatch and absence of well-developed fluid-attenuated inversion recovery changes of acute diffusion lesions; Cho et al., 2008). The DWI and CTP Assessment in the Triage of Wake-Up and Late Presenting Strokes Undergoing Neurointervention trial is evaluating whether MR perfusion or CTP based endovascular treatment in patients with wake-up and late presenting strokes is as safe and effective as standard endovascular treatment performed within 8 h of symptom onset (phase I) and leads to improved outcomes when compared with best medical treatment (phase II, randomized controlled trial; Nogueira et al., 2009a,2009b).

SENTIS

Partial aortic occlusion may increase cerebral perfusion by increasing cerebral collateral recruitment and salvage of penumbra tissue. SENTIS (Safety and Efficacy of NeuroFlo for Treatment of Ischemic Stroke) was a randomized, controlled multicenter trial designed to demonstrate the safety and efficacy of the NeuroFlo treatment compared to standard medical care in ischemic stroke patients presenting within 10 h of onset (Shuaib et al., 2010). Preliminary results of the trial were presented at the 2010 International Stroke Conference and did not show benefit to the device.

Conclusion

Acute revascularization is associated with improved outcomes in ischemic stroke patients. It is unclear which method of IA intervention, if any, is ideal. Promising approaches in acute stroke treatment are likely a combination of IV and endovascular revascularization efforts, combining early treatment initiation with direct clot manipulation and/or PTA/stenting (Gupta et al., 2006). Further randomized studies comparing medical therapy, IV and endovascular treatments are essential, and their implementation will require the wide support and enthusiasm from the neurologic, neuroradiologic and neurosurgical stroke communities.

Conflict of Interest Statement

The authors declare that the research was conducted in the absence of any commercial or financial relationships that could be construed as a potential conflict of interest.

Appendix

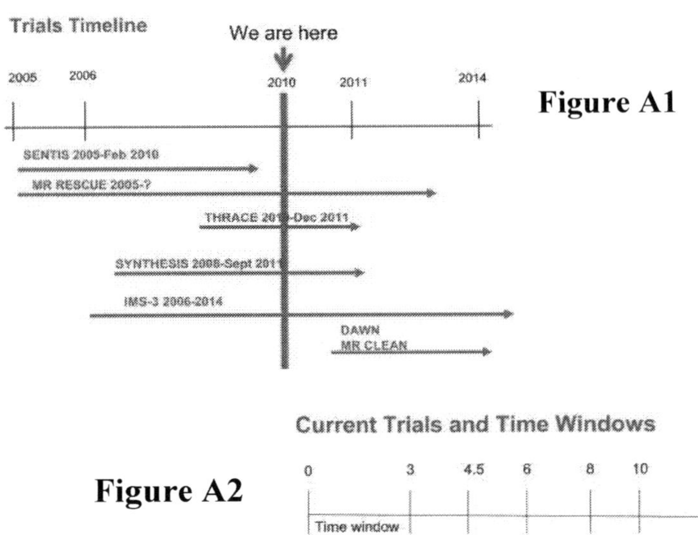

Trials Timeline
We are here
2005
2006
2010
2011
2014
SENTIS 2005-Feb 2010
MR RESCUE 2005-?
IMS-3 2006-2014
DAWN
MR CLEAN

Figure A1

Figure A2

Current Trials and Time Windows
0
3
4.5
6
8
10
indeterminate
Time window
IMS III: IV vs IV/IA
SYNTHESIS: IV vs IA
MR CLEAN: standard vs IA
MR RESCUE: standard vs IA
SENTIS: standard vs IA aortic balloon
DAWN: standard vs IA

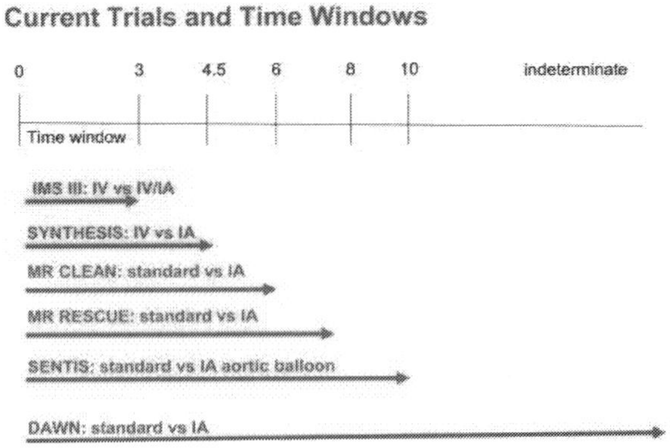

Footnotes

www.clinicaltrials.gov

References

1. Abou-Chebl A., Bazjer C. T., Krieger D. W., Furlan A. J., Yadav J. S. (2005). Multimodal therapy for the treatment of severe ischemic stroke combining GPIIb/IIa antagonists and angioplasty after failure of thrombolysis. Stroke 36, 2286–228810.1161/01.STR.0000179043.73314.4f [PubMed] [Cross Ref]

2. Alexandrov A. V., Grotta J. C. (2002). Arterial reocclusion in stroke patients treated with intravenous tissue plasminogen activator. Neurology 59, 862–867 [PubMed]

3. Bose A., Henkes H., Alfke K., Reith W., Mayer T. E., Berlis A., Branca V., Sit S. P. for the Penumbra Phase 1 Stroke Trial Investigators (2008). The penumbra system: a mechanical device for the treatment of acute stroke due to thromboembolism. AJNR Am. J. Neuroradiol. 29, 1409–141310.3174/ajnr.A1110 [PubMed] [Cross Ref]

4. Brekenfeld C., Schroth G., Mattle H. P., Do D. D., Remonda L., Mordasini P., Arnold M., Nedeltchev K., Meier N., Gralla J. (2009). Stent placement in acute cerebral artery occlusion: use of a self-expandable intracranial stent for acute stroke treatment. Stroke 40, 847–85210.1161/STROKEAHA.108.533810 [PubMed] [Cross Ref]

5. Cho A. H., Sohn S., Han M. K., Lee D. H., Kim J. S., Choi C. G., Sohn C. H., Kwon S. U., Suh D. C., Kim S. J., Bae H. J., Kang D. W. (2008). Safety and efficacy of MRI-based thrombolysis in unclear-onset stroke. Cerebrovasc. Dis. 25, 572–579 [PubMed]

6. Ciccone A., Valvassori L., Ponzio M., Ballabio E., Gasparotti R., Sessa M., Scomazzoni F., Tiraboschi P., Sterzi R. the SYNTHESIS Investigators (2010). Intra-arterial or intravenous thrombolysis for acute ischemic stroke? The SYNTHESIS pilot trial. J. Neurointerv. Surg. 2, 74–79 [PubMed]

7. Davis S. M., Donnan G. A. (2009). 4.5 hours. The new time window for tissue plasminogen activator in stroke. Stroke 40, 2266–226710.1161/STROKEAHA.108.544171 [PubMed] [Cross Ref]

8. del Zoppo G. J., Poeck K., Pessin M. S., Wolpert S. M., Furlan A. J.,

Ferbert A., Alberts M. J., Zivin J. A., Wechsler L., Busse O. (1992). Recombinant tissue plasminogen activator in acute thrombotic and embolic stroke. Ann. Neurol. 32, 78–86 [PubMed]

9. Furlan A., Higashida R., Wechsler L., Gent M., Rowley H., Kase C., Pessin M., Ahuja A., Callahan F., Clark W. M., Silver F., Rivera F. (1999). Intra-arterial pro-urokinase for acute ischemic stroke. The PROACT II study: a randomized controlled trial. Prolyse in acute cerebral thromboembolism. JAMA 282, 2003– 201110.1001/jama.282.21.2003 [PubMed] [Cross Ref]

10. Grotta J. C., Welch K. M., Fagan S. C., Lu M., Frankel M. R., Brott T., Levine S. R., Lyden Pd. (2001). Clinical deterioration following improvement in the NINDS rtPA stroke trial. Stroke 32, 661–668 [PubMed]

11. Gupta R., Vora N. A., Horowitz M. B., Tayal A. H., Hammer M. D., Uchino K., Levy E. I., Wechsler L. R., Jovin T. G. (2006). Multimodal reperfusion therapy for acute ischemic stroke: factors predicting vessel recanalization. Stroke 37, 986–99010.1161/01.STR.0000209303.02474.27 [PubMed] [Cross Ref]

12. Hacke W., Kaste M., Bluhmki E., Brozman M., Dávalos A., Guidetti D., Larrue V., Lees K. R., Medeghri Z., Machnig T., Schneider D., von Kummer R., Wahlgren N., Toni D. (2008). Thrombolysis with alteplase 3 to 4.5 hours after acute ischemic stroke. N. Engl. J. Med. 359, 1317-1329. [PubMed]

13. Halloran J. I., Bekavac I. (2004). Unsuccessful tissue plasminogen activator treatment of acute stroke caused by a calcific embolus. J. Neuroimaging 14, 385–387 [PubMed]

14. Henkes H., Reinartz J., Lowens S., Miloslavski E., Roth C., Reith W., Kuhne D. (2006). A device for fast mechanical clot retrieval from intracranial arteries (Phenox Clot Retriever). Neurocrit. Care 5, 134–140 [PubMed]

15. Jovin T. G., Gupta R., Uchino K., Jungreis C. A., Wechsler L. R., Hammer M. D., Tayal A., Horowitz M. B. (2005). Emergent stenting of extracranial internal carotid artery occlusion in acute stroke has a high revascularization rate. Stroke 36, 2426–243010.1161/01.STR.0000185924.22918.51 [PubMed] [Cross Ref]

16. Kase C. S., Furlan A. J., Wechsler L. R., Higashida R. T., Rowley H. A., Hart R. G., Molinari G. F., Frederick L. S., Roberts H. C., Gebel J. M., Sila C. A., Schulz G. A., Roberts R. S., Gent M. (2001). Cerebral

hemorrhage after intra-arterial thrombolysis for ischemic stroke: the PROACT II trial. Neurology 57, 1603–1610 [PubMed]

17. Kelly M. E., Furlan A. J., Fiorella D. (2008). Recanalization of an acute middle cerebral artery occlusion using a self-expanding, reconstrainable, intracranial microstent as a temporary endovascular bypass. Stroke 39, 1770–177310.1161/STROKEAHA.107.506212 [PubMed] [Cross Ref]

18. Kerber C. W., Wanke I., Bernard J., Woo H. H., Liu M. W., Nelson P. K. (2007). Rapid intracranial clot removal with a new device: the alligator retriever. AJNR Am. J. Neuroradiol. 28, 860–863 [PubMed]

19. Khatri P., Hill M. D., Palesch Y. Y., Spilker J., Jauch E. C., Carrozzella J. A., Demchuk A. M., Martin R., Mauldin P., Dillon C., Ryckborst K. J., Janis S., Tomsick T. A., Broderick J. P. for the IMS III Investigators (2008) . Methodology of the Interventional Management of Stroke III Trial. Int. J. Stroke 130–137 [PMC free article] [PubMed]

20. Kleindorfer D., Lindsell C. J., Brass L., Koroshetz W., Broderick J. P. (2008). National US estimates of recombinant tissue plasminogen activator use: ICD-9 codes substantially underestimate. Stroke 39, 924–92810.1161/STROKEAHA.107.490375 [PubMed] [Cross Ref]

21. Leary M. C., Saver J. L., Gobin Y. P., Jahan R., Duckwiler G. R., Vinuela F., Kidwell C. S., Frazee J., Starkman S. (2003). Beyond tissue plasminogen activator: mechanical intervention in acute stroke. Ann. Emerg. Med. 41, 838–846 [PubMed]

22. Lee K. Y., Han S. W., Kim S. H., Nam H. S., Ahn S. W., Kim D. J., Seo S. H., Kim D. I., Heo J. H. (2007). Early recanalization after intravenous administration of recombinant tissue plasminogen activator as assessed by pre- and post-thrombolytic angiography in acute ischemic stroke patients. Stroke 38, 192–19310.1161/01.STR.0000251788.03914.00 [PubMed] [Cross Ref]

23. Levy E. I., Kim S. H., Bendok B. R., Boulos A. S., Xavier A. R., Yahia A. M., Qureshi A. I., Guterman L. R., Hopkins L. N. (2004). "Interventional neuroradiolgic therapy," in Stroke: Pathophysiology, Diagnosis, and Management, 4th Edn. eds Mohr J. P., Choi D. W., Grotta J. C., Weir B., Wolf P. A., editors. (Philadelphia: Churchill Livingstone;).

24. Liebig T., Reinartz J., Hannes R., Miloslavski E., Henkes H. (2008). Comparative in vitro study of five mechanical embolectomy systems: effectivenss of clot removal and risk of distal embolization.

Neuroradiology 50, 43–5210.1007/s00234-007-0297-y [PubMed] [Cross Ref]

25. Macleod M. R., Davis S. M., Mitchell P. J., Gerraty R. P., Hankey F. G., Stewart-Wynne E. G., Rosen D., McNeil J. J., Bladin C. F., Chambers B. R., Herkes G. K., Young D., Donnan G. A. (2005). Results of a multicentre, randomized controlled trial of intra-arterial urokinase in the treatment of acute posterior circulation ischaemic stroke. Cerebrovasc. Dis. 20, 12–17 [PubMed]

26. Malik A. M., Aleu A., Lin R., Ranawat N. S., Vora N. A., Jankowitz B., Jumaa M., Zaidi S. F., Kanaan H., Kostov D. B., Reddy V. K., Hammer M. D., Uchino K., Gupta R., Wechsler L. R., Horowitz M. B., Jovin T. G. (2010). Incidence, predictors, and outcomes of intracranial vessel perforation during endovascular therapy for acute ischemic stroke. Stroke 41, 108

27. Mocco J., Hanel R. A., Sharma J., Hauck E. F., Snyder K. V., Natarajan S. K., Llinfante I., Siddiqui A. H., Hopkins L. N., Boulos A. S., Levy E. I. (2010). Use of a vascular reconstruction device to salvage acute ischemic occlusions refractory to traditional endovascular recanalization methods. J. Neurosurg. 112, 557– 562.10.3171/2009.8.JNS09231 [PubMed] [Cross Ref]

28. Nakano S., Iseda T., Yoneyama T., Kawano H., Wakisaka S. (2002). Direct percutaneous transluminal angioplasty for acute middle cerebral artery trunk occlusion: an alternative option to intra-arterial thrombolysis. Stroke 33, 2872– 287610.1161/01.STR.0000038985.26269.F2 [PubMed] [Cross Ref]

29. Nguyen T. N., Lanthier S., Roy D. (2008). Iatrogenic arterial perforation during acute stroke interventions. AJNR Am. J. Neuroradiol. 29, 974–97510.3174/ajnr.A0958 [PubMed] [Cross Ref]

30. Nogueira R. G., Liebeskind D. S., Sung G., Duckwiler G., Smith W. S., MERCI; Multi MERCI Writing Committee (2009a). Predictors of good clinical outcomes, mortality, and successful revascularization in patients with acute ischemic stroke undergoing thrombectomy: pooled analysis of the Mechanical Embolus Removal in Cerebral Ischemia (MERCI) and Multi MERCI Trials. Stroke 40, 3777–3783 [PubMed]

31. Nogueira R. G., Schwamm L. H., Hirsch J. A. (2009b). Endovascular approaches to acute stroke, Part 1: drugs, devices, and data. AJNR Am. J. Neuroradiol. 30, 649–66110.3174/ajnr.A1486 [PubMed] [Cross Ref]

32. Nogueira R. G., Yoo A. J., Buonanno F. S., Hirsch J. A. (2009c).

Endovascular approaches to acute stroke, Part 2: a comprehensive review of studies and trials. AJNR Am. J. Neuroradiol. 30, 859–87510.3174/ajnr.A1604 [PubMed] [Cross Ref]

33. Ogawa A., Mori E., Minematsu K., Taki W., Takahashi A., Nemoto S., Miyamoto S., Sasaki M., Inoue T. for the MELT Japan Study Group (2007). Randomized trial of intraarterial infusion of urokinase within 6 hours of middle cerebral artery stroke. Stroke 38, 2633–263910.1161/STROKEAHA.107.488551 [PubMed] [Cross Ref]

34. Parsons M. W., Miteff F., Bateman G. A., Spratt N., Loiselle A., Attia J., Levi C. R. (2009). Acute ischemic stroke: imaging-guided tenecteplase treatment in an extended time window. Neurology 72, 915–92110.1212/01.wnl.0000344168.05315.9d [PubMed] [Cross Ref]

35. Penumbra Pivotal Stroke Trial Investigators (2009a). Penumbra Pivotal Stroke Trial. Safety and effectiveness of a new generation of mechanical devices for clot removal in intracranial large vessel occlusive disease. Stroke 40, 2761–2768 [PubMed]

36. Penumbra POST (2009b). A Multicenter Real World Look at Penumbra System Results. Boca Raton, FL; Society of Neurointerventional Surgery, Oral Presentation

37. Qureshi A. I., Harris-Lane P., Kirmani J. F., Janjua N., Divani A. A., Mohammad Y. M., Suarez J. I., Montgomery M. O. (2006). Intra-arterial reteplase and intravenous abciximab in patients with acute ischemic stroke: an open-label, dose-ranging, phase I study. Neurosurgery 59, 789–79610.1227/01.NEU.0000232862.06246.3D [PubMed] [Cross Ref]

38. Rha J. H., Saver J. L. (2007). The impact of recanalization on ischemic stroke outcome: a meta-analysis. Stroke 38, 967–97310.1161/01.STR.0000258112.14918.24 [PubMed] [Cross Ref]

39. Saver J. L. (2006). Does the Merci Retriever work? Stroke 37, 1340–134110.1161/01.STR.0000217436.17531.fa [PubMed] [Cross Ref]

40. Shuaib A., Schellinger P., Shownkeen H., Rutledge N., Molina C., Concha M., Bernardini G. L. (2010). "SENTIS: a multi-center randomized trial evaluating cerebral perfusion augmentation via partial aortic occlusion in acute ischemic stroke," in International Stroke Conference, San Antonio, TX

41. Smith W. S., Sung G., Saver J., Budzik R., Duckwiler G., Liebeskind D. S., Lutsep H. L., Rymer M. M., Higashida R. T., Starkman S., Gobin Y. P. and for the Multi MERCI Investigators (2008). Mechanical thrombectomy for acute ischemic stroke: final results of the multi

MERCI trial. Stroke 39, 1205–121210.1161/STROKEAHA.107.497115 [PubMed] [Cross Ref]

42. The National Institute of Neurological Disorders and Strokert-PA Stroke Study Group(1995) Tissue plasminogen activator for acute ischemic stroke. N. Engl. J. Med. 333, 1581–1587 [PubMed]

43. Tomsick T. (2007). TIMI, TIBI, TICI: I came, I saw, I got confused. AJNR Am. J. Neuroradiol. 28, 382–384 [PubMed]

44. Ueda T., Sakaki S., Nochide I., Kumon Y., Kohno K., Ohta S. (1998). Angioplasty after intra-arterial thrombolysis for acute occlusion of intracranial arteries. Stroke 29, 2568–2574 [PubMed]

45. Wechsler L. R. (2006). Does the Merci retriever work? Against. Stroke 37, 1341–134210.1161/01.STR.0000217361.50952.5c [PubMed] [Cross Ref]

46. Wolpert S. M., Bruckmann H., Greenlee R., Wechsler L., Pessin M. S., del Zoppo G. J. (1993). Neuroradiologic evaluation of patients with acute stroke treated with recombinant tissue plasminogen activator: the rtPA Acute Stroke Study Group. AJNR Am. J. Neuroradiol. 14, 3–13 [PubMed]

47. Zaidat O. O., Wolfe T., Hussain S. I., Lynch J. R., Gupta R., Delap J., Torbey M. T., Fitzsimmons B. F. (2008). Interventional acute ischemic stroke therapy with intracranial self-expanding stent. Stroke 39, 2392–239510.1161/STROKEAHA.107.510966 [PubMed] [Cross Ref]

Formats:

Related citations in PubMed

Trial design and reporting standards for intra-arterial cerebral thrombolysis for acute ischemic stroke.[Stroke. 2003]

Mechanical approaches combined with intra-arterial pharmacological therapy are associated with higher recanalization rates than either intervention alone in revascularization of acute carotid terminus occlusion.[Stroke. 2009]

Lin R, Vora N, Zaidi S, Aleu A, Jankowitz B, Thomas A, Gupta R, Horowitz M, Kim S, Reddy V, et al. Stroke. 2009 Jun; 40(6):2092-7. Epub 2009 Apr 23.

Acute stroke therapy 1981-2009.[Klin Neuroradiol. 2009]

Eckert B. Klin Neuroradiol. 2009 Mar; 19(1):8-19. Epub 2009 May 15.

Endovascular treatment strategies for acute ischemic stroke.[Int J Stroke. 2011]

Ellis JA, Youngerman BE, Higashida RT, Altschul D, Meyers PM. Int J Stroke. 2011 Dec; 6(6):511-22.

Endovascular treatment of acute ischemic stroke: the end or just the beginning?[Neurosurg Focus. 2014]

Mokin M, Khalessi AA, Mocco J, Lanzino G, Dumont TM, Hanel RA, Lopes DK, Fessler RD 2nd, Ringer AJ, Bendok BR, et al. Neurosurg Focus. 2014 Jan; 36(1):E5

Cited by other articles in PMC

Successful recanalization with multimodality endovascular interventional therapy in acute ischemic stroke[World Journal of Clinical Cases : WJCC. 201...]

Jongsathapongpan A, Raumthanthong A, Muengtaweepongsa S. World Journal of Clinical Cases : WJCC. 2014/03/16 00:00; 2(3)78-85

Recent Activity

Intra-Arterial Treatment Methods in Acute Stroke Therapy

Intra-Arterial Treatment Methods in Acute Stroke
Therapy Frontiers in Neurology. 2011; 2()
Endovascular abdominal aortic aneurysm repair
Endovascular abdominal aortic aneurysm repair
Postgraduate Medical Journal. 2007 Jan; 83(975)21
Technical features of the INCRAFT™ AAA Stent Graft System.
Technical features of the INCRAFT™ AAA Stent Graft System.
J Cardiovasc Surg (Torino). 2014 Oct ;55(5):705-15. Epub 2014 Jul 16 .
Interventional approaches to deep vein thrombosis.
Interventional approaches to deep vein thrombosis.

Am J Hematol. 2012 May ;87 Suppl 1:S113-8. doi: 10.1002/ajh.23145. Epub 2012 Mar 3 .
PubMed
New Tools and Techniques in Interventional Radiology: Ultrasound-Enhanced Thromb...
New Tools and Techniques in Interventional Radiology: Ultrasound-Enhanced Thrombolysis: EKOS EndoWave Infusion Catheter System Seminars in Interventional Radiology. 2008 Mar; 25(1)37

See more...

Tissue plasminogen activator for acute ischemic stroke. The National Institute of Neurological Disorders and Stroke rt-PA Stroke Study Group[N Engl J Med. 1995]

N Engl J Med. 1995 Dec 14; 333(24):1581-7.

Thrombolysis with alteplase 3 to 4.5 hours after acute ischemic stroke.[N Engl J Med. 2008]

Hacke W, Kaste M, Bluhmki E, Brozman M, Dávalos A, Guidetti D, Larrue V, Lees KR, Medeghri Z, Machnig T, Schneider D, von Kummer R, Wahlgren N, Toni D, ECASS Investigators

N Engl J Med. 2008 Sep 25; 359(13):1317-29.

4.5 hours: the new time window for tissue plasminogen activator in stroke[Stroke. 2009]

Davis SM, Donnan GA Stroke.

2009 Jun; 40(6):2266-7.

Recombinant tissue plasminogen activator in acute thrombotic and embolic stroke.[Ann Neurol. 1992]

del Zoppo GJ, Poeck K, Pessin MS, Wolpert SM, Furlan AJ, Ferbert A, Alberts MJ, Zivin JA, Wechsler L, Busse O

Ann Neurol. 1992 Jul; 32(1):78-86.

Tissue plasminogen activator for acute ischemic stroke. The National Institute of Neurological Disorders and Stroke rt-PA Stroke Study Group.[N Engl J Med. 1995]

N Engl J Med. 1995 Dec 14; 333(24):1581-7.

Early recanalization after intravenous administration of recombinant tissue plasminogen activator as assessed by pre- and post-thrombolytic angiography in acute ischemic stroke patients[Stroke. 2007]

Lee KY, Han SW, Kim SH, Nam HS, Ahn SW, Kim DJ, Seo SH, Kim DI, Heo JH

Stroke. 2007 Jan; 38(1):192-3.

Clinical deterioration following improvement in the NINDS rt-PA Stroke Trial.[Stroke. 2001]

Grotta JC, Welch KM, Fagan SC, Lu M, Frankel MR, Brott T, Levine SR, Lyden PD

Stroke. 2001 Mar; 32(3):661-8.

Arterial reocclusion in stroke patients treated with intravenous tissue plasminogen activator[Neurology. 2002]

Alexandrov AV, Grotta JC Neurology.

2002 Sep 24; 59(6):862-7.

National US estimates of recombinant tissue plasminogen activator use: ICD-9 codes substantially underestimate[Stroke. 2008]

Kleindorfer D, Lindsell CJ, Brass L, Koroshetz W, Broderick JP

Stroke. 2008 Mar; 39(3):924-8.

Thrombolysis with alteplase 3 to 4.5 hours after acute ischemic stroke[N Engl J Med. 2008]

Hacke W, Kaste M, Bluhmki E, Brozman M, Dávalos A, Guidetti D, Larrue V, Lees KR, Medeghri Z, Machnig T, Schneider D, von Kummer R, Wahlgren N, Toni D, ECASS Investigators

N Engl J Med. 2008 Sep 25; 359(13):1317-29.

See more ...

Neuroradiologic evaluation of patients with acute stroke treated with recombinant tissue plasminogen activator. The rt-PA Acute Stroke Study Group.[AJNR Am J Neuroradiol. 1993]

Wolpert SM, Bruckmann H, Greenlee R, Wechsler L, Pessin MS, del Zoppo GJ

AJNR Am J Neuroradiol. 1993 Jan-Feb; 14(1):3-13.

Review TIMI, TIBI, TICI: I came, I saw, I got confused.[AJNR Am J Neuroradiol. 2007]

Tomsick T

AJNR Am J Neuroradiol. 2007 Feb; 28(2):382-4.

The impact of recanalization on ischemic stroke outcome: a meta-analysis.[Stroke. 2007]

Rha JH, Saver JL

Stroke. 2007 Mar; 38(3):967-73.

Predictors of good clinical outcomes, mortality, and successful revascularization in patients with acute ischemic stroke undergoing thrombectomy: pooled analysis of the Mechanical Embolus Removal in Cerebral Ischemia (MERCI) and Multi MERCI Trials[Stroke. 2009]

Nogueira RG, Liebeskind DS, Sung G, Duckwiler G, Smith WS, MERCI, Multi MERCI Writing Committee

Stroke. 2009 Dec; 40(12):3777-83.

See more ...

Predictors of good clinical outcomes, mortality, and successful revascularization in patients with acute ischemic stroke undergoing thrombectomy: pooled analysis of the Mechanical Embolus Removal in Cerebral Ischemia (MERCI) and Multi MERCI Trials.[Stroke. 2009]

Nogueira RG, Liebeskind DS, Sung G, Duckwiler G, Smith WS, MERCI, Multi MERCI Writing Committee

Stroke. 2009 Dec; 40(12):3777-83.

Predictors of good clinical outcomes, mortality, and successful revascularization in patients with acute ischemic stroke undergoing thrombectomy: pooled analysis of the Mechanical Embolus Removal in Cerebral Ischemia (MERCI) and Multi MERCI Trials.[Stroke. 2009]

Nogueira RG, Liebeskind DS, Sung G, Duckwiler G, Smith WS, MERCI, Multi MERCI Writing Committee

Stroke. 2009 Dec; 40(12):3777-83.

Review Endovascular approaches to acute stroke, part 1: Drugs, devices, and data.[AJNR Am J Neuroradiol. 2009]

Nogueira RG, Schwamm LH, Hirsch JA

AJNR Am J Neuroradiol. 2009 Apr; 30(4):649-61.

Intra-arterial reteplase and intravenous abciximab in patients with acute ischemic stroke: an open-label, dose-ranging, phase I study.[Neurosurgery. 2006]

Qureshi AI, Harris-Lane P, Kirmani JF, Janjua N, Divani AA,

Mohammad YM, Suarez JI, Montgomery MO

Neurosurgery. 2006 Oct; 59(4):789-96; discussion 796-7.

Review Endovascular approaches to acute stroke, part 1: Drugs, devices, and data.[AJNR Am J Neuroradiol. 2009]

Nogueira RG, Schwamm LH, Hirsch JA

AJNR Am J Neuroradiol. 2009 Apr; 30(4):649-61.

Acute ischemic stroke: imaging-guided tenecteplase treatment in an extended time window.[Neurology. 2009]

Parsons MW, Miteff F, Bateman GA, Spratt N, Loiselle A, Attia J, Levi CR

Neurology. 2009 Mar 10; 72(10):915-21.

Review Endovascular approaches to acute stroke, part 1: Drugs, devices, and data.[AJNR Am J Neuroradiol. 2009]

Nogueira RG, Schwamm LH, Hirsch JA

AJNR Am J Neuroradiol. 2009 Apr; 30(4):649-61.

Intra-arterial prourokinase for acute ischemic stroke. The PROACT II study: a randomized controlled trial. Prolyse in Acute Cerebral Thromboembolism.[JAMA. 1999]

Furlan A, Higashida R, Wechsler L, Gent M, Rowley H, Kase C, Pessin M, Ahuja A, Callahan F, Clark WM, Silver F, Rivera F

JAMA. 1999 Dec 1; 282(21):2003-11.

Cerebral hemorrhage after intra-arterial thrombolysis for ischemic stroke: the PROACT II trial.[Neurology. 2001]

Kase CS, Furlan AJ, Wechsler LR, Higashida RT, Rowley HA, Hart RG, Molinari GF, Frederick LS, Roberts HC, Gebel JM, Sila CA, Schulz GA, Roberts RS, Gent M

Neurology. 2001 Nov 13; 57(9):1603-10.

Randomized trial of intraarterial infusion of urokinase within 6 hours of middle cerebral artery stroke: the middle cerebral artery embolism local fibrinolytic intervention trial (MELT) Japan.[Stroke. 2007]

Ogawa A, Mori E, Minematsu K, Taki W, Takahashi A, Nemoto S, Miyamoto S, Sasaki M, Inoue T, MELT Japan Study Group

Stroke. 2007 Oct; 38(10):2633-9.

Results of a multicentre, randomised controlled trial of intra-arterial urokinase in the treatment of acute posterior circulation ischaemic stroke.[Cerebrovasc Dis. 2005]

Macleod MR, Davis SM, Mitchell PJ, Gerraty RP, Fitt G, Hankey GJ, Stewart -Wynne EG, Rosen D, McNeil JJ, Bladin CF, Chambers BR, Herkes GK, Young D, Donnan GA

Cerebrovasc Dis. 2005; 20(1):12-7.

Early recanalization after intravenous administration of recombinant tissue plasminogen activator as assessed by pre- and post-thrombolytic angiography in acute ischemic stroke patients.[Stroke. 2007]

Lee KY, Han SW, Kim SH, Nam HS, Ahn SW, Kim DJ, Seo SH, Kim DI, Heo JH

Stroke. 2007 Jan; 38(1):192-3.

Predictors of good clinical outcomes, mortality, and successful revascularization in patients with acute ischemic stroke undergoing thrombectomy: pooled analysis of the Mechanical Embolus Removal in Cerebral Ischemia (MERCI) and Multi MERCI Trials.[Stroke. 2009]

Nogueira RG, Liebeskind DS, Sung G, Duckwiler G, Smith WS, MERCI, Multi MERCI Writing Committee

Stroke. 2009 Dec; 40(12):3777-83.

Unsuccessful tissue plasminogen activator treatment of acute stroke caused by a calcific embolus.[J Neuroimaging. 2004]

Halloran JI, Bekavac I

J Neuroimaging. 2004 Oct; 14(4):385-7.

Iatrogenic arterial perforation during acute stroke interventions.[AJNR Am J Neuroradiol. 2008]

Nguyen TN, Lanthier S, Roy D

AJNR Am J Neuroradiol. 2008 May; 29(5):974-5.

Mechanical thrombectomy for acute ischemic stroke: final results of the Multi MERCI trial.[Stroke. 2008]

Smith WS, Sung G, Saver J, Budzik R, Duckwiler G, Liebeskind DS, Lutsep HL, Rymer MM, Higashida RT, Starkman S, Gobin YP, Multi MERCI Investigators, Frei D, Grobelny T, Hellinger F, Huddle D, Kidwell C, Koroshetz W, Marks M, Nesbit G, Silverman IE

Stroke. 2008 Apr; 39(4):1205-12.

A device for fast mechanical clot retrieval from intracranial arteries (Phenox clot retriever).[Neurocrit Care. 2006]

Henkes H, Reinartz J, Lowens S, Miloslavski E, Roth C, Reith W, Kühne D

Neurocrit Care. 2006; 5(2):134-40.

Comparative in vitro study of five mechanical embolectomy systems: effectiveness of clot removal and risk of distal embolization.[Neuroradiology. 2008]

Liebig T, Reinartz J, Hannes R, Miloslavski E, Henkes

H Neuroradiology. 2008 Jan; 50(1):43-52.

Rapid intracranial clot removal with a new device: the alligator retriever.[AJNR Am J Neuroradiol. 2007]

Kerber CW, Wanke I, Bernard J Jr, Woo HH, Liu MW, Nelson

PK AJNR Am J Neuroradiol. 2007 May; 28(5):860-3.

A device for fast mechanical clot retrieval from intracranial arteries (Phenox clot retriever).[Neurocrit Care. 2006]

Henkes H, Reinartz J, Lowens S, Miloslavski E, Roth C, Reith W, Kühne D

Neurocrit Care. 2006; 5(2):134-40.

The Penumbra System: a mechanical device for the treatment of acute stroke due to thromboembolism.[AJNR Am J Neuroradiol. 2008]

Bose A, Henkes H, Alfke K, Reith W, Mayer TE, Berlis A, Branca V, Sit SP, Penumbra Phase 1 Stroke Trial Investigators

AJNR Am J Neuroradiol. 2008 Aug; 29(7):1409-13.

The penumbra pivotal stroke trial: safety and effectiveness of a new generation of mechanical devices for clot removal in intracranial large vessel occlusive disease.[Stroke. 2009]

Penumbra Pivotal Stroke Trial Investigators

Stroke. 2009 Aug; 40(8):2761-8.

Angioplasty after intra-arterial thrombolysis for acute occlusion of intracranial arteries.[Stroke. 1998]

Ueda T, Sakaki S, Nochide I, Kumon Y, Kohno K, Ohta S

Stroke. 1998 Dec; 29(12):2568-74.

Direct percutaneous transluminal angioplasty for acute middle cerebral artery trunk occlusion: an alternative option to intra-arterial thrombolysis.[Stroke. 2002]

Nakano S, Iseda T, Yoneyama T, Kawano H, Wakisaka S

Stroke. 2002 Dec; 33(12):2872-6.

Predictors of good clinical outcomes, mortality, and successful revascularization in patients with acute ischemic stroke undergoing thrombectomy: pooled analysis of the Mechanical Embolus Removal in Cerebral Ischemia (MERCI) and Multi MERCI Trials.[Stroke. 2009]

Nogueira RG, Liebeskind DS, Sung G, Duckwiler G, Smith WS, MERCI, Multi MERCI Writing Committee

Stroke. 2009 Dec; 40(12):3777-83.

Review Beyond tissue plasminogen activator: mechanical intervention in acute stroke.[Ann Emerg Med. 2003]

Leary MC, Saver JL, Gobin YP, Jahan R, Duckwiler GR, Vinuela F, Kidwell CS, Frazee J, Starkman S

Ann Emerg Med. 2003 Jun; 41(6):838-46.

Multimodal therapy for the treatment of severe ischemic stroke combining GPIIb/IIIa antagonists and angioplasty after failure of thrombolysis.[Stroke. 2005]

Abou-Chebl A, Bajzer CT, Krieger DW, Furlan AJ, Yadav JS

Stroke. 2005 Oct; 36(10):2286-8.

Recanalization of an acute middle cerebral artery occlusion using a self-expanding, reconstrainable, intracranial microstent as a temporary endovascular bypass.[Stroke. 2008]

Kelly ME, Furlan AJ, Fiorella D

Stroke. 2008 Jun; 39(6):1770-3.

Interventional acute ischemic stroke therapy with intracranial self-expanding stent.[Stroke. 2008]

Zaidat OO, Wolfe T, Hussain SI, Lynch JR, Gupta R, Delap J, Torbey MT, Fitzsimmons BF

Stroke. 2008 Aug; 39(8):2392-5.

Stent placement in acute cerebral artery occlusion: use of a self-expandable intracranial stent for acute stroke treatment.[Stroke. 2009]
Brekenfeld C, Schroth G, Mattle HP, Do DD, Remonda L, Mordasini P, Arnold M, Nedeltchev K, Meier N, Gralla J
Stroke. 2009 Mar; 40(3):847-52.

Use of a vascular reconstruction device to salvage acute ischemic occlusions refractory to traditional endovascular recanalization methods.[J Neurosurg. 2010]
Mocco J, Hanel RA, Sharma J, Hauck EF, Snyder KV, Natarajan SK, Linfante I, Siddiqui AH, Hopkins LN, Boulos AS, Levy EI
J Neurosurg. 2010 Mar; 112(3):557-62.

Emergent stenting of extracranial internal carotid artery occlusion in acute stroke has a high revascularization rate.[Stroke. 2005]
Jovin TG, Gupta R, Uchino K, Jungreis CA, Wechsler LR, Hammer MD, Tayal A, Horowitz MB
Stroke. 2005 Nov; 36(11):2426-30.

Review TIMI, TIBI, TICI: I came, I saw, I got confused.[AJNR Am J Neuroradiol. 2007]
Tomsick T
AJNR Am J Neuroradiol. 2007 Feb; 28(2):382-4. Does
the Merci Retriever work? Against.[Stroke. 2006]
Wechsler LR
Stroke. 2006 May; 37(5):1341-2; discussion 1342-3.

Does the Merci Retriever work? For.[Stroke. 2006]
Saver JL
Stroke. 2006 May; 37(5):1340-1; discussion 1342-3.

Methodology of the Interventional Management of Stroke III Trial.[Int J Stroke. 2008]
Khatri P, Hill MD, Palesch YY, Spilker J, Jauch EC, Carrozzella JA, Demchuk AM, Martin R, Mauldin P, Dillon C, Ryckborst KJ, Janis S, Tomsick TA, Broderick JP, Interventional Management of Stroke III Investigators
Int J Stroke. 2008 May; 3(2):130-7.

Thrombolysis with alteplase 3 to 4.5 hours after acute ischemic stroke.[N Engl J Med. 2008]

Hacke W, Kaste M, Bluhmki E, Brozman M, Dávalos A, Guidetti D, Larrue V, Lees KR, Medeghri Z, Machnig T, Schneider D, von Kummer R, Wahlgren N, Toni D, ECASS Investigators

N Engl J Med. 2008 Sep 25; 359(13):1317-29.

Intra-arterial or intravenous thrombolysis for acute ischemic stroke? The SYNTHESIS pilot trial.[J Neurointerv Surg. 2010]

Ciccone A, Valvassori L, Ponzio M, Ballabio E, Gasparotti R, Sessa M, Scomazzoni F, Tiraboschi P, Sterzi R, SYNTHESIS Investigators

J Neurointerv Surg. 2010 Mar; 2(1):74-9.

Safety and efficacy of MRI-based thrombolysis in unclear-onset stroke. A preliminary report.[Cerebrovasc Dis. 2008]

Cho AH, Sohn SI, Han MK, Lee DH, Kim JS, Choi CG, Sohn CH, Kwon SU, Suh DC, Kim SJ, Bae HJ, Kang DW

Cerebrovasc Dis. 2008; 25(6):572-9.

Predictors of good clinical outcomes, mortality, and successful revascularization in patients with acute ischemic stroke undergoing thrombectomy: pooled analysis of the Mechanical Embolus Removal in Cerebral Ischemia (MERCI) and Multi MERCI Trials.[Stroke. 2009]

Nogueira RG, Liebeskind DS, Sung G, Duckwiler G, Smith WS, MERCI, Multi MERCI Writing Committee

Stroke. 2009 Dec; 40(12):3777-83.

Review Endovascular approaches to acute stroke, part 1: Drugs, devices, and data.[AJNR Am J Neuroradiol. 2009]

Nogueira RG, Schwamm LH, Hirsch JA

AJNR Am J Neuroradiol. 2009 Apr; 30(4):649-61.

Multimodal reperfusion therapy for acute ischemic stroke: factors predicting vessel recanalization.[Stroke. 2006]

Gupta R, Vora NA, Horowitz MB, Tayal AH, Hammer MD, Uchino K, Levy EI, Wechsler LR, Jovin TG

Stroke. 2006 Apr; 37(4):986-90.

Tissue plasminogen activator for acute ischemic stroke. The National Institute of Neurological Disorders and Stroke rt-PA Stroke Study

Group.

N Engl J Med. 1995 Dec 14; 333(24):1581-7.

Thrombolysis with alteplase 3 to 4.5 hours after acute ischemic stroke.

Hacke W, Kaste M, Bluhmki E, Brozman M, Dávalos A, Guidetti D, Larrue V, Lees KR, Medeghri Z, Machnig T, Schneider D, von Kummer R, Wahlgren N, Toni D, ECASS Investigators

N Engl J Med. 2008 Sep 25; 359(13):1317-29.

4.5 hours: the new time window for tissue plasminogen activator in stroke.

Davis SM, Donnan GA Stroke.

2009 Jun; 40(6):2266-7.

Recombinant tissue plasminogen activator in acute thrombotic and embolic stroke.

del Zoppo GJ, Poeck K, Pessin MS, Wolpert SM, Furlan AJ, Ferbert A, Alberts MJ, Zivin JA, Wechsler L, Busse O

Ann Neurol. 1992 Jul; 32(1):78-86.

Early recanalization after intravenous administration of recombinant tissue plasminogen activator as assessed by pre- and post-thrombolytic angiography in acute ischemic stroke patients.

Lee KY, Han SW, Kim SH, Nam HS, Ahn SW, Kim DJ, Seo SH, Kim DI, Heo JH

Stroke. 2007 Jan; 38(1):192-3.

Clinical deterioration following improvement in the NINDS rt-PA Stroke Trial.

Grotta JC, Welch KM, Fagan SC, Lu M, Frankel MR, Brott T, Levine SR, Lyden PD

Stroke. 2001 Mar; 32(3):661-8.

Arterial reocclusion in stroke patients treated with intravenous tissue plasminogen activator.

Alexandrov AV, Grotta JC Neurology.

2002 Sep 24; 59(6):862-7.

National US estimates of recombinant tissue plasminogen activator use: ICD-9 codes substantially underestimate.

Kleindorfer D, Lindsell CJ, Brass L, Koroshetz W, Broderick JP

Stroke. 2008 Mar; 39(3):924-8.

Neuroradiologic evaluation of patients with acute stroke treated with recombinant tissue plasminogen activator. The rt-PA Acute Stroke Study Group.

Wolpert SM, Bruckmann H, Greenlee R, Wechsler L, Pessin MS, del Zoppo GJ

AJNR Am J Neuroradiol. 1993 Jan-Feb; 14(1):3-13.

Review TIMI, TIBI, TICI: I came, I saw, I got confused.

Tomsick T

AJNR Am J Neuroradiol. 2007 Feb; 28(2):382-4.

The impact of recanalization on ischemic stroke outcome: a meta-analysis.

Rha JH, Saver JL

Stroke. 2007 Mar; 38(3):967-73.

Predictors of good clinical outcomes, mortality, and successful revascularization in patients with acute ischemic stroke undergoing thrombectomy: pooled analysis of the Mechanical Embolus Removal in Cerebral Ischemia (MERCI) and Multi MERCI Trials.

Nogueira RG, Liebeskind DS, Sung G, Duckwiler G, Smith WS, MERCI, Multi MERCI Writing Committee

Stroke. 2009 Dec; 40(12):3777-83.

Review Endovascular approaches to acute stroke, part 1: Drugs, devices, and data.

Nogueira RG, Schwamm LH, Hirsch JA

AJNR Am J Neuroradiol. 2009 Apr; 30(4):649-61.

Intra-arterial reteplase and intravenous abciximab in patients with acute ischemic stroke: an open-label, dose-ranging, phase I study.

Qureshi AI, Harris-Lane P, Kirmani JF, Janjua N, Divani AA, Mohammad YM, Suarez JI, Montgomery MO

Neurosurgery. 2006 Oct; 59(4):789-96; discussion 796-7.

Acute ischemic stroke: imaging-guided tenecteplase treatment in an extended time window.

Parsons MW, Miteff F, Bateman GA, Spratt N, Loiselle A, Attia J, Levi CR

Neurology. 2009 Mar 10; 72(10):915-21.

Intra-arterial prourokinase for acute ischemic stroke. The PROACT II study: a randomized controlled trial. Prolyse in Acute Cerebral Thromboembolism.

Furlan A, Higashida R, Wechsler L, Gent M, Rowley H, Kase C, Pessin M, Ahuja A, Callahan F, Clark WM, Silver F, Rivera F

JAMA. 1999 Dec 1; 282(21):2003-11.

Cerebral hemorrhage after intra-arterial thrombolysis for ischemic stroke: the PROACT II trial.

Kase CS, Furlan AJ, Wechsler LR, Higashida RT, Rowley HA, Hart RG, Molinari GF, Frederick LS, Roberts HC, Gebel JM, Sila CA, Schulz GA, Roberts RS, Gent M

Neurology. 2001 Nov 13; 57(9):1603-10.

Randomized trial of intraarterial infusion of urokinase within 6 hours of middle cerebral artery stroke: the middle cerebral artery embolism local fibrinolytic intervention trial (MELT) Japan.

Ogawa A, Mori E, Minematsu K, Taki W, Takahashi A, Nemoto S, Miyamoto S, Sasaki M, Inoue T, MELT Japan Study Group

Stroke. 2007 Oct; 38(10):2633-9.

Results of a multicentre, randomised controlled trial of intra-arterial urokinase in the treatment of acute posterior circulation ischaemic stroke.

Macleod MR, Davis SM, Mitchell PJ, Gerraty RP, Fitt G, Hankey GJ, Stewart -Wynne EG, Rosen D, McNeil JJ, Bladin CF, Chambers BR, Herkes GK, Young D, Donnan GA

Cerebrovasc Dis. 2005; 20(1):12-7.

Levy E. I., Kim S. H., Bendok B. R., Boulos A. S., Xavier A. R., Yahia A. M., Qureshi A. I., Guterman L. R., Hopkins L. N. (2004). "Interventional neuroradiolgic therapy," in Stroke: Pathophysiology, Diagnosis, and Management, 4th Edn. eds Mohr J. P., Choi D. W., Grotta J. C., Weir B., Wolf P. A., editors. (Philadelphia: Churchill Livingstone;).

Unsuccessful tissue plasminogen activator treatment of acute stroke caused by a calcific embolus.

Halloran JI, Bekavac I

J Neuroimaging. 2004 Oct; 14(4):385-7.

Iatrogenic arterial perforation during acute stroke interventions.

Nguyen TN, Lanthier S, Roy D

AJNR Am J Neuroradiol. 2008 May; 29(5):974-5.

Malik A. M., Aleu A., Lin R., Ranawat N. S., Vora N. A., Jankowitz B., Jumaa M., Zaidi S. F., Kanaan H., Kostov D. B., Reddy V. K., Hammer M. D., Uchino K., Gupta R., Wechsler L. R., Horowitz M. B., Jovin T. G. (2010). Incidence, predictors, and outcomes of intracranial vessel perforation during endovascular therapy for acute ischemic stroke. Stroke 41, 108

Mechanical thrombectomy for acute ischemic stroke: final results of the Multi MERCI trial.

Smith WS, Sung G, Saver J, Budzik R, Duckwiler G, Liebeskind DS, Lutsep HL, Rymer MM, Higashida RT, Starkman S, Gobin YP, Multi MERCI Investigators, Frei D, Grobelny T, Hellinger F, Huddle D, Kidwell C, Koroshetz W, Marks M, Nesbit G, Silverman IE

Stroke. 2008 Apr; 39(4):1205-12.

A device for fast mechanical clot retrieval from intracranial arteries (Phenox clot retriever).

Henkes H, Reinartz J, Lowens S, Miloslavski E, Roth C, Reith W, Kühne D

Neurocrit Care. 2006; 5(2):134-40.

Comparative in vitro study of five mechanical embolectomy systems: effectiveness of clot removal and risk of distal embolization.

Liebig T, Reinartz J, Hannes R, Miloslavski E, Henkes H

Neuroradiology. 2008 Jan; 50(1):43-52.

Rapid intracranial clot removal with a new device: the alligator retriever. Kerber CW, Wanke I, Bernard J Jr, Woo HH, Liu MW, Nelson PK AJNR Am J Neuroradiol. 2007 May; 28(5):860-3.

The Penumbra System: a mechanical device for the treatment of acute stroke due to thromboembolism.

Bose A, Henkes H, Alfke K, Reith W, Mayer TE, Berlis A, Branca V, Sit

SP, Penumbra Phase 1 Stroke Trial Investigators

AJNR Am J Neuroradiol. 2008 Aug; 29(7):1409-13.

The penumbra pivotal stroke trial: safety and effectiveness of a new generation of mechanical devices for clot removal in intracranial large vessel occlusive disease.

Penumbra Pivotal Stroke Trial Investigators

Stroke. 2009 Aug; 40(8):2761-8.

Penumbra POST (2009b). A Multicenter Real World Look at Penumbra System Results. Boca Raton, FL; Society of Neurointerventional Surgery, Oral Presentation

Angioplasty after intra-arterial thrombolysis for acute occlusion of intracranial arteries.

Ueda T, Sakaki S, Nochide I, Kumon Y, Kohno K, Ohta

S Stroke. 1998 Dec; 29(12):2568-74.

Direct percutaneous transluminal angioplasty for acute middle cerebral artery trunk occlusion: an alternative option to intra-arterial thrombolysis.

Nakano S, Iseda T, Yoneyama T, Kawano H, Wakisaka

S Stroke. 2002 Dec; 33(12):2872-6.

Review Beyond tissue plasminogen activator: mechanical intervention in acute stroke.

Leary MC, Saver JL, Gobin YP, Jahan R, Duckwiler GR, Vinuela F, Kidwell CS, Frazee J, Starkman S

Ann Emerg Med. 2003 Jun; 41(6):838-46.

Multimodal therapy for the treatment of severe ischemic stroke combining GPIIb/IIIa antagonists and angioplasty after failure of thrombolysis.

Abou-Chebl A, Bajzer CT, Krieger DW, Furlan AJ, Yadav JS

Stroke. 2005 Oct; 36(10):2286-8.

Recanalization of an acute middle cerebral artery occlusion using a self-expanding, reconstrainable, intracranial microstent as a temporary endovascular bypass.

Kelly ME, Furlan AJ, Fiorella D

Stroke. 2008 Jun; 39(6):1770-3.

Interventional acute ischemic stroke therapy with intracranial self-expanding stent.
Zaidat OO, Wolfe T, Hussain SI, Lynch JR, Gupta R, Delap J, Torbey MT, Fitzsimmons BF
Stroke. 2008 Aug; 39(8):2392-5.

Stent placement in acute cerebral artery occlusion: use of a self-expandable intracranial stent for acute stroke treatment.
Brekenfeld C, Schroth G, Mattle HP, Do DD, Remonda L, Mordasini P, Arnold M, Nedeltchev K, Meier N, Gralla J
Stroke. 2009 Mar; 40(3):847-52.

Use of a vascular reconstruction device to salvage acute ischemic occlusions refractory to traditional endovascular recanalization methods.
Mocco J, Hanel RA, Sharma J, Hauck EF, Snyder KV, Natarajan SK, Linfante I, Siddiqui AH, Hopkins LN, Boulos AS, Levy EI
J Neurosurg. 2010 Mar; 112(3):557-62.

Emergent stenting of extracranial internal carotid artery occlusion in acute stroke has a high revascularization rate.
Jovin TG, Gupta R, Uchino K, Jungreis CA, Wechsler LR, Hammer MD, Tayal A, Horowitz MB
Stroke. 2005 Nov; 36(11):2426-30. Does

the Merci Retriever work? Against.

Wechsler LR

Stroke. 2006 May; 37(5):1341-2; discussion 1342-3.

Does the Merci Retriever work? For.

Saver JL

Stroke. 2006 May; 37(5):1340-1; discussion 1342-3. Methodology

of the Interventional Management of Stroke III Trial.

Khatri P, Hill MD, Palesch YY, Spilker J, Jauch EC, Carrozzella JA, Demchuk AM, Martin R, Mauldin P, Dillon C, Ryckborst KJ, Janis S, Tomsick TA, Broderick JP, Interventional Management of Stroke III Investigators

Int J Stroke. 2008 May; 3(2):130-7.

Intra-arterial or intravenous thrombolysis for acute ischemic stroke? The SYNTHESIS pilot trial.

Ciccone A, Valvassori L, Ponzio M, Ballabio E, Gasparotti R, Sessa M, Scomazzoni F, Tiraboschi P, Sterzi R, SYNTHESIS Investigators

J Neurointerv Surg. 2010 Mar; 2(1):74-9.

Safety and efficacy of MRI-based thrombolysis in unclear-onset stroke. A preliminary report.

Cho AH, Sohn SI, Han MK, Lee DH, Kim JS, Choi CG, Sohn CH, Kwon SU, Suh DC, Kim SJ, Bae HJ, Kang DW

Cerebrovasc Dis. 2008; 25(6):572-9.

Shuaib A., Schellinger P., Shownkeen H., Rutledge N., Molina C., Concha M., Bernardini G. L. (2010). "SENTIS: a multi -center randomized trial evaluating cerebral perfusion augmentation via partial aortic occlusion in acute ischemic stroke," in International Stroke Conference, San Antonio, TX

Multimodal reperfusion therapy for acute ischemic stroke: factors predicting vessel recanalization.

Gupta R, Vora NA, Horowitz MB, Tayal AH, Hammer MD, Uchino K, Levy EI, Wechsler LR, Jovin TG

Index

A

B

C

D

E

F

G

I

L

N

O

P

Q

R

S

T

U

V

W

X

Y

Z

Made in the USA
Middletown, DE
04 January 2025

67063872R00316